How Drowned Was My Valley

How Drowned Was My Valley

Geoffrey Bond

With notes by

Roger Stringer

WEAVER
—PRESS—

Published in 2022 by
Weaver Press, Box A1922, Avondale, Harare.
<www.weaverpresszimbabwe.com>

ISBN
978-1-77922-422-4 (paperback)
978-1-77922-423-1 (e-book)

Cover design by Rhona Chapman, Harare

The cover photo, from a slide taken by Geoffrey Bond, shows part of the river in the middle Zambezi Valley that was flooded when Lake Kariba was formed.

Typeset by TextPertise, Harare

For my long-suffering family,

Marina, Ivan and Nina

Writing a book is a horrible, exhausting struggle,
like a long bout of some painful illness.

George Orwell

I see no point in reading.

Louis XIV

Contents

How Drowned Was My Valley

Appendices

Foreword

Geoff Bond mapped the Zambezi Valley before it was flooded to form Lake Kariba. A far cry from his upbringing in England, he loved the harsh African bush and spent many weeks there exploring in his trusty Land-Rover, Susan, along with Million, his right-hand man. The events described in this book shaped his life and career until he was appointed as the founder Professor of the Geology department of what is now the University of Zimbabwe.

The manuscript was commissioned by Louis Bolze, who had founded Books of Rhodesia and, in 1978/79, had begun publishing two series by people who had significantly contributed to the country's development: 'Men of Our Time' and 'Down Memory Lane'.

On the completion of his writing, my father apparently felt that the material had become irrelevant now that Zimbabwe had attained independence from Great Britain in 1980. Then came his untimely death in 1983.

The typescript had been read by a British geologist visiting Zimbabwe, who deemed it 'unworthy of publishing' with its lack of accompanying photographs. However, having stayed safely in a drawer for over 37 years, the manuscript reappeared in 2020, a good year to realize in hindsight that it should go to print, especially while there were still people who had known him.

I will always be indebted to Charles and Lesley Castelin who were given a rare opportunity to see the typescript, declaring it 'a cracking read' and no doubt of interest to a wider audience. So started the journey and the idea to add the appendices and the illustrations. Geoff Bond was a gentleman, modest and honest, making no mention of the numerous accolades he had collected during his career. This information has been gathered for posterity from personal tributes by those who knew and greatly respected him. These contributions appear in an appendix.

The manuscript has been preserved as my father wrote it, with footnotes and photographs added by Roger Stringer, who had the daunting task of editing a work by a deceased author. He has incorporated the appendices in such a way as to not detract from the original idea of the book.

Geoff Bond's wicked sense of humour is evident in his writing. He loved playing with words, composing 'doggerel' in his spare time. I can just imagine him wondering what the title of his book might be, when, with a twinkle in his eye, he thought of Richard Llewellyn's famous book, *How Green Was My Valley*, and came up with the title, *How Drowned Was My Valley*.

Nina Bond Crowhurst

Acknowledgements

A project of this magnitude requires a team. It has been an honour to have Roger Stringer at the helm, guiding the whole process and leaving no stone unturned. Rhona Chapman has had an incredible eye for detail, giving much thought to the cover and the photographic side. We will forever be indebted to you for your invaluable contributions.

Thank you to Lee Bond for helping to transfer the book from the original typescript into an editable format – a major task – and to Warren Crowhurst, for then helping with the initial proof-reading before the editing process could even begin.

Thank you to Charles and Lesley Castelin, for keeping a keen eye on the project, and to Jenna van de Ruit who helped me get back on track.

The co-operation of the staff of the Geology Department at the University of Zimbabwe is also much appreciated, who held the field notebooks and other material that have been invaluable in putting the book together.

Finally to my family, who have been so supportive. Thank you Jason, Rosa, Nick and all those who have given encouragement and input along the way.

To my mum, thank you for keeping the typescript safely stored all these years.

And, last but not least, to my husband Russell for all your support and tireless enthusiasm right from the beginning. Thank you!

Nina Bond Crowhurst

When it is not possible for a copy-editor to ask questions of an author about his work, it becomes necessary to seek the assistance of others. This can be even more difficult when the work was written more than forty years earlier and there are very few people who will have known him in his younger years.

The 'tributes' that Charles Castelin initially arranged to get from Geoffrey Bond's former students, of whom he was one, and that appear in Appendix 1, added more information about the 1960s. But there were still others to find, and the recollections of some of Geoffrey Bond's colleagues now complement them; the contributions of all are greatly appreciated.

Through the University of Zimbabwe's Geology department, we were able to access many of Geoffrey Bond's field notebooks from between 1946 and 1969 and even his slides and photographs from those years; we have chosen some of those to illustrate his story. Rhona Chapman has worked her magic in making old pictures look as good as possible.

In addition, Geoffrey Bond made reference to many people, places and events that may have been recognized at the time he was writing but would be much less familiar today. I felt that it would enhance his story if I could at least provide some clues as to who or what they were, hence the footnotes, all of which I have added. I could not have done all this without the assistance of a number of people, both in Zimbabwe and around the world, but any errors or inaccuracies in the footnotes are entirely my responsibility.

I am pleased to have been able to include some information from people he mentions working with, in particular from Terence Coffin-Grey, thanks to his daughter, Cathy Peacock; from Jeremy Burden, with whom he worked in Malawi; and from Philip Allsworth-Jones, with whom he communicated about his work in Nigeria.

I would like to thank Andrew Leitch for the photograph of Seascale preparatory school, where he was the last head boy. They had recently started a Facebook 'nostalgia page', as 2022 marked fifty years since the school closed. Dr Anthony Reeve, Hon. Archivist of the St Beghian Society, was very helpful in finding some information about Geoffrey Bond's time at St Bees school and providing photographs. Colin Carlin of *Abercornucopia* provided some useful information about the Gamwell sisters in Zambia.

Friends and colleagues were always available to help with information where clarification was needed, and the book could not have been completed in this way without them. I must especially thank Tim Broderick, whose knowledge of the geology and archaeology of Zimbabwe is immense. Others who have helped include Rob Burrett, for information about the Bulawayo Museum; Fr Edward Murphy SJ, for information about Fr Terörde's grave; Chris Magadza, Brian Marshall and Athol McLachlan, for information about the Lake Kariba Research Station; Sean Burke, for images from the Rhodesian Study Circle website; as well as 'bits and pieces' from Mark Hyde, Jim Latham, Doug McClymont, Joyce Makwenda, Dominic Muntanga, Ray Roberts, Dawn Siemers, Harald Solberg, Robin Taylor and Jonathan Waters.

Finally, it should be noted that minimal changes have been made to the text of the original typescript, especially in terms of changes in the names of places, etc., after Independence. Therefore the reader will find, for example, Salisbury used rather than Harare for the name of the capital city. This has been done in order to retain the integrity of what is effectively a 'historical' manuscript written in the late 1970s.

Roger Stringer

Preface

Books about people can be put into groups, under well-known labels. Biographies are written by professional writers to further their own reputations, under the guise of cataloguing the achievements of someone alleged to have been of some significance during his lifetime. People who are really significant generally produce memoirs. Autobiographies may be written by people who have little or no significance but think they ought to have. If they can find a publisher, they may even achieve some.

To be asked by a publisher to write one's own biography is an odd event and, in my case, it produced some strange reactions. In the first place it was gratifying. I have my share of vanity, and it certainly did my ego good. Less satisfying reactions followed. It indicated, certainly, that at least one person thought I had done something worth writing about, but it implied that I had reached a stage when I could not expect to last much longer, at least in the productive capacity. Another reaction was that, as I had never written a book, I would have to think how it was to be done from a base of no experience; another was that it would be a considerable labour.

As might have been expected, vanity overcame apprehension and I decided to 'have a go'. It is extremely unlikely to be a best-seller, or even to venture outside the borders of the country that I have regarded as home for forty years and has been so good to me all that time; but it does offer the opportunity to put my gratitude on record in a permanent form. It will consist almost entirely of what my irreverent family call 'anecdotage', but what else is there to put in one's own biography?

I am more at home in the impersonal style of writing, which is the fashion in scientific papers, than in a style halfway between that and spoken English. I will not go as far as Humpty Dumpty and make the words mean what I like, and pay them extra on Saturdays, but I have always taken liberties with strict English grammar. It's my only language and I feel free to bend it somewhat, as the situation demands.

I do not claim to have had a career, which implies planning. As will be seen, I have never had a plan to follow. What has happened has been a series of rather disconnected lucky chances, over which I seem to have had little control. Perhaps a career is better defined as something to look back on towards its end rather than something to look forward to from near its beginning. There is nothing unique about what follows, except that it all happened to me. It consists partly of stories from the bush, telling of things that happened while there were still almost unexplored bits left, where the rivers were shown on the maps as dotted lines, if there were any maps. Modern maps from ubiquitous air photography have spoilt all that; the aeroplane, the satellite and the camera have snooped all over my happy and unspoilt Zambezi Valley, where there are now so many roads that I get lost without a map.

I have tried to isolate, evaluate and describe the people in my early life who influenced me most as I grew up, as well as those who later taught me what geology was all about. What I have been able to contribute to the development of the piece of Gondwanaland that has now become Zimbabwe has been through the science of geology, but the circumstances under which the work was done have been described more fully than the scientific results, though I hope those results have had some significance.

The whole thing is meant as nothing more than a pastime for those people lucky enough to have time on their hands to pass.

Geoff Bond

– 1 –

In the Beginning ...

It all started at about 6.00p.m. on Monday, 27 May 1912 in a house in that part of Brighton known as Hove. My mother was thirty-seven at the time and, as I was her first child, I had a rather difficult arrival. In gratitude to her doctor she chose his Christian name for me. It happens to be one that did not occur on either side of the family, and to be one that I have never liked; even now I find myself reluctant to say it aloud.

If maternal family tradition had been followed, I would have been James. Perhaps I was lucky to have escaped that name because of the exploits of a certain secret agent of world renown, known also as 007. The surname has sometimes helped to smooth my passage through immigration formalities in various parts of the world, when officers have made jokes about James, thinking they were being original. And my family were rung up three Saturday nights in succession, very late, by a joker at a party asking to speak to James.

The reason why I was born in Brighton was that my father, Denys, was at the time a master at Brighton College, and I was christened in the chapel. If I had arrived in school holidays, it would have been at our cottage in the depths of the West Yorkshire Dales country, a village called Arncliffe on a tributary of the River Wharfe, the Skirfare. Until I came to Zimbabwe, this was my home, and I have always resented the fact that I was not born a Yorkshireman.

One cannot choose one's birthplace any more than one can choose one's parents, but genetics and early environmental influence both have a profound effect on subsequent events. My parents were ill-suited. Both my grandfathers were Church of England parsons but of very different circumstances, and from opposite ends of Britain. I have never learned how my parents met, but I suspect it was in Lincoln, a sort of halfway house.

My father was born and brought up in Somerset and went to Sherbourne School and from there to Cambridge, where he took Modern Languages (French and German) and launched out on a teaching career in public schools. He played tennis for Cambridge and earned a half-blue, which was all it was thought to be worth in those days.

When World War I broke out, he joined the 'Wavy Navy', the Royal Naval Volunteer Reserve (RNVR), and was posted to Dartmouth, to the Royal Naval College as an 'instructor', which merely meant that he continued to teach French and German, this time to naval cadets in the fourteen-year-plus age group. He was so short-sighted that he was totally unfit for active service, but he held the rank of lieutenant.

So we went to live in Dartmouth, and I think my first memories date from that sleepy Devonshire town. Our first home was called Tower Cottage and it had quite a large garden. I remember two things about this place. One is a white wooden gate at the foot of the staircase; I am not sure whether this was designed to stop me going upstairs or to keep me upstairs. The other memory is of a cadet called, at that time, Louis Battenberg. With others, he sometimes came to rather formal Sunday afternoon teas organized by my mother. He is the only one I remember because he played so nicely with me. Sometimes we played bears with the hearth rug, and, even more exciting, he built towering constructions with my building bricks, and then took out most of the basement courses without their falling down. This left an indelible mark on my three-and-a-half-year-old mind.

His family soon changed its name from Battenberg to a literal translation from the German: Mountbatten. He died, at the hands of Irish terrorists, as Admiral of the Fleet, Earl Mountbatten of Burma.[1] I still have a letter that he wrote to my mother, thanking her most politely for a Sunday tea and signed Louis Battenberg.

This had a sequel in Salisbury. I have had quite close contact with Louis Mountbatten School, and it occurred to me some years ago to offer a photostat of this very early letter to the headmaster to add to their considerable collection of Mountbatten material, for he had always kept in touch with the school. I had not been in contact with Lord Mountbatten since his time as a cadet at Dartmouth, but I wrote to him to say what I had done and waited to see whether he would answer a rebel Rhodesian.

The result was an answer almost by return on his own signature. The letter gave the impression that he was by then a rather lonely widower and quite glad to go back into his early youth. What was, perhaps, more surprising was his plea for a photostat copy of that simple letter of thanks. He said that he had practically nothing in his own archives from that period of his career. Naturally, I sent him a copy and at the same time asked if he had been one of the twenty-four cadets that my mother had invited for one particularly ill-

[1] 'IRA bombs kill Mountbatten and 17 soldiers', *The Guardian*, 28 August 1979.

fated tea party. Eggs were a rarity even in a small town like Dartmouth, which was surrounded by farming country, but for this day she had wangled two dozen, which were duly boiled and served up for afternoon tea. Every single one was hard-set and contained a chicken almost at the point of hatching. The resulting consternation on the faces of two-dozen naval cadets when they each discovered this while they were on their very best behaviour can well be imagined. History does not relate how they coped with the situation. Lord Mountbatten was not one of them.

Once again, I had a reply immediately, but we had no further correspondence before his murder. I attended the memorial service at Louis Mountbatten School with a feeling of real personal sorrow.

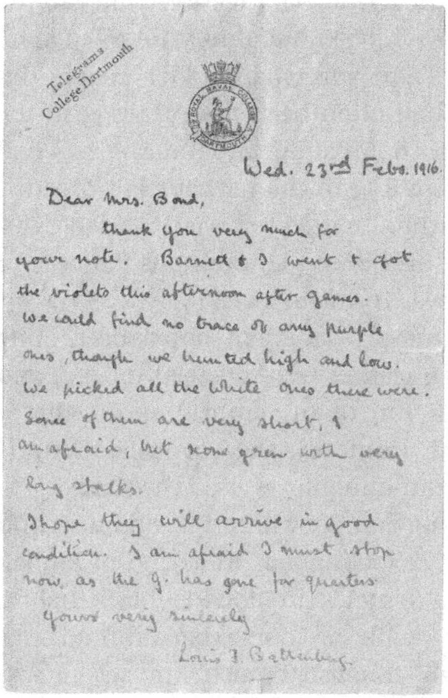

Telegrams
College Dartmouth

Wed. 23rd Febr. 1916.

Dear Mrs. Bond,

thank you very much for your note. Barnett & I went & got the violets this afternoon after games. We could find no trace of any purple ones, though we hunted high and low. We picked all the white ones there were. Some of them are very short, I am afraid, but some grew with very long stalks.

I hope they will arrive in good condition. I am afraid I must stop now, as the G. has gone for quarters.

Yours very sincerely

Louis F. Battenberg.

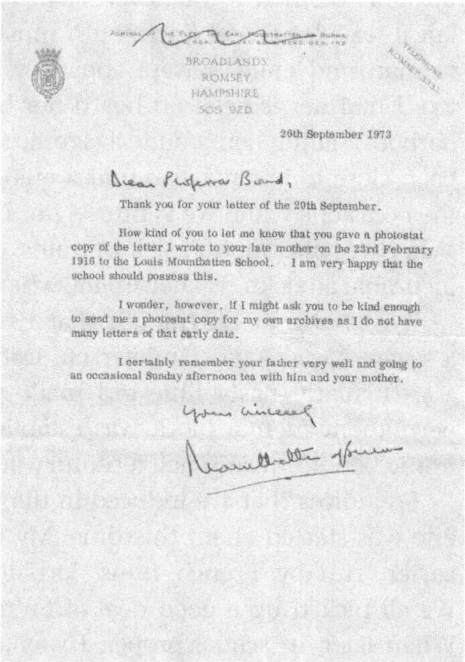

BROADLANDS
ROMSEY
HAMPSHIRE
SO5 9ZD

TELEPHONE
ROMSEY 2333

26th September 1973

Dear Professor Bond,

Thank you for your letter of the 20th September.

How kind of you to let me know that you gave a photostat copy of the letter I wrote to your late mother on the 23rd February 1916 to the Louis Mountbatten School. I am very happy that the school should possess this.

I wonder, however, if I might ask you to be kind enough to send me a photostat copy for my own archives as I do not have many letters of that early date.

I certainly remember your father very well and going to an occasional Sunday afternoon tea with him and your mother.

Yours sincerely

Mountbatten of Burma

Dartmouth had a good deal of influence on me. I learnt to swim in the college baths under Chief Petty Officer Clark, from whom I also acquired a basic vocabulary of naval invective, delivered in a strong Devonshire accent. This was improved while helping the local farmers, who, in return, added a few choice phrases from their own special vocabulary. I have never lost this facility and, indeed, it was considerably enriched while I spent five years in the Sergeants' Mess in the RAF during World War II. Only the accent has changed.

Another, more practical, seed was sown at Dartmouth - an interest in mechanical things, and particularly in internal combustion engines. Our last home in Dartmouth was right in the town, not far from a garage where I spent endless hours with a mechanic called Dixon, watching him mend motor cars, including early Daimlers, which seemed to have masses of beautifully polished copper tubing under the bonnet.

One of Dartmouth College's prized possessions was Nelson's old wooden battleship, HMS *Britannia*. She was permanently moored in the Dart, more or less opposite the college, dismasted but resplendently painted black and white. I had a great love for her, but she was obsolete, even for training purposes. The Powers That Be decreed that she be towed away and broken up. This seemed to me, even at the age of six, to be a terrible thing to do, but it was done, and Dartmouth must have turned out almost to every man, woman and child. I wept copiously, as I am sure many older people did too. I had never been on board her but had often been taken sailing in the harbour, sailing being quite a significant part of seamanship training, so I had been close to her massive hull. I enjoyed sailing in the harbour but dreaded the occasions when we ventured out into the Channel as I was invariably seasick. I was also motion sick in trains, cars and buses, an ominous foretaste of incurable airsickness later in life, when I was trying to learn to fly.

It was also at Dartmouth that I became a permanent non-dancer, quite a severe social handicap later on. Dancing classes were decreed for me, and a particularly nasty little girl made such fun of my clumsy efforts that a permanent mental block was established. Ironically, she was called Nina - ironic because nearly half a century later our daughter was so christened.

Prejudices that are induced in the young can persist for life, and a second one was started about this time. My first schooling was in a kind of kindergarten run by French nuns, forbidding females in flowing black habits. We all picked up a good deal of French - with, of course, beautiful accents. When later, in school proper, I was taught 'English French', my accent was mercilessly ridiculed by the rest of the class, and to this day I cannot bring myself to utter a word of French. The laughter of the young is a cruel weapon, perhaps even more than the thoughtless laughter of parents at their own children's slips.

I did not enjoy kindergarten, and I did not like the Sisters. My parents seldom beat me, having rather advanced psychological ideas on child-rearing. As I was an only child, all their experiments were concentrated on me, but I got punished for coming home one afternoon and announcing that 'Sister Benetild is a black bastard'. I suspect it was not so much about what I said as who I had said it about.

Anti-German feeling was actively fostered in Britain in World War I and was intense. While we were closely connected with the Royal Naval College, where there were one or two royal princelings as cadets, I very unpatriotically got German measles. Queen Mary, like a previous Queen of England, was not amused. I do not know whether she knew that my father had, on one occasion, had to rebuke one of the princelings for his bad spelling. The princely answer was, 'I'm sorry, sir, but all my people spell atrociously.'

At the time of the Armistice, 11 November 1918, my father still held the rank of lieutenant in the RNVR. The German fleet surrendered at Scapa Flow and he was sent up as an interpreter for this historic occasion, yet he never seems to have talked about it, certainly not to me. Perhaps this was not surprising, as we were never very close.

My parents separated, not suddenly but rather gradually, beginning when I was about seven or eight. By then I was living with my mother in her cottage at Arncliffe, in the Dales of the West Riding of Yorkshire, and which I regarded as my real home. My father continued to teach in various schools but used to come to Arncliffe for parts of the holidays, more for trout fishing than family. He was, to me, a colourless, shadowy figure and our only mutual interest was in fishing; it was from him that I learnt the rudiments of fly fishing. This had an added advantage of being a solitary sport, which allowed me to disappear all day on my own, out of reach of parental authority.

In that part of Yorkshire we small boys practised three forms of trout fishing: 'fly' when the river was down; 'worm when there was a big brown flood after heavy rain, quite a common thing in the Pennines; and 'groping' or 'tickling' when the weather was hot and the river exceptionally low.

The catchment of the Skirfare and its tributaries or 'becks' included sandstone tops to the hills with heather and grouse moors, and a lot of hill peat, overlying limestones, which made up the valley flanks and floors. In such a setting much of the river flow is underground, and it is an area riddled with caves, both large and small. The peaty moors provide run-off which is dark coloured, but the normal low flow is gin-clear, and trout are very hard to catch when the river is in that state. But who can blame small boys stripping off for a bathe in hot weather?

We always selected pools with moderate-sized boulders in them, under which the fish retreated when we bathed. They were not difficult to corner and, with skill, fingers can be worked into their gills. Their necks were then quickly broken and the corpse hidden in the nearest rabbit hole or under a rock on the bank. It was all highly illegal, which added spice to it, but very productive when fish could not be got by rod and line. The standard fine for 'groping' was £1 per fish, but none of us was ever caught because we kept a very keen watch for the local bobby or anyone who might be likely to report us to him.

Worm fishing in a flood would also be frowned on by the purist trout fisherman and it is illegal in some countries, but it was widely done in the Dales and it is a good deal more skilful than the purist imagines. It is also not without risk because the floods are fearsome. My mother must have had a lot of anxious hours wondering if I had fallen in and drowned. I had a few anxious moments too.

Fly fishing on the Skirfare could be done either downstream 'wet' or upstream 'semi-dry' but the tackle sounds very old-fashioned now. Gut was easy to get but the preference was for horsehair, even though it had so little tensile strength: it never got soggy like gut, and was far less visible in the water. It was rather special stuff because mares' hair was usually rotten but geldings' hair was said to have long flat areas that reflect light. The only usable hair was from grey stallions, whose tail hair was round in cross-section and reasonably strong. Neither was it brittle, so it tied well. To catch a trout over the eight-inch limit on such tackle in fast-running water was quite an art and developed a very delicate touch. I wonder how it compares with modern monofilament nylon, which is so much stronger, as to refractive index and therefore visibility in the water.

Arncliffe was quite a famous trout-fishing village, and the local pub, The Falcon, had rights for its visitors over about ten miles of water, except on Sundays.[2] No one fished on Sunday, however tempting the weather and the condition of the river. The pub, like all others in the region, was shut on Sundays, licences being for only six days a week. I expect both types of Sabbath observance have lapsed since my time. The visitors used to come armed with expensive tackle, boxes full of different patterns of flies, special fishing clothes and hip-length waders. They usually caught very few fish, for two reasons. Because they had such a variety of flies, they spent an inordinate amount of time sitting on the bank changing them. Tackle will not catch fish unless it is in the water, and even then only when there are fish where it is cast. The casual visitor did not know the river either.

On days when the river was 'off', I spent a lot of time watching trout, learning their habits and especially their favourite places. I did not have boxes full of flies, but I did know the river; and I could almost think like a trout. The results were often gratifying. There was no limit to the number of fish that could be taken by an individual rod, but no fish below eight inches was

[2] The Falcon Inn later featured in the British television series, Emmerdale Farm, when it was first broadcast in 1972. *The Falcon Inn*, <http://thefalconinnarncliffe.co.uk>.

allowed. These had to be put back. From discussions with fisheries people much later in life, it has become obvious that this was a mistaken policy if its aim was to promote the quality of the fishing, particularly if anglers measure success by the size of individual fish. The old fisherman's grace indicates that traditionally this is so.

Lord give me grace to catch a fish
So large that even I,
When boasting of it afterwards,
May have no need to lie. (Anon)

The policy should have been to take out as many small fish as possible so that the limited food supply would go to the big ones and make them even bigger.

Nevertheless, some big brown trout came out of that river. There was a monster of eight pounds, speared with a hay fork during a drought when the river was in danger of drying up completely. Another monster, fourteen pounds, was left stranded in a meadow when the flash flood after a cloudburst finally subsided.

I watched trout for hours, watching their feeding habits and the peck order established by the bigger fish, who lay at the head of the pools in quiet water just out of the main stream, saving their energy but with a good view of all food coming downstream. So perhaps it was not surprising that, barefoot and with the crudest tackle, I caught fish when expensively equipped visitors did not. They were often visibly put out by this contrast. Some of them never seemed to realize that sitting on the bank changing flies was a poor substitute for fishing in the right places at the right time and that presentation of the fly and personal stealth were the keys to success. Lord Grey of Fallodon was absolutely right when he defined a day of fly fishing as twelve hours of creeping and crawling.[3]

I can evaluate my mother so much more clearly now than then. She was an arch snob, though she would have denied it hotly if she had ever been accused of snobbery. But it was plain to me, even as a child, and I rebelled against it. She was powerless to prevent me from mixing with my own age group in the village, for there were no other boys of what she regarded as her class. Their dialect became my language, my accent indistinguishable from theirs, and my summer-time appearance differed only because I preferred to go barefoot and they wore clogs. In winter, so did I. But my accent vanished on the loathsome journey back to boarding school, which I hated except for games.

[3] Sir Edward Grey, *Fly Fishing* (London: Dent, 1899).

For a long time, we lived in a rented cottage in the village. There were only two families of landed gentry, who owned all the farms except for small patches of land belonging to the Church. My mother regarded these two families as her class equals because she had been brought up in a land-owning family in Scotland. I was never clear how this had come about because her father was a Church of England parson and she was born in Wakefield while he was vicar of one of the parishes. Somehow, he owned a property in Kincardineshire called Drumtochty, with a large country house, grouse moor, trout stream, and a large staff of gamekeepers and grooms to look after the stables. I gather it is now a school.[4]

Her attitudes were moulded rigidly in the Victorian mode, but her early country-house environment accounted for her excellence at billiards (not snooker: that was very low grade) and with a rifle. For some reason, shotguns were not for the girls in Scotland. I would have thought cigarette smoking would have been frowned on even more, but her story was that, at the age of eighteen, she had had hay fever very badly. Nothing was known in those days (about 1890) about allergies, and in desperation her doctor suggested trying cigarettes. They had no effect on the hay fever, but she remained a heavy smoker until she died at the age of 87.

Her knowledge of manners and etiquette was profound, and I was in constant trouble for transgressing the rules of eating. I learned from experience that the logical way of dealing with food was always the wrong way according to the book. How to decapitate a boiled egg is a good example. Basically, I am left-handed, but left-handedness at the table was frowned on. So I had to battle right-handedly. The proper way is to take the egg spoon in the right hand, gently tap round the appropriate part of the blunt end of the egg, then insert the tip of the spoon and lever off the lid. The logical way for me would have been to take a sharp knife in my left hand and, with one shrewd swipe, cleanly take off the top. I still cringe if I see this done or, worse still, the top peeled off with fingers. Salt had to be applied in a neat heap on the right-hand side of the plate. Pepper was taboo. What she would have thought of the American habit of cutting the egg in half and scooping or pouring the contents either on to toast or into a glass, I shudder to think.

As for turning the fork concave side up so that peas stay on in reasonable numbers, that was logical but totally forbidden. Butter knives, sugar spoons

[4] During the Second World War, Drumtochty Castle was bought by the Norwegian government-in-exile as a boarding school for Norwegian children who fled the Nazi invasion of their homeland. Scott Begbie, 'How Drumtochty Castle became a safe haven and school for Norwegian children fleeing the Nazis', *Press and Journal*, 30 April 2021.

In 1947 Elizabeth and Robert Langlands bought Drumtochty Castle and started a preparatory school for boys. 'Neither Robert nor Elizabeth was businesslike, and Drumtochty was engulfed by financial troubles and had to close in 1971.' Obituary: 'Elizabeth Langlands', *The Scotsman*, 26 November 2002.

and such things were essential for a gentleman, even when eating alone. No short cuts were allowed. If sugar was added to tea, only one gentle circuit of the teaspoon was allowed. The result, of course, was that most of the sugar stayed undissolved at the bottom of the cup. Very illogical, but it was strictly 'correct' to do it this way.

The fact that we only rented a cottage went against the grain, and eventually she became a landowner by buying an old, disused cotton mill, a row of empty double-storey cottages that had been used by workers before the mill was closed in about 1850, and a few other cottages around the village green. A disused dam and a field of about ten acres was also included in the deal. I think it crippled her financially, but she was now landed gentry again. I was about eleven when this happened, and I wept bitterly. All the antique silver, which even at that age I enjoyed, had gone into the deal, and I foresaw penury for us both.

However, she proceeded to turn the old mill into a peculiarly personal kind of house. It had stood empty for decades, a giant rectangular, five-storey, windowless, sandstone shell, with a low-pitched roof made of great slabs of micaceous flagstone that measured four feet by two feet at the eaves and were nearly two inches think. These immensely heavy roofs were the normal finish for houses and barns in that part of Yorkshire, and they were supported by equally massive timber trusses made of imported Baltic timber. Each slab had two neat holes picked through it and was held in place by pegs made from sheep's ribs. The walls of this prison-like building were more than three feet thick.

When we took it over, the waterwheel and some of the metal shafting were still in position. The wheel was a gigantic overshot affair, twenty-eight feet in diameter and six feet wide, and the tail race a dank, underground, stone-lined cavern into which I never dared to venture. It must have originally led into the river about 500 yards away, but the exit had long since caved. The bearings of the cast-iron axle of the wheel were massive sandstone blocks, presumably lubricated in some way, possibly with tallow, and the friction and wear must have been considerable.

I think the local paper had always regarded my mother as probably slightly mad, and this was surely confirmed when it became known that she was going to live in the old mill: it was well known that it was haunted, and not a soul in the village had ever, or would ever, set foot inside it. I had explored it from top to bottom, but none of my small friends could ever be persuaded to come with me. Admittedly, much of the old flooring was rotten and there was always the chance of falling through, but this merely added to the spice of the exploration.

"The Old Mill, Arncliffe"

This was the shell of what my indomitable parent was determined to make into a house, which, of course, she did, by taking off the roof, knocking off three storeys, replacing the roof and partitioning one storey into a 'flat'. We never had any ghostly visitors, but the roof did make odd creakings, especially when it had a thick covering of snow in winter.

In spite of becoming a landowner, my mother's relationship with the others was peculiar, to say the least, and was conditioned by her upbringing. One was an ex-captain from World War I who retained his rank and had made his money in textiles. He was regarded as *nouveau riche* and was treated with condescension. The other family consisted of three spinster sisters, the Misses Hammonds, who had inherited their land; this was respectable, however it had been acquired in the first place. This was especially so as one of them was 'the old woman in the red petticoat' of Charles Kingsley's *Water Babies*, which he wrote while staying in Malham Tarn, only five miles from Arncliffe.

My mother was much influenced by psychology, having read Freud, Jung and Adler avidly and whose teaching she obviously tried to apply in bringing up her only young single-handed. Her passions in life were breeding Alsatians, closely followed by music, particularly the period of music dominated by the three composers born in 1685: Bach, Handel and Scarlatti. This rubbed off on me, but I grew to loathe dogs, especially Alsatians. I have retained both feelings throughout my life.

She was a product of mid-Victorian rigidly stratified society and firmly adhered to its principles and its tidy way of compartmentalizing people. She could not claim to belong to the very top bracket, the aristocracy, since there were no titles in the family, but she rubbed shoulders easily with them and, for instance, knew the Dalhousie family quite well. She was staunchly 'upper class', in which there were idle rich and gentlemen who earned respectable livings in the Church, the army and navy, and as schoolmasters in the Rugby mould, and it was one of these she married (*not* 'teacher' or 'school teacher', please; this was reserved for the government-school staff). Oddly enough, medical men were marginal, depending on their family background.

Alice Rosamund Bond (known as Puff)

Anything Scottish, which she always called Scotch, was enthusiastically accepted, especially bagpipes, tribal dress and customs, grouse shooting and deer stalking. I rebelled very early against tribal dress for small boys and have been heavily biased against anything 'Scotch' ever since, including their accent. Why should this be acceptable when cockney, northern and West Country accents are frowned on?

Such was the mother who shaped my prejudices, some positively, some negatively; of my father there is little to say. His influence was practically restricted to trout-fishing and to his initial genetic contribution. Environmentally, he contributed remarkably little.

His father was also a Church of England parson and was at one time Archdeacon of Lincoln. Consequently, we had 'Lincoln Imps'[5] on brass toasting-forks and door-knockers wherever we went. Having two ordained grandfathers may well have influenced my subsequent religious beliefs, but probably not in the way they would have wished. But he was no match for my mother, and they parted when I was very young, after which he moved

[5] 'A grotesque carving having weird and prominent ears and nursing the right leg crossed over the left, in the Angel Choir of Lincoln Cathedral. He is said to have been turned to stone by the angels for misbehaving in the Angel Choir.' Ivor H. Evans, *Brewer's Dictionary of Phrase and Fable* (London: Cassell, 14th edn., 1989). Now 'the mischievous, sinister but somehow likeable mascot of the city'. Cory Santos, 'Tracking the mysterious origins of the Lincoln Imp', *The Lincolnite*, 19 April 2013.

from one school appointment to another in an aimless drift, gradually taking refuge in more and more whisky until he was a eventually an unemployable alcoholic.

Eventually he was taken care of by his sister (wife of the Vicar of Louth in Lincolnshire) where he died suddenly from a coronary thrombosis at the age of fifty-six. All I inherited on his death was a silver half-hunter,[6] which now lies on my desk and keeps remarkably good time in spite of being nearly 100 years old, and a pair of leather gauntlets, which have long since vanished.

Such was my genetic and early environmental make-up, and it is hard to say which had greater influence. Certainly, the wide-open spaces of the Yorkshire Dales left a permanent inclination to spend as much time as possible in similarly wild regions. As it was largely limestone country, I became aware of fossils at a very early age, and this was strengthened by a family, father and two sons, a few years older than I, who spent all their holidays in the neighbourhood and were regarded by the locals as highly peculiar.[7] One was a budding botanist and consequently they ate a variety of fungi, which the locals regarded with utmost suspicion. But they survived. The other was a budding geologist who went around with his little hammer and a map. They often slept out on the moors in the depths of winter in thick snow and freezing temperatures.

What happened to the botanist I am not sure, but I think he became a don at either Oxford or Cambridge. The geologist became famous for his exploration of the ice cap in Greenland and for his geological work on the coast of that frozen land. He was a member of the 1933 expedition to Mount Everest and climbed without oxygen to more than 28,000 feet. With his companion, Wyn-Harris, they found an ice-axe that had belonged to either Mallory or Irvine on an earlier expedition, who may have successfully climbed the mountain, but vanished on the descent.[8] He was later appointed Professor of Geology at Oxford and a Fellow of the Royal Society. His speciality was in ultra-basic rocks of the kind that make up the Great Dyke of Zimbabwe and he spent two weeks with me in the late 1950s, seeing our Dyke first hand. His early influence certainly nudged me in the direction that I eventually took, and we remained in at least intermittent contact until his death.

[6] A pocket watch that has a lid with a small round window so that the time can be seen without opening the lid.

[7] These were the Wagers. The father was the headmaster of Hebden Bridge Grammar School.

[8] This was Laurence Wager: W.A. Deer, 'Laurence Rickard Wager, 1904–1965'. *Biographical Memoirs of Fellows of the Royal Society* (1967), 13: 358–85.

A strong push towards geology took place on one of my long train journeys to boarding school on the Cumberland coast, when I was about eleven or twelve and was still at prep school at Seascale. The rigid confinement of boarding school irked me more than a little after the freedom of my home in the Dales, where mealtimes were flexible and bedtime indefinite: meals on demand when I was hungry and bedtime was when I was tired. Arncliffe was (and still is) sixteen miles from a railway station, and in the earlier days this journey was done by pony and trap and took all day; later it was by primitive motor car, but that made no difference. It was sixteen miles of gloom followed by four hours of misery in the train, a prelude to thirteen weeks of what, to me, was a prison sentence that I was to serve intermittently for twelve years.

> There is, on the whole, nothing on earth intended for innocent people so horrible as a school. To begin with, it is a prison. But it is in some respects more cruel than a prison. In a prison, for instance, you are not forced to read books written by the warders and the governor.

A quotation from Bernard Shaw's 'Treatise on Parents and Children', with which I agree; but society has yet to throw up a workable alternative.[9]

On this particular journey I was bundled into a compartment with a small elderly man with a beard, who must have been aware fairly quickly of the halo of gloom surrounding me. He began to talk about the country we were going through, about the shapes of the hills, what the rocks were and how they were formed, and the fossils they contained. He kept me fascinated for the full four hours, no mean effort on his part to hold a small boy's attention for so long. He asked my name and address, and in due course a book arrived through the post. It was called *A Popular Geology* and the author was William Platt, the small bearded man who had so beguiled me on the journey.[10]

This was a significant landmark, and that book was my first textbook on geology. It was another ten years before the next major influence happened, and though it is chronologically out of order, it should be brought in here. By that time I was teaching in a prep school in Hessle, near Hull, owned by one Arthur England, who at least fed and housed me in term time and paid me the princely sum of £90 per annum to teach a variety of subjects for which I was totally unqualified.[11]

My only other resident colleague was a man a few years older, Fred Bagley, who had a degree in geology from Oxford and, being a graduate, was

[9] Bernard Shaw, *Misalliance: With a Treatise on Parents & Children* (London: Constable, 1922).

[10] William Platt, *A Popular Geology* (London, The Sheldon Press, 1924).

[11] This was Woodleigh School. See *Memories of Woodleigh School from 1928 to 1987*, a collection of reminiscences of Woodleigh School compiled by Joanna Passmore and Mary Robertson, daughters of the late Arthur R. England and his late wife Barbara.

paid about three times my salary. He lent me his copy of Lake and Rastall's *Textbook of Geology* and this really fired me.[12] Together, we visited all the local quarries and collected fossils.

Bagley plus book and fieldwork awoke the dormant interest first instilled by William Platt and his book, and from then on geology became an absorbing hobby, ranking at least equal with cricket and trout-fishing. Equally important, if I could only get properly qualified, and that meant a university degree, I might even earn my living by turning my hobby into a career. It became something worth aiming for and towards which to work, and it turned a completely rudderless hand-to-mouth existence into an ambition. But a great deal had to be done, and a few lucky breaks had to happen before the yawning chasm between ambition and achievement could be bridged.

The years of school dragged on, first prep school at Seascale and then public school at St Bees a few miles north up the Cumberland coast. Both schools were in the strict English boarding-school tradition. This entailed various essential elements. Boys must be kept busy, from waking up in the morning to going to sleep at night. They must never have enough bedding in winter to keep warm, there must be no facilities that will adequately dry wet clothes,

[12] *Lake and Rastall's Textbook of Geology*, revised by R.H. Rastall (London: E. Arnold, 5th edn., 1941).

and they must be perpetually hungry. And, perhaps most important of all, their existence must be one of the strictest monastic seclusion, the only feminine presence being that of the matron and any mistresses employed to teach. Naturally, these must be unattractive and at least middle-aged. Exercise to the point of near exhaustion must be taken daily. Young *Homo sapiens sapiens* is a tough animal; under those conditions he had to be able to survive, and it was remarkable how few succumbed.

In the course of a lifetime, and particularly in its early years, there are just a few people who, with the wisdom of hindsight, can be singled out as having had a lasting influence and who have significantly shaped one's attitudes and performance. There was one such person at Seascale. He was a young Englishman, obviously of Scandinavian ancestry since his name was Phil Nansen. He had all the qualities of a good, instinctive teacher, and though he was a graduate, he had no teaching qualification in the modern sense. He was universally liked and looked up to, and this was not only because of his very obvious talents in sport. We copied him in so many ways - how he brushed his hair, how he kicked a ball, how he batted and bowled, and even how he wrote. And we learnt from him even in unpopular subjects like Maths and Latin because he held one's attention and made us laugh while we worked.

His innate good manners must also have rubbed off by example. I doubt if he realized what an influence he was having on plastic young minds. If he had known, he would probably have lost much of his effectiveness, as he would have become self-conscious and self-questioning. As it was, he was brilliant. I wonder what became of him, and how long he endured the grind of teaching small boys and being paid peanuts for his efforts.

By one of those strange quirks of fortune, there is another man in Zimbabwe who remembers him. In about 1970 I was host to a party in Fort Victoria held by the National Museums and Monuments Commission after a Board of Trustees meeting. The last guest to arrive was Colonel George Hartley, later Speaker of the Legislative Assembly in Zimbabwe Rhodesia.[13] As I met him at the door, he said he had been wanting to ask me a personal question for years. This was a bit disconcerting and I wondered what he was going to ask. His question was, simply, had I been at Seascale prep school between 1926 and 1930! We had indeed gone through that school on the west coast of Cumberland together and had then diverged to different schools; but our paths had crossed once more, forty years later, six thousand miles to the south-south-east. He had good archival material to prove it, including photo-

[13] Lieutenant-Colonel George Holland Hartley of the Rhodesian African Rifles was appointed OBE in the Queen's New Year's Honours list on 1 January 1955. *The London Gazette*, 31 December 1954, supplement 40366, 9.

graphs and an autograph album in which I had drawn a cartoon of him behind the stumps as the school wicket-keeper, the year I was captain.

St Bees, my so-called public school, a few miles north of Seascale on the Cumberland coast, was small and old, one of the earlier grammar schools.[14] It was founded in 1583 by Archbishop Grindal and, to its inmates, most of the buildings might well have been dated from its foundation. Discomfort and inconvenience were their keynotes, and the motto over the main entrance to the quad, *Ingredere ut proficias* ('Enter to become proficient' is a loose translation), seemed to me totally inappropriate; 'Enter and be damned' would have been more apt. I hated it as much as I had hated Seascale but with a greater sense of resignation. After all, by then I had only four or five years left of my sentence to serve, with possible remission for a year if I got through O and A as quickly as possible.

Once again, however, there was one man who made his mark on me, but he was so different from Phil Nansen that I am surprised that he could have had any impact at all. He was known as 'Kick John' Boulter because he had a stiff knee from an unsuccessful cartilage operation. He was middle-aged, forbidding in appearance, sarcastic in his dealings with boys, and free with the cane – a combination of which, in theory, ought to have damned him, or at least made him an object of furtive ridicule throughout the school, but to me, at least, he had 'something'. His two absorbing interests were religion and Beethoven, the first of no interest to me even then, but his gramophone recitals for select groups of senior boys stirred up dormant leanings towards classical music, and Beethoven's *Eroica* symphony in particular made a deep impression.[15]

When my parents lived in Brighton, my mother owned a Steinway concert grand piano, and after I arrived I must have spent hours on the floor while

[14] He joined the school on 17 September 1926 on a Junior Scholarship and left in July 1930. Edmund Grindal was born in the village and went on to become the Archbishop of Canterbury: 'The History', *St Bees School*, <https://www.stbeesschool.co.uk/about-us/the-history>. The 'old boys' have their own society, the St Beghian Society, and have published its history: *The Story of St. Bees: 1583–1939* (London: Buck & Wootton, 1939).

[15] 'John Boulter was a long-serving and influential master at the school. He started in 1913 and retired in 1945 as Headmaster. He was asked to become Headmaster at short notice in 1939 and, with a much depleted staff, saw it through the War years. He was Second Master (i.e. Deputy Head) for a long time in the 1930s. He taught Classics, Geography and most games. Despite his knee injury he was a talented and all-round sportsman and apparently an excellent coach. He seems to have been immersed in most aspects of school and village life, and a former colleague wrote thus of him: "Despite his somewhat abrupt manner, he had a warm heart and was held in high regard by all who worked with him."'.

my mother played Bach preludes and fugues and exercises by Handel, enshrined in an ancient leather-bound book (which I still have) consisting mainly of themes and vast numbers of variations. This exposure resulted in a form of brainwashing that has lasted ever since. The St Bees school choir produced Handel's *Messiah* when I was sixteen. I sang in the choir as a bass, and still have my copy of the Ebenezer Prout edition, with 'G. Bond II' on the flyleaf. There happened to be another Bond in the school who was my senior and who was, therefore, Bond major while I was Bond minor. He later played full back for England.[16]

While there were people at school who had a profound effect on me in the positive sense, there were other influences that were entirely negative. One was literature for O level and the other was compulsory religion.[17] For O level in English Literature, I had *Macbeth* as a set book during term time and *Pride and Prejudice* as a holiday task. I took particular exception to being expected to do school work in my holidays. It was a totally unwarranted intrusion of my assumed prerogative of having no connection with school once the train left St Bees station until I reluctantly stepped off it again on the first day of the next spell of incarceration. My pride and prejudice precluded me from doing anything with Jane Austen's classic beyond hiding it in a drawer until I got back on the train at Skipton station and made the dreary four-hour journey back to prison.

During the journey I skimmed through some of it and remained unimpressed. Three hours of that journey was spent locked in the train lavatory with Jane Austen, who at least helped to pass the time while I sat on the seat with *Pride and Prejudice* on my knees: my colleagues had discovered a cunning way of fastening the door from the outside, and they opened it only at St Bees station. No examination was set on it and, as far as I can recall, it was never mentioned in class, which merely reinforced my feeling that the whole exercise was a complete farce. Not so with *Macbeth*. We were expected to know it virtually by heart. There must be legions of boys, and girls, who have suffered under this sort of introduction to Shakespeare, and who probably agree with Ivor Brown:

> Like most schoolboys I had been sickened of Shakespeare by education. ... All too well did I know and was able to repeat on paper what the Rialto was: I could define an argosy to any teacher's delight and could be profound about Arden, Ducdame,

[16] 'Bond major' was William Ronald Bond from Windermere, who was at the school from 1923 to 1928. He subsequently played in two rugby union 'trial' games for England, once in 1942 and again in 1947.

[17] Although he writes about 'O levels' and 'A levels' here, they were not brought in until 1951, and the equivalent examinations then were the School Certificate and the Higher School Certificate.

> and the Significance of Jaques. ... By the age of eighteen I was allergic to Shakespeare almost beyond hope of therapy.[18]

Substitute a blasted heath for the Rialto, and other characters from *Macbeth* for those quoted by Ivor Brown, and the situation exactly fits. I have remained beyond hope of therapy ever since. It must have been my kind of boy in my kind of school who prompted Irwin Edman to remark that 'education is the process of casting false pearls before real swine'.

Luckily, my O level results allowed a choice between the Arts and the Sciences, though I do not remember being consulted. The choice was made for me and I entered the Lower Sixth to do Maths, Physics and Chemistry. There was no Biology option, but I had to do French as a subsidiary, very subsidiary, subject. Science teaching was formal and rather crude, but our labs were well equipped, and school took on a slightly less distasteful aspect. We were treated much more like grown-ups. Maths was different because I had been badly frightened by it preparing for O level and had a distinct maths phobia which has lasted ever since. Frank Rhodes taught Physics much better than P.G. (Piggy) Gow taught Chemistry. Both had sufficient individual peculiarities to be sources of amusement, but their teaching stuck.

Wiggie Evans suffered from alopecia, and was bald to the point of no beard, eyebrows or eyelashes, never mind hair on the top of his head. He wore a ginger wig, which was very obvious and slipped about on his shiny cranium in an uncontrollable manner. This amused us and he seemed not to care how he looked. For his age, which was considerable but indefinite, a ginger wig without a single grey hair was singularly incongruous. His teaching reinforced my maths phobia but got me through A level twice.[19]

And so the educational die was cast, that I should leave the Arts for the Sciences, that first major dichotomy which halved the options at, in my case, the tender age of fifteen. Before this point, the future is like a filter funnel, an open conical end in which life and school begin, narrowing progressively until the long, thin stem confines the victim relentlessly within it for the rest of his life. This is known as your 'career'. Only the accident of breakage permits escape with all the hazards of being cast adrift into a hostile world, outside the confines of a defined pathway.

[18] Ivor Brown, *Shakespeare* (London: Collins, 1949), 197.

[19] In December 1927 (when he would have been 15) he sat the School Certificate exam and passed with seven 'Credits'. He then stayed on and took the Higher School Certificate in July 1930 (when he was 18) and also passed that. Information in the school magazines shows that he was regularly achieving 'Distinctions' in school 'internal' exams.

An allergy to Shakespeare and other classical literature was paralleled by an allergy to orthodox religion induced by the imposition of compulsory religion of a particular brand. Since the school had been founded by an Archbishop of the Established Church, there was no choice. Memory does not recall what happened to any boy who was, by upbringing, a Jew or Roman Catholic. There must have been some, as there seemed to be no restriction. They certainly did not get the dispensation given to them on the church parades in the RAF. On such occasions they had to parade and march to whatever makeshift 'church' was in use on that particular station. It was generally an empty hangar, stiflingly hot in summer and arctic in winter. On arrival they got the command, 'Fall out the Roman Catholics and Jews', whereupon they could get into the shade or the sun, according to season, and loaf and smoke until the service was over. The more intrepid souls left and sneaked off to their billets.

Religion at St Bees was well organized but highly inconvenient. Breakfast was squeezed in between pre-breakfast prep and first period of school but had to be ingested at speed because chapel also had to be crammed in before first period. The school chapel was impressive and one of the older parts of the school. There was a real pipe organ, which had to be blown by hand, and needed two boys on the bellows for each service. This was an unpopular fatigue because the bellows were operated from behind the organ. There was no need to blow when the organ was not being played, but the pressure had

The chapel, St Bees School

Photos courtesy of the St Beghian Society

to be put on in time for the organ to be fully operational as soon as it was needed. This required a good knowledge of the order of service, for an organ with no wind is as bad as a pub with no beer, or a hydroelectric plant with no water. To embarrass the organist was to invite rapid and painful retribution. On the other hand, to blow when the organist was not playing was a waste of effort and tended to produce creaking and bumping noises, which lowered the tone of the service. Altogether it was an unenviable labour, to be avoided at all costs. I never blew for the organ. Singing in the choir was the surest ticket for exemption and I happened to enjoy singing. The other chore that I dreaded was reading the lesson. Again, being in the choir was the best way of avoiding this nerve-racking performance. I never read a lesson.

However, two chapel services daily except Saturday, when there was only one morning service, and two full services complete with sermons every Sunday, combined with periods of 'Divinity' each week of term, induced a lasting allergy to orthodox religion. It had started long before public-school days because of my clerical grandparents and the consequent upbringing of my parents along lines of unquestioning orthodoxy.

My science teaching, particularly in Sixth Form, had instilled in me the need to subject all evidence to the acid test of doubt. Religion demands unquestioning faith, which I found distinctly unpalatable and illogical. I was perfectly prepared to accept the Bible as history, albeit that it was compiled from dubious and defective written evidence, combined with a good deal of oral tradition; that it had been translated from originals in Hebrew and Greek and in consequence might be expected to contain obscurities and even exaggerations, as well as mistranslations.

One of my clerical great uncles, Henry Bramley, had been a biblical scholar and wrecked one of my early mental pictures by assuring me that Joseph's coat of many colours was a mistranslation. It should have been Joseph's coat with long sleeves, which would have destroyed not only my childish mental picture, but the whole foundation for a recently highly successful 'musical'.[20]

The boredom of chapel, and the feeling that I was being forced to accept a particular version of orthodox religion before I was able to judge its merits for myself, made me distinctly 'anti'. Why should any child blindly accept a faith he has never had the opportunity to compare with any other version of that faith, or indeed with any other major or even minor religions of the world? Comparative religious studies were not on offer, I suspect quite deliberately. Neither was scepticism encouraged or even countenanced. I rebelled, and have remained rebellious, but I have never been an atheist. The marvellous order

[20] The first full production of *Joseph and the Amazing Technicolour Dreamcoat*, with lyrics by Tim Rice and music by Andrew Lloyd Webber, was recorded in 1974.

of the universe almost compels belief in an overall design, which implies a designer. But no orthodox religion of which I am aware is big enough to comprehend the things that I need in it as a geologist. Genesis Chapter I is marvellously correct in its relative timescale. As an explanation of the origin of things for a primitive peasant community, it is ideal. There is no need to fiddle around to try and fit it to modern sophisticated society, or the recently determined geochronology based on radioactive decay series.

Agnostic I may be, but agnosticism is a perfectly respectable state of mind: the state of being unconvinced by the evidence, but still open to conviction, and unable to accept dogma or doctrine by an act of faith. I am still unconvinced, and I am still open to conviction. Much biblical history has been substantiated by archaeology, but I am still dubious about the theistic concepts and their interminable interpretations. As literature in its Authorized and Revised versions, the Bible is supreme. Modern translations may be more accurate, but are either wishy-washy attempts to be popular or just plain dull.

On the other hand, much religiously inspired music is inherently beautiful and needs no 'faith' for pure enjoyment. Bach may have, of necessity, been a musical sausage machine, but the products are enduring and, to me, enchanting.

H P D Sykes, M O W Ainsworth, G Bond, R D Carruthers, G D Addey-Jibb, M M Evans, S G Coade

J D Younghusband, G A Walker (V Capt.), G W C Meikle (Capt.), T A Hanson, N Scorah

St Bees Ist XI cricket team, 1929/30

F A Rhodes, J S Boulter, D Bond, H Howes, N R S Cameron, H O Roberts, R C Barkworth

E A Bell, R B Brown, A B Cowburn

E K Quick, S A Weare

St Bees Masters cricket team, 1926/27

Photos courtesy of the St Beghian Society

– 2 –

Dropout, Dole and Depression

The term to 'drop out' had not been invented in 1930, but in retrospect it is apt to describe my years between school and my first real job with Rayleigh G. Strutt at his private prep school in Putney.

I had always hated school, though in the later years at St Bees I came to regard it as some kind of necessary evil but one to be terminated at the earliest possible moment. Because I had taken O levels at a rather precocious age, it was possible to take A levels twice before I was much over eighteen, and the authorities duly inflicted this on me. Perhaps this was just as well, as Britain and the rest of the world were in the grip of the greatest economic depression that had ever happened. But by the end of the summer term in 1930, with A levels safely behind me, and the end of the school year likewise, I was horrified to learn that I was expected to go back to St Bees for yet another year of grinding away at Physics, Chemistry and Maths. The excuse given was that I should sit some kind of scholarship exam in the following June, which would lead to a science degree, probably in Chemistry, at Cambridge, my father's old university.

I had reached a major crossroads, though it was not really obvious at the time. The thought of spending the rest of my life in some smelly chemical laboratory did not appeal. On reflection, too, the thought came to mind that really all the school was interested in was for me to play one more season in the 1st XI cricket team, in which I had been quite useful as an opening bat. I rebelled, and bluntly informed my long-suffering mother that I refused to go back to school. Secretly, I had the feeling that, as far as cricket was concerned, I could do much better by joining Skipton Cricket Club and trying to get into their Yorkshire League side.

The young need to be mobile, but mobility was no problem: I already had an ancient but quite reliable motorcycle. Similarly, living was no problem: there was plenty of room at our house in Arncliffe. Money was a problem, but with the blind faith and optimism of youth, I was pretty certain that I could find some kind of job, notwithstanding the length of the unemployment queues. So I left school.

Job-hunting did not really begin in earnest for some time. Life was comfortable at home, a succession of motorcycles ensured mobility, though looking back, it was an uncomfortable form of transport in that climate. It seemed to rain for most of the so-called summer, and snow was common in the Dales in winter.

I quickly graduated to a really fast 350cc overhead camshaft Velocette. This was tuned and re-tuned in my own makeshift workshop until it was fit for competitions. Velocettes were winning just about everything at the time, and I did moderately well at local meetings on grass tracks, freak hill climbs, scrambles and reliability trials. However, if it was a handicap event, the mere fact of being astride a 'Velo' was enough to make one the scratch rider. But it was fun.

The works in Birmingham were liberal with technical advice and kept the price of spares very low to 'works-assisted riders'. Fuel, chains, tyres, oil and plugs were supplied free by the makers, and the sport was therefore relatively cheap. Prizes were equally cheap, but they were more usually in the form of small cups and other trophies than as cash. We all converged on whatever meeting was on, riding our machines. Lamps, silencers and other extras were removed in the pits, the races took place, and all the bits reassembled at the end so that the machines could be ridden home.

It says much for the reliability of these early motorcycles that not once did I have any kind of mechanical failure that made my mount unrideable. Lamps were, however, a problem. I could not afford electric lights and had to rely on a carbide generator amidships, with rubber tubes fore and aft to rather inefficient lamps. In winter, the water tended to freeze in the generator and the whole antiquated system was thoroughly unreliable. I met carbide lamps again when I began to explore small gold mines in Matabeleland.

Cricket worked out quite well with the Skipton league side, but work proved to be very difficult to find. The Depression persisted, and all I could find was a succession of temporary jobs on the local farms; extra hands were always needed in hay time and lambing time. The hours were long and the pay was poor, but meals were provided. As a small boy I had always done both in school holidays just for fun, and knew all the complicated techniques of hay-making - including using a scythe on slopes too steep to mow by horse-drawn machine. Tractors were unknown in the Dales; it was all horse traction.

Lambing time was fascinating; the birth of mammals has always appealed to me as an awesome and wonderful process. Lambing time in the Dales is in late March and most of April, when the weather can be quite arctic and snow is not uncommon. Sheep are brought down into the valleys for shelter, and an all-night vigil is kept on the ewes. As lambs are born, so mother and offspring are put into the shelter of barns, at least until the next morning. The night shift was always given to the hired extras for obvious reasons. One plodded round the field with a storm lantern, finding new lambs, putting them in shelter, and assisting with difficult births. My hands were small compared with the massive paws of the average farm labourer and had always been in demand for the really difficult cases. With some Vaseline, I could get my hand and arm right into the uterus, turn the lamb around, find the front legs, and gently assist in its delivery. It was surprising how often this was successful, and how often both lamb and ewe survived our crude and unhygienic obstetric methods.

Even so, motherless lambs had to be fed. Traditionally two methods of feeding were used. If a ewe had lost a lamb, it was skinned and the skin put on the motherless lamb. This method of 'mothering' worked well.[1] If it failed, or if for some other reason bottle-feeding had to be resorted to, it was done in the farmhouse kitchen by the womenfolk, using cows' milk. Again, there was a traditional way of doing it. The milk was not put in a pan and warmed on the fire – the poker was put in the fire until it was red hot. It was then quenched in the milk. No one could explain why it was done that way – I suspect that it successfully pasteurized the milk without boiling it and prevented fatal infections – but it was always done this way.

One singularly unpleasant job that was wished on to me by my dog-loving mother was as a very junior kennel boy at a well-known boarding kennel just outside Leeds. It consisted mostly of cleaning out smelly kennels, feeding the occupants on smelly horse meat from the knackers' yards, and interminably exercising dogs on leads, making sure that if there was more than one at a time (to save mileage) they were reasonably compatible. I stuck it for about six months before giving notice. This experience, added to my mother's kennel of Alsatians, cured me of ever wanting a dog of my own.

Hay time, lambing time, odd jobs mending primitive radios, shovelling snow for the local council in winter, and any other job I could find made up the time when I was not trout-fishing, playing cricket, or fixing and tuning my precious Velocette. Snow-cutting was important in winter, as Arncliffe was frequently cut off by deep drifts on the road, which was winding and confined between high, dry-stone walls. As a left-handed shoveller, my pay was a halfpenny an hour more than for right-handers. Since only about eight

[1] 'If the ewe has lost her own lamb(s), skinning the dead lamb to create a 'jumper' for the foster lamb is a very effective technique, providing the tail is left on and the skin covers the anus of the foster lamb, and is reasonably well fitting.' FAI, *Fostering and Orphan Lambs*, <https://www.faifarms.com/wp-content/uploads/2016/03/Fostering-Orphans.pdf>.

per cent of Europeans are left-handed, it had a scarcity value. A left-handed and right-handed shoveller can be put to work in the same working end. It is the only situation known to me when left-handedness pays in hard cash.

When I could find no kind of work I tried to get on the dole. This was difficult unless one had a card with the required number of employment stamps on it, but I managed it twice. These lean years left a mark on me for life. They gradually generated in me a deep sense of insecurity, which I still have. I don't think I am mean by nature, but some Scottish blood, a Yorkshire upbringing, and those years of unemployment have made me rather careful. However, when eventually I became a poverty-stricken university student, this attitude of mind at least allowed me to be poor without a sense of novelty.

The next major impact, and it was a decisive one, came after this rather lengthy spell, from the ages of eighteen to twenty-two, as what would now be described as a dropout. Necessity, in the form of a mother whose patience became exhausted on my twenty-first birthday, demanded that I should make a supreme effort to find a job and become more or less self-supporting. Since I had absolutely no qualifications, and England was still in the Depression, this posed a considerable problem.

It was suggested that a private preparatory school might be a possibility, since they were always poor and almost preferred unqualified young men: they could pay them the merest pittance, house and feed them after a fashion during school terms, and disclaim all responsibility for them during the holidays. In theory, slavery had not existed in Britain for centuries and child labour had been abolished, but my own experience, first as a prep-school boy and later as a master, suggested that both still existed under the thinnest of disguises.

My mother remembered the name and address of a firm of schoolmaster brokers, Messrs Gabbitas and Thring, and I wrote to them to see if they still existed.[2] Presumably, my father had been on their books for a long time and

[2] 'Established nearly 150 years ago in 1873, Gabbitas, originally called Gabbitas and Thring, began life as a pioneering agency recruiting teachers for top schools both here in the UK and overseas. A few years later they added tutors to their range of activities. Amongst the more illustrious tutor alumni there are several well-known authors and composers including Evelyn Waugh, H.G. Wells, C.S. Lewis and Edward Elgar plus many other outstanding figures from the arts and sciences. Over the years, Gabbitas has worked with crowned heads of Europe, celebrities from stage and screen together with families from all walks of life looking to find the best education solutions for their children, using the independent sector.' 'Gabbitas – supporting the education journey', *Gabbitas*, <https://gabbitas.com/supporting-independent-education-journey>.

had been provided with lists of vacancies for which to apply in his long slide down the scholastic ladder. They replied, and I was duly added to their books. Only one vacancy was passed on to me. It was in a small school in Putney, the headmaster of which was Rayleigh G. Strutt. I applied, giving at least equal weight to my sporting and academic records. An answer came promptly, inviting me to go down in July for an interview at his expense. I went; a trip to London would be quite an adventure in itself.

I arrived at the school, a converted three-storey suburban house with quite a large garden, to find cricket practice in full swing in the nets on the back lawn. The Head was in charge, and I was commanded to get my pads on and perform. I was then allowed to bowl and coach. The performance must have lasted two hours. It was plain that Strutt was a most accomplished, keen and knowledgeable cricketer with the ability to inspire small boys and get results.[3] At the end of nets, he said casually, 'I suppose you can start in September?' 'But, sir,' I said, 'we haven't talked about work yet.' 'Oh,' he replied, 'we can do that when you get here.' The job was mine, at a salary of £50 per annum, living in during term and presumably fending for myself in the holidays.

Rayleigh G. Strutt was a large man in early middle age. His background was a good public school, followed by four years in the army during World War I, which prevented him going straight to Oxford. By the end of the war, finance was at a low ebb and Oxford was beyond his reach, but by dint of odd jobs and scrimping and saving, and by part-time private study, he managed to get to the Royal College of Music, where he took his Bachelor of Music degree and Associateship of both the Royal College of Music and the Royal College of Organists. More scraping and saving got him to Oxford, and eventually a Master of Arts. He was a relation of Lord Rayleigh, the famous physicist, whose family name was Strutt, and descended from a Strutt who had been an artist in the days before photography, and who had worked as an illustrator for such magazines as the *Illustrated London News*.[4]

He had started this prep school some years before on almost no resources, but by the time he hired me, he had begun to do quite well. There were seventy to eighty boys from 'good' families, with only a handful of boarders; the rest were local day boys. The staff were an odd assortment, mostly young, mainly male, and all just as lacking in any formal qualifications as myself.

[3] He wrote a book, first published in the 1940s, on schoolboy cricket: Rayleigh G. Strutt, *Schoolboy Cricket: The Boys' and Masters' Guide* (London: Stanley Paul, 4th edn, 1964).

[4] Lord Rayleigh, in full John William Strutt, 3rd Baron Rayleigh of Terling Place (1842–1919). 'English physical scientist who made fundamental discoveries in the fields of acoustics and optics that are basic to the theory of wave propagation in fluids. He received the Nobel Prize for Physics in 1904 for his successful isolation of argon, an inert atmospheric gas.' Lindsay, R. Bruce. 'Lord Rayleigh', *Britannica*, <https://www.britannica.com/biography/John-William-Strutt-3rd-Baron-Rayleigh>.

They were a colourless lot on the whole. But Strutt made up for that by being almost larger than life. He was a big man in every way, and his early example gave me the inspiration and direction I had so far lacked. I was easily persuaded to sing in his choir in St Stephen's, Rochester Row, in Westminster, twice on Sundays and one evening practice during the week. His period of music was my period of music, the choir was really good and had boy trebles and altos, stiffened by a professional tenor and bass.

He had just missed a Blue at Oxford as a wicket-keeper batsman. He had also invented, and had constructed, a mechanical bowling machine, working on the lines of a swinging spring-loaded arm. It could be made to bowl almost any kind of delivery from a simple 'donkey drop' to a fast and venomous leg-break. It was known in the school as 'J.C. White' after the Somerset and England bowler, 'Jack' White, and its main merit was that, once set, it would bowl the same delivery time after time.

He was also an enthusiastic chess player and revived my interest in the game. We used a peg board to play while we walked to the station on the way to church on Sundays, in the Tube, and even in the vestry. We could not quite manage it during the sermon, as my choir stall was on the opposite side of the church from the organ.

After one year with Strutt, I changed jobs and went to Hessle, where I met Fred Bagley and was reinfected with geology. A year later I went back to Strutt, who suggested to me that Chelsea Polytechnic was the place to go for evening classes, and I spent three years there doing Geology.[5] In that time, I worked through the entire London degree syllabus for the fun of it, all for the remarkable fee of 30/- per annum. For this bargain price I had three lectures a week and three practical periods of two hours each under Dr W.F. Fleet, who ran the whole department single-handed for a day college and an evening college.[6] He lectured in all branches of geology and ran all the practicals. Ask a modern university lecturer to undertake even a fraction of such a load and he would have a purple fit. In addition, he ran field days at weekends.

[5] The site on the corner of Manresa Road and King's Road, Chelsea, had been earmarked for a college as early as 1890, was opened as South West Polytechnic in 1895, and became the Chelsea Polytechnic in 1922. In 1985 the college merged with nearby Queen Elizabeth College and soon thereafter the merged college was itself amalgamated into King's College London. 'Chelsea College of Science and Technology', *Wikipedia*, <https://en.wikipedia.org/wiki/Chelsea_College_of_Science_and_Technology>,

[6] An obituary of William Frederick Fleet (1890–1966) appears in *Proceedings of the Geologists Association* (1968), 79(2): 262–6.

Strutt plus Fleet was a compelling combination. At the end of this Chelsea episode, Fleet suggested to me that I should enter for a Royal Scholarship exam. My first question was, 'How much does it cost to enter?' The answer was that there was no entry fee but generally only one scholarship was awarded each year, tenable at the Imperial College of Science and Technology in South Kensington, and that it was open country-wide. The only restriction was that the candidate must not have done more than one year as a full-time undergraduate at Imperial College.

These Royal Scholarships stemmed from the profits of the Crystal Palace Exhibition of 1851 and were highly prized and quite adequately funded to allow the student four full years at Imperial College, leading to a Bachelor's degree.

I entered and, in the fullness of time, wrote the exam in Geology and forgot all about it. There seemed only a remote chance of success, but if I was successful, the bridge between ambition and achievement would have been built. Weeks later, in September 1937, the bridge materialized. I was awarded the coveted Royal Scholarship and could now look forward to four years of opportunity to turn my hobby into my profession.

This was the most measurable impact that has ever been made on me. I got my degree in Geology with First Class Honours from an institution that ranked very highly in the British hierarchy of such bodies of learning, working under the very cream of the country's geological brains. It was literally the chance of a lifetime at a time in my life when, without it, I would have been condemned to struggling mediocrity for ever more.

– 3 –

University

The Imperial College of Science and Technology is situated in South Kensington among the museums and art galleries.[1] This constellation of learning includes the Victoria and Albert Museum, the Natural History section of the British Museum,[2] the Geological Museum, the Science Museum and the old Imperial Institute. These are connected to South Kensington station, which is served by both the Inner Circle and Piccadilly lines of London's Underground system, by a long tunnel, for pedestrians only, which has exits to all these institutions.[3] Since they are all favourite places to which parents send their offspring on wet weekends, this tunnel can be packed with small boys and girls on such occasions.

In my first year at Imperial College I spent a lot of weekend time absorbing geology in the galleries of the Geological Museum, and I soon became more than ever convinced that Darwin was right. The noises made by young *Homo sapiens sapiens*[4] in a confined tunnel and in high spirits are remarkably like those heard at feeding time in the monkey house at the London Zoo, consisting mainly of non-verbal shrieks and howls. Not that even then did I believe that modern man was descended from any existing ape or monkey but from a common ancestor away back in geological time. Our line, branching off at the point of dichotomy, specialized in brain and became verbal; the other branch specialized in brawn and remained non-verbal.

[1] Hannah Gay, *The History of Imperial College London, 1907–2007: Higher Education and Research in Science, Technology and Medicine* (London: Imperial College Press, 2007).

[2] Now known as the Natural History Museum, it was officially known as British Museum (Natural History) until 1992, although it had been legally separated from the British Museum in 1963. The Geological Museum merged with the Natural History Museum in 1985.

[3] 'The loop was referred to as the Inner Circle in Parliamentary papers as early as 1864, but the Circle line name was not used officially until 1936. It was circular, or more accurately, elliptical, until 2009. ... Due to its origins as a service rather than a separate line, all of the stations on the Circle line are served by at least one other Tube line.' 'The Circle Line', *London Transport Museum*, <https://www.ltmuseum.co.uk/collections/stories/transport/circle-line>.

[4] '*Homo sapiens sapiens*, in anthropology and palaeontology, the subspecies of *Homo sapiens* that consists of the only living members of genus *Homo*, modern human beings. Traditionally, this subspecies designation was used by palaeontologists and anthropologists to separate modern human beings from more-archaic members of *Homo sapiens*.' John P. Rafferty, 'Homo sapiens sapiens', *Britannica*, 2 July 2020, <https://www.britannica.com/topic/Homo-sapiens-sapiens>.

There is a theory that the development of the young of any species from conception to maturity - ontogeny - repeats, in a telescoped form, the evolution of the species itself, or its phylogeny. Hence the saying that ontogeny repeats phylogeny.[5] The young howlers and shriekers in that tunnel were, of course, verbal, but there was a strong lower-primate flavour about their non-verbal efforts. I still believe that Darwin was right, but I have the benefit of all the East African and South African fossil material that has been discovered in the last fifty years. This was totally lacking in Darwin's time but, with uncanny instinct and foresight, he came to the same conclusion.

The Imperial College of Science and Technology is a collection of buildings of various ages and shapes and sizes scattered around in this area of great learning, which must have more Fellows of the Royal Society in its golden mile of science than any other area of similar size in the world. It really is a tremendous concentration of brains of the highest order.

I.C. (for short) comprises three constituent colleges: the Royal College of Science, the Royal School of Mines and the City and Guilds Institute. Colloquially, they are abbreviated to RCS, RSM and C&G. The whole is incorporated in that sprawling monstrosity, the University of London, but I.C. has an identity within London University and regards itself as distinct and superior. I belonged to I.C. first and to London University as a poor second. Even the three individual constituent colleges had their own identity, which generated a good deal of inter-college rivalry in sport, particularly rowing, which was done on the Thames. From the three rowing clubs, I.C. could pick a powerful eight who invariably did well in the annual Head of the River Race on the Thames. Frequently they won it.[6]

[5] Ontogeny is the developmental history of an organism within its own lifetime, as distinct from phylogeny, which refers to the evolutionary history of a species. 'Ontogeny', *Wikipedia*, <https://en.wikipedia.org/wiki/Ontogeny>.

'The phrase "ontogeny recapitulates phylogeny" was coined by Ernst Haeckel in 1866 and for many decades was accepted as natural law. Haeckel meant it in the strict sense: that an organism, in the course of its development, goes through all the stages of those forms of life from which it has evolved. Modern biology now rejects this dogmatic perspective. Though recognizing that human beings evolved from fish and reptiles, biologists cannot discern in our development any stages that correspond precisely to those of a fish or a reptile.' 'Tool Module: The Connection between Ontogeny and Phylogeny', *The Brain*, <https://thebrain.mcgill.ca/flash/capsules/outil_bleu12.html>.

[6] 'The Head of the River Race (HORR) is an against-the-clock ("processional") rowing race held annually on the River Thames in London, England between eights. ... The race was founded in 1925 by the rowing coach Steve Fairbairn who was a great believer in the importance of distance training over the winter. 'Head of the River Race', *Wikipedia*, <https://en.wikipedia.org/wiki/Head_of_the_River_Race>.

A vast amount of rebuilding has been done in post-war years, and men of my generation (there were then virtually no girls) are lost in the new concrete jungles, which seem to us to have far less character and appeal than the old buildings. Functionally they are far better.

There was, for example, the old Huxley building next door to the massive Victoria and Albert Museum and dating from the same period.[7] It housed the original Royal College of Science, the first of the three Royal Colleges that now make up I.C. By modern standards of university building, nothing could have been designed and built that was less suitable for its purpose. The exterior was ornate, the interior dark, dirty and labyrinthine.

In my time it housed only the department of Mathematics, which for me has always been a dark, labyrinthine and mysterious subject. How such a building coped with Chemistry, Physics, Botany, Zoology and Geology is an unsolved riddle. But they must have been there, and notable discoveries were made in these unlikely surroundings - including the first quartz fibres for suspensions in delicate instruments. It is said that the first one was made in a passage, with a miniature bow, using a quartz rod as an arrow. This was heated at one point with an oxyacetylene torch and, when it melted, the arrow was released, so pulling out an almost microscopically fine thread of fused quartz. How it could be found on the grubby floor of a dark passage is not recorded.

The famous Professor Huxley taught science there. Investigating a cupboard in the basement yielded at least circumstantial evidence of his presence. I found a dusty cardboard box, heavy, and packed with geological specimens wrapped in Victorian newspapers. The specimens were mostly

[7] It was named after Thomas Henry Huxley (1825–1895), grandfather of Aldous Huxley. 'Staff papers deposited in the [Imperial College] Archives include those of some of the most significant names of the nineteenth-century scientific world [who] include T.H. Huxley – Biologist, Palaeontologist, School of Mines and Royal College of Science, "Darwin's Bulldog".' 'Archive Catalogues', *Imperial College London*, <https://www.imperial.ac.uk/admin-services/acru/collegearchives/archivecatalogues>. He was known as 'Darwin's Bulldog' for his advocacy of Charles Darwin's theory of evolution.

fossil shells from a period in which I had a particular interest. One specimen looked strangely familiar. It was a bivalve,[8] concavo-convex, and separated into three pieces. One was the upper valve, and the second the lower, and the third the cast of the space between them in which the soft parts had lived 250 million years ago. It showed all the scars of the muscles and some of the other internal organs, and the more I looked at it, the more familiar it seemed.

I went to the library and took out the monograph in which this group of shells had been described and figured in mid-Victorian times. Sure enough, among the beautiful hand-drawn engravings, was the very shell I had unearthed. Every detail tallied. The specimen was numbered and was said to be in the Geological Survey collections, by then housed in the Geological Survey Museum on the other side of the road.[9]

So the next step was to put the specimen in my pocket, cross the road, and ask to see specimen number so and so. The member of staff who came to the enquiries desk to see me took the number and disappeared. After a long time, he reappeared with a look of horror on his face. The specimen I had asked to see was missing. The loss of a 'figured specimen' is a major scientific calamity which can seriously interfere with research. I asked if there was any evidence that the specimen was out on loan, and a further search of the records produced the information that this particular specimen had been borrowed, decades earlier, by Professor Huxley, but obviously he had never returned it, which was, indeed, inexcusable. I produced the specimen from my pocket, with the tell-tale Survey Collection number still clearly on it, and the tell-tale red circle of paper, the size of a flake of confetti, that was the hallmark of a 'figured specimen'. Together we looked up the mid-Victorian monograph again and compared figure with specimen. It was unmistakably the same and was promptly returned to its rightful place in the Survey's collection. Huxley was a great scientist, but even Homer nodded.[10]

[8] 'A type of mollusc, such as an oyster, that has its body inside two connected shells.' 'bivalve', *Cambridge Dictionary*, <https://dictionary.cambridge.org/dictionary/english/bivalve>.

[9] 'Begun in 1929, the new Survey building (with "Geological Survey Museum" carved over the Exhibition Road entrance) was almost complete by early 1933 at a cost of £220,000, when it was taken over for the summer for the Government's International Economic and Monetary Conference. During 1934, Survey geologists were involved in preparing labels, maps, diagrams, dioramas and models for the new museum, which was opened by the Duke of York on 3 July 1935 in front of an audience of 1000.' 'Geological Museum', *Natural History Museum*, <https://www.nhm.ac.uk/CalmView/Record.aspx?src=CalmView.Catalog&id=GM>.

[10] *Quandoque bonus dormitat Homer* [Sometimes even good old Homer nods]. Horace, *De Arte Poertica*. 'That "Homer sometimes nods" is an expression that means that even the wisest and most gifted of men, like Homer, make mistakes.' Morton S. Freeman, *A New Dictionary of Eponyms* (Oxford: Oxford University Press, 1997).

The other buildings were distinctly newer, but of varied character and more in keeping with their functions. The Physics and Chemistry departments were in the wings of a massive grey, stone, three-storeyed building, a formidable home for formidable disciplines. The Royal School of Mines was of the same architecture and had an equally forbidding exterior, but was, because it houses both the Mining and Geology departments,[11] full of fascinating wall-cases with specimens of rocks, minerals and fossils. The spaces between were hung with coloured geological maps from various parts of the world. The corridors were an education in themselves. There was a cubic yard of polished kimberlite, the primary host of diamond, from the big hole in Kimberley itself. Firmly attached to the top surface, in a metal case, was a diamond, about 1/5 carat uncut, to illustrate the average volume of rock which had to be treated to recover this amount of diamond.

The laboratories in which Part I undergraduates worked were in a newer wing, which stuck out at right angles at the back of the main building. This was away from traffic noise but its windows opened on to the back of the next-door building, which was the Royal College of Music. All their practice rooms were at the back of their building. This was perfectly acceptable in winter, when all the windows of both buildings were tightly shut to retain what little heat their antiquated radiator systems provided, but in summer the weird mixture of sounds from those practice rooms was almost beyond endurance. Every kind of instrument was added to every type of human voice, all doing different things in different keys, and ranging from interminable scales to grand opera. Summer did not happen very often. The botanists and zoologists lived in newer buildings surrounding a quad with two hard tennis courts, and this complex included the Students Union, with the bar as its focal point. The Administration also lurked somewhere in the same complex.

Botany looked on to the steps of the Albert Hall, adorned with female statues in a state of near toplessness. They were periodically found to be clothed in genuine female underwear, which scandalized the residents of the block of flats on the other side of the steps: these flats contained a colony of retired army, navy and air force officers. Complaints were first made to Kensington Borough Council offices, who disclaimed all responsibility for the statuary. Neither the police nor the fire brigade could be persuaded to take action. The statues were, in fact, the responsibility of the electricity authorities, who had cunningly concealed a substation in a dug-out below them. Naturally, the whole process took at least a day to operate. This was particularly pleasing if the decorations included certain items of old-fashioned bedroom furniture neatly inverted on the ladies' heads.

[11] Geology is now part of the department of Earth Science and Engineering.

The City and Guilds Institute was architecturally anomalous, and inside was filled with the massive machinery so beloved by engineers. Its only interest for me was the beautiful display of polished Shap granite round the main doorways.[12] When the building was demolished soon after I came to the university in Salisbury, I managed to get a few small specimens of this beautiful building-stone from the college authorities.

The three constituent colleges continue to emphasize their identity by awarding their own degrees, Associate of each, in addition to the London degrees. The holders of such Associateships are, therefore, distinct from the run-of-the-mill London graduates.

This then, was the physical environment into which a Royal Scholarship admitted me, but if I thought this distinction would mark me out among the new intake, I was to be disillusioned. It was a truly hand-picked bunch. Differentiation from the common herd did not occur until my second year, and then in a totally non-academic way.

In my first year I was so afraid of failing – which meant the end, for there was no repeating a year – that I shunned all sport in case I got crocked, and this even included cricket. Having got into the second year (of the four-year degree), I relaxed this rule and joined the Imperial College Cricket Club. After the first net practice, I was put into the First XI as opening bat, and in an early game, against Reading University, I made a hundred. What I did not know was that from the professor down to the technicians, the staff of the Geology department were keen on cricket almost to a man. This century brought instant recognition, and from then on I was 'somebody' and was recognized and greeted by the staff in quite a new way. To reinforce my new status, I did it again the next year. My ducks seemed to be overlooked.

I. C. had virtually no residential accommodation, nor was there any system of licensed or recommended lodgings. Students fended for themselves. Chelsea was a cheap area, so that was my choice. It was also dingy, but full of old four- to six-storey houses let off as rooms. They were generally owned and run by elderly spinsters or widows, who advertised on cards outside stationers and other shops. There were hundreds of them.

[12] 'Shap granite is described by geologists as a coarse-grained granite, formed by the cooling of a large body of igneous rock, called a pluton, which was intruded into the pre-existing Cumbrian rocks. The granite is very distinctive and easily identified by the large crystals (phenocrysts) of pink orthoclase feldspar contained within its matrix.' <https://teessidepsychogeography.wordpress.com/2017/05/30/shap-granite>.

I chose one in Oakley Street, just off King's Road. No. 2 Oakley Street was owned by Mrs Spanner, a dragon with a build so thin that she would have been transparent but for her never-changing widows' weeds. She was hard on top but soft underneath, and she charged me 12/6 rent per week for a third-floor room at the back. It was silent as a grave, had one electric light and a four-inch gas ring. There was no heater, but a piece of steel plate on the gas ring did quite well, except that the meter was greedy and I was poor. For water, I had to go down two flights of stairs to a cold tap in the only bathroom that I could find. Mrs Spanner lived in a solitary troglodytic state in the basement. Cold baths were tolerable in summer but unbearable in the two winter terms. Chelsea public baths were, however, just round the corner, where a superb hot bath with clean towel and soap could be had for 6*d*.[13] No. 2 Oakley Street was twenty-five minutes' walk from the Geology department.

Most of my exploration of London as a student was done on foot. Much later, most of my exploration of the middle Zambezi was also done on foot. I came to know both very well indeed. To eke out my £100 p.a. scholarship I did two or three evenings a week as either a waiter or washer-upper in a small café in Kings Road. At least there was a free and substantial meal. Since it appeared that my dropout days were finally over, my mother relented and I spent vacations with her at Arncliffe.

So, in October 1937, I had become a full-time undergraduate of the Royal College of Science, in the Imperial College of Science and Technology, that august seat of learning to which my Royal Scholarship had given me entry.

The first year and a half was tough. It was nearly seven years since I had left school, and I had to do Physics, Chemistry, Maths, Botany, Zoology and Geology. Maths I feared most of all, having been so badly frightened by all my previous teachers, but here I was lucky. The first year was taught by Professor Hyman Levy, who had the gift of making matters understandable, if not exactly enjoyable. Classes were over a hundred, and one never actually met the Professor, but he got me through. It was about thirty years later that I met him and thanked him. He was, in fact, the father of Professor Lawrence Levy, who by then was my colleague, as Professor of Surgery in the Godfrey Huggins School of Medicine, at the University of Rhodesia.[14]

[13] The public baths are now a listed building. See <https://historicengland.org.uk/listing/the-list/list-entry/1080716>.

[14] Laurence Levy: <https://en.wikipedia.org/wiki/Laurence_Levy>.

By the end of my second year, the war clouds were gathering over Europe and the world, but we all plodded steadily on, hoping that cataclysm could be averted. Munich gave us false hope, and the business of learning continued.

In the department of Geology, I came under the influence of two more remarkable men, utterly different in looks and character but equally brilliant in their teaching: Professors Brammall and Read.[15] I had come into geology mainly through fossils. Both these men were hard-rock petrologists and mineralogists. I wanted to learn only about fossils. They soon showed me that I had to become proficient in all branches of geology; specialization could come later. This philosophy was so palpably right that it was my guiding principle when the time came to set up my own department in Salisbury; their influence was enduring.

Brammall looked like Lloyd George, without making an effort to do so. He did not care what he looked like. He lived for hard-rock geology (particularly the Dartmoor granites), music, painting and drawing. His sketches of features to be seen under the microscope were quick, accurate and revealing. As students (there were only four of us), we were expected to learn the technique so that we could illustrate our practical notes. We did tolerably well. His lectures were dramatic and volatile, generally in shirt sleeves and butterfly collar, and the mental picture of him darting backwards and forwards behind the lecture bench, sketching in coloured chalks on the board, acting the behaviour of fast and slow waves, and occasionally breaking into snatches of song, is still vivid.

Read was totally different. He was a big man and had done years of hard fieldwork in Scotland while doing mapping for the Geological Survey of Great Britain, based in the Edinburgh Office.[16] He was either a Kentish man, or a Man of Kent, I can never remember the subtle distinction.[17] Essentially a country man, he eventually retired to his native county to write textbooks on

[15] Brammell is Alfred Brammell. A mineral was named for him: 'Brammallite is a sodium rich analogue of illite. First described in 1943 for an occurrence in Llandybie, Carmarthenshire, Wales, it was named for British geologist and mineralogist Alfred Brammall (1879–1954).' 'Brammallite', *Wikipedia*, <https://en.wikipedia.org/wiki/Brammallite>.

Read is Herbert Harold Read. John Sutton, 'Herbert Harold Read, 1889–1970', *Biographical Memoirs of Fellows of the Royal Society*, 16: 479–97.

[16] Now known as the British Geological Survey, it is the world's oldest national geological survey, founded in 1835. It changed its name several times through its history and had been renamed the Geological Survey of Great Britain in 1905. In 1966 it was combined with the Overseas Geological Surveys and renamed the Institute of Geological Sciences, but was renamed the British Geological Survey on 1 January 1984. 'Our history: About BGS', *British Geological Survey*, <https://www.bgs.ac.uk/about-bgs/our-work/our-history>.

[17] 'If you are born on the east side of the Medway, you may call yourself a Man of Kent. If you were born to the west, a Kentish Man.' 'Kent – a county of two halves: A brief history.' *Kent: Family History Society*, <https://www.kfhs.org.uk/kent-history>.

geology, keep bees and watch cricket. He kept bees at his Kent home while he was Professor at Imperial College. The Royal School of Mines building had a flat roof and Read decided that he would bring one of his hives up for the spring and summer. The RSM is within easy bee range of Kensington Gardens and Hyde Park and he reasoned that the bee population of these well-kept public gardens was well below their carrying capacity. His logic, as usual, was right: that hive broke the English record for a season's honey production in its first year.

Read's lectures were quiet affairs compared with Brammall's histrionics, but they were absorbing just the same. He reasoned that even undergraduates could read textbooks, and he never lectured from the book. He drew on his own vast store of field and laboratory experience, and his lectures were personal embroidery on the textbook and the syllabus, a lesson I tried to follow many years later.

In the field (and we often went on excursions with him), his powers of observation were phenomenal on all branches of geology, except perhaps fossils, about which he pretended to know nothing. He tried to sit beside me at meetings of the Geological Society in Burlington House, Piccadilly, whenever papers on fossils were being presented. The meetings were in a lecture theatre set out rather like a miniature House of Commons, two tiers of benches facing each other across a space with tables for maps and exhibits. The President sat with the Secretary and Treasurer at one end and the speaker performed at the other. This room had seen history made in geological science for over a hundred years. The Piltdown skull had been described in it,[18] and all the great names from the past had performed there. The few great battles and clashes of personalities had also taken place in it, though, on the whole, geologists take their differences of opinion very calmly. It was a room with an atmosphere and had the ancient ballot boxes where still one black ball was enough to block election to Fellowship.[19] It was also badly ventilated and, during the reading of scientific papers, as quiet as the tomb. Read took a wicked delight in turning to me during papers on fossils and saying in a loud stage whisper, 'What *is* he talking about?' In those days and in those august surroundings I was easily embarrassed.

[18] 'In 1912 Charles Dawson, an amateur archaeologist, claimed to have discovered the "missing link" between ape and man. He had found part of a human-like skull in Pleistocene gravel beds near Piltdown village in Sussex.' 'Piltdown Man', *Natural History Museum*, <https://www.nhm.ac.uk/our-science/departments-and-staff/library-and-archives/collections/piltdown-man.html>.

[19] 'A large supply of black and white balls is provided for voters. Each voter audibly casts a single ball into the ballot box under cover of the box, or of a combination of a cloth and the box itself, so that observers can see who votes but not how they are voting. When all voting is complete, the box is opened and the balls displayed: all present can immediately see the result, without any means of knowing which members are objecting.' 'Blackballing', *Wikipedia*, <https://en.wikipedia.org/wiki/Blackballing>.

Read had a knack of saying remarkably profound things straight off the cuff, and it is those sayings that have stuck so long in my mind. I have found myself quoting them to my own students quite spontaneously and, I regret, often without acknowledgement. In my years as an undergraduate under him there were only four students and Read treated us as though we were a very special and select group. It did us all good and made us feel important, but I suspect, in retrospect, that he had the gift of making each group feel this way, another lesson which I hope I was able to absorb and apply to my own students.

World War II cut my undergraduate years into two parts, three years before my service in the RAF and, five and a half years later, a fourth year. Read was still there to welcome me back and pick up where we had left off. It was soon clear to him that I intended to return to Bulawayo after graduating to join the staff of the National Museum as Keeper of Geology. He was much put out. Why did I want to go and bury myself in the bush? I argued that there was an enormous, almost virgin, field for research in Rhodesia, and that a museum job would give me great opportunities. He was genuinely concerned because he had already earmarked a job for me on his staff. This would inevitably have led to narrow specialization in some minor aspect of British geology, and Britain had already been crawled over by scores of geologists for nearly a century and a half. Looking for minor openings and gaps in knowledge in such circumstances seemed a poor substitute for the great open spaces of Africa, but he was unconvinced, and assured me that I was making a cardinal error.

Years later, the Geological Society of South Africa invited Read to give their prestige lecture, the Alex du Toit Memorial lecture, and he spent eight weeks being shown around South African geology. I did not see him during that visit, but out of the blue came a letter from him after his return to London. In it was his complete change of heart. He expressed himself strongly that my decision to 'bury myself in the bush' had been entirely justified, and was backed up by the remark that, if he had been a young man in my place, he would have done exactly the same thing.

Because of my demobilization date (around Christmas 1945), I had to do my last year's work at I.C. in two terms instead of three, which was tough. Read knew this, and he also knew that my scholarship had remained unchanged in face value but had decreased in real terms because of inflation, and that ex-Servicemen's gratuities were pretty stingy. Mine was £90, plus sports jacket and slacks, a pair of shoes and a raincoat, which, even when it was new, leaked. I spurned the hat which completed my civilian outfit. Out of the kindness of his heart, he suggested that I could earn a little money by being a lecturer in Palaeontology at Queen Mary College in the Mile End Road each Friday. This offer was gratefully accepted, but only Read at I.C. knew where I was on Fridays and only the head of department at Queen Mary College knew that I was only an undergraduate for the rest of the week.[20]

I lectured to students up to their final B.Sc. year and conducted practicals and learnt an awful lot in the process. Being by then much older than the average student, it was possible to get away with the situation. Again, the experience was to stand me in good stead later. All in all, Read had a deep and lasting influence on me, the last at pre-graduate level, in that handful of people who moulded me and guided me when it was most needed.

My last meeting with him was in 1963, while I was on a 'contact' visit from my embryo department of Geology in Salisbury. He had retired by then but kept a room in the School of Mines. Three of us had an hour's talk with him. We were all his old students and all of the same vintage. The talk was quite inconsequential, but at the end of it he said it had been, for him, historic. I asked him why. He replied simply, 'You are all my old students, and you are all full professors.'

I can see his point of view. I never saw him again. I hope that for the sake of their own students, the Read imprint was as strong on the other two as it was on me.

[20] East London College became Queen Mary College, named after Mary of Teck, wife of King George V, in 1934. After several mergers, its public name was changed to Queen Mary University of London in 2000. 'Our history', *Queen Mary University of London*, <https://www.qmul.ac.uk/about/history>.

R.A.F. Form 2520A
AIRMAN
ROYAL AIR FORCE
SERVICE AND RELEASE BOOK
Rank
Service Number
IDENTITY CARD FOR R.A.F. AND W.A.A.F. PERSONNEL (All Ranks)
This is to certify that No. 1378345
BOND G.
(Rank and full name)
whose description is contained herein is serving in the Royal Air Force
Women's Auxiliary Air Force and is stationed at
BLACKPOOL
Changes of rank or station are to be certified below
Date
Rank
Station
Signature and rank of officer certifying
Description
Height 5' 10"
Colour of eyes GREY
Build MEDIUM
Colour of hair SANDY
Date of birth 27th MAY 1919
No. 618220
Signature of holder
Signature of Issuing Officer
Rank F/O.
Station BLACKPOOL
SERVICE PARTICULARS
Service Number 1378345
Rank A/F/Sgt
Air Crew Category and/or R.A.F. trade
Air Crew Badges awarded (if any)
Overseas Service
R.A.F. Character
Proficiency A
B
Decorations, Medals, Clasps, Mention in Despatches, Commendations, etc.
Defence Medal
Educational and Vocational Training Courses and Results
DESCRIPTION
Date of Birth 27.5.19
Height 5' 10"
Marks and Scars
Specimen Signature of Airman
(Last day of service in unit)
Particulars of his Service are shown
Brief statement of any special aptitudes or qualities of employment for which recommended
Signature of Officer Commanding
Date 13.12.1945

– 4 –

The Tale of the Missing 's'

I arrived in Africa as a misprint – or perhaps it would be more accurate to say as one four-hundredth part of a misprint. During the War, the RAF used Southern Rhodesia as one of its Empire Training Centres for aircrew, and it is said that the following signal was sent to the Air Ministry from the Rhodesian Air Training Group HQ in Salisbury:

Send 400 Aircrew for Oxfords

As a result, 400 pupil pilots were sent overseas from England forthwith, to the considerable embarrassment of the staff of the local RAF stations because the signal should have read:

Send 400 Airscrews for Oxfords

This story may or may not be true, but it was widely believed in Rhodesia at the time. Certainly, the term 'airscrew' went out of use soon afterwards in favour of the unambiguous but less high-sounding word 'propeller'.

For one member of this band of four hundred uprooted recruits, this posting was an unexpected fulfilment of an early childhood dream that in later years had almost been forgotten. The desire to go to Africa had been planted in a six- or seven-year-old boy by a middle-aged mother who read aloud supremely well. One book that she read so well in the cold climate of our village in the Yorkshire Dales was Col. Patterson's account of the two man-eating lions of Tsavo in East Africa, which actually brought the construction of a railway line to a complete stop. It is a piece of straightforward narrative writing. There is no pretence at literary skill or embellishment. It is completely true. No fictional hunting tale could be so exciting, but it cannot have been easy to read it aloud. I still keep a copy of *The Man-Eaters of Tsavo*, and my own, limited, experience of a family of lions with a taste for human flesh has made me appreciate Patterson's achievements.[1]

[1] J.H. Patterson, *The Man-Eaters of Tsavo, and Other East African Adventures* (London: Macmillan, 1907).

School in the Lake District, holidays in the isolation of the Dales and a scientific training all conspired together, inducing an urge to see at first hand places which a life in England can never satisfy. An overseas posting in January 1941 was, therefore, not unwelcome, but, of course, security dictated that none of us should know where we were going. Liverpool was the port of embarkation, an old Dutch East India Company ship of some 20,000 tons on her first trooping run was the transport, and the destination was anybody's guess.

Our convoy was an imposing sight as we set out into the Atlantic. There were about a dozen fast liners and a few fast cargo boats escorted by three battleships, including *Nelson*,[2] six cruisers and at least twelve destroyers. We had all been issued with tropical kit, but this might have been camouflage, for we sailed steadily north-westwards, zigzagging, at some sixteen knots.

Rumour spread that the previous convoy had met the *Scharnhorst* and, through lack of protection, had been severely mauled.[3] No doubt the authorities would have welcomed a second encounter, and it was reassuring to look round the horizon and see so much of the Royal Navy. Nothing happened.

One morning, some days out from Liverpool, the sun came up on the wrong side. The mess-deck navigators pronounced that the convoy had turned on to a general south-easterly course and that our destination could not be Canada or the States but might be southern Africa. The Pole Star sank lower every night, the temperature increased and flying fish appeared. Our tropical kit ceased to be regarded as camouflage and was worn in spite of its ghastly shape. Excessive sunburn was officially declared 'self-inflicted injury', and rather amateurish lectures on the relationships to be observed between white troops and black women were given by pink-and-white young British officers who looked as though they had never been further south than Land's End.

One or two liners and most of our private navy were missing one morning, leaving us with a slightly naked feeling, and the pundits opined that we were well on our way to Freetown in Sierra Leone. They must have been intensely gratified when a Sunderland flying boat appeared, and soon afterwards a long indistinct blur on the horizon turned into our first sight of land since leaving England. Freetown may be bearable for a civilian in peacetime, but from a troop ship's deck it is not an attractive place. It was, for most of us,

[2] Kennedy Hickman, 'World War II: HMS *Nelson*', *ThoughtCo.*, 7 August 2019, <https://www.thoughtco.com/battleship-hms-nelson-2361541>.

[3] 'German battleship *Scharnhorst*', *Wikipedia*, <https://en.wikipedia.org/wiki/German_battleship_Scharnhorst>.

our first sight of Africa. It would have been excusable to hope that it would be the last. No one was allowed ashore. The ventilating system broke down, port-holes and watertight doors were closed.

Fruit was abundantly on offer all day long from all manner of small native craft, manned by dark-skinned strangers who swam like fish and dived for silver (but only for silver) coins thrown in the sea. In that heat, fruit would have been – next to long, cool drinks in tall glasses with tinkling ice-cubes – about the most desirable thing imaginable. An official decree, however, forbade all fruit from over the side and, apart from normal serious penalties for disobedience, forecast unpleasant-sounding diseases for lawbreakers. As it had not been designed to carry passengers in the heat of the tropics, the ship seemed like an oven, and in any case was grossly overcrowded. In retrospect, I doubt if I have ever had to endure anything like it since, except deep underground in poorly ventilated mines.

So this was Africa, an almost landlocked harbour, with low, swampy shores and palm trees, hot and humid. It hardly stirred my dormant memories of Col. Patterson's dry Nyika thornbush.

For a week we looked at Africa, almost encircled by it. Africa could be seen and heard, for the locals never gave up hope with their fruit, but it could not be trodden. There was nothing to do but speculate about what lay inside this stinking rim of the continent. Then the convoy moved out through the narrow jaws of the harbour. The Pole Star sank to the horizon immediately astern and that singularly disappointing constellation of the Southern Cross came up out of the sea ahead. The mess-deck navigators declared that we had crossed the equator, but for security reasons Father Neptune remained below and no certificates were issued.

The lazy routine rolled on. The same young officers, browner now, gave lectures on tropical hygiene with the same lack of assurance. Well away from any threat of enemy aircraft, they also gave lectures on aircraft recognition, with obvious first-hand knowledge of the subject. We still had *Nelson* to remind us of the war on the other half of the world. The weather became cooler and we changed back into blue. I changed jobs, from mess orderly, to member of the crew of a six-inch naval gun mounted on the stern. This was made in 1906. Where do they store these ancient fowling pieces in the years between wars? Do they still work? As we did all our drill without even the dignity of dummy shells, the question can never be answered. Thirty-five days after leaving Liverpool, we changed suddenly to an easterly course, and, on the thirty-sixth day, Table Mountain came up out of the sea.

I am going to give a rather full account of my first impressions of southern Africa, not because there is anything new about them but because of the impact they made on me, for it was largely these first impressions that determined where I would spend my working life. In this sense they are vital and essential for all that follows. That posting, even if it was not after all the result of a misprint, was one of the lucky chances that shaped my whole life.

The hard pavement of the Victoria Dock in Table Bay was the first solid feel of Africa under my feet, and it was good. The air was cool, Table Mountain was a lovely sight in the soft sunlight, and Cape Town looked a fine city. But a train was waiting in the dock siding, and we slid slowly past the end of Adderley Street and on over the Cape Flats.

The landscape of the Cape is a strange mixture. The Cape Flats are low sand-dune country, almost at sea level, with very meagre vegetation, which somehow manages to survive without anything that, by normal standards, can be called soil. This low plain is surrounded by beautiful, jagged, rocky mountains several thousand feet high, up which there is no apparent track that the railway can use to reach the inner part of the country. There could be no more beautiful country on which land-hungry mortals could stand after thirty-six days in a troopship.

The line reached the end of the Flats and found a deep valley in the face of the ranges that divide the coast from the high plateau of the interior. The southern half of Africa is like an inverted soup plate. There is a narrow coastal plain, a great escarpment running down the east coast, round the Cape and up the west coast, and the flat interior plateau, much of which is 4,000 to 5,000 feet above sea level. The escarpment guarded the secrets of the interior long after the coastal plain was known.

These Cape ranges are twice as high as my familiar Lake District mountains and, like them, are full value for their height since they rise from a plain that is almost at sea level. The altitude was given with the name of each railway station. Even at the foot of the Hex River Pass, which breaches the barrier, the height was a mere few hundred feet above sea level, yet the mountains that press in from all sides are several thousand feet high.

It was dark with a full moon as the train began to climb, one engine pulling and one pushing from behind. I was intensely curious to see how this steep, winding track climbed through the pass on to the plateau. The train got slower and slower, and eventually crawled onto a level section where there is a station cut out of the side of a mountain. It stopped. The station's nameboard showed about 3,000 feet but it was still in the mountains.

The last sitting for dinner, about the seventh, was served. The train remained at rest. People walked on the track beside the train in the moonlight. Rumour, always ready to come to the troops' assistance, had it that our engine had broken down. It seemed unlikely that the train would move for some hours. Resignedly, I slept.

When I woke, the sun was up and the magic transformation had taken place. Last night, the line lay among dark mountains; the valley was green and full of prosperous-looking farms, with tall trees and neat vineyards. This morning, the horizon was nearly flat, with a few table-topped or sharp pointed hills. The station boards gave heights of over 4,000 feet. The earth was baked and bare, here and there with clumps of juicy-looking green bush. The farmhouses were miles apart. The only trees were languid gums with great trunks, giving some shade to the homesteads.

The glare was intense, even at breakfast time, and the only signs of life were a few brown sheep on the parched earth. This was the Karoo, which is called desert but is some of the finest sheep country in the world. It is typical of so much of the flat part of the inverted soup plate beyond the fringing escarpment. It is a hard country, where life is a constant battle against lack of water, and it was conquered only by pioneers who were prepared to live hard and keep up the battle against Africa. Even today, Africa hits back hard. A road neglected for a season or two is quickly obliterated. It is either overgrown by bush or scoured out to a deep scar by the rains. Africa seems to resent settlement and civilization. Given the chance, she soon rubs out the marks. Termites eat out the frames of abandoned houses, and the bush covers the remains. Even abandoned motor cars, left beside the lonely roads, gradually disappear until all that is left is the rusting remains of the body-panelling.

Only constant effort keeps civilization going in the interior. When one becomes too old to keep up the battle, it is time to retire to the softer air and lower altitudes of the coastal plains. This is now obviously an overstatement, for the cities on the plateau are well founded and permanent, but there is still a great deal of Africa that is young man's country.

The Karoo intrigued me by its vastness and comparative emptiness. Distance like this cannot be found in England. The train passed through the same type of country for more than twenty-four hours. Kimberley came and went, and one desert was exchanged for the fringes of another, Karoo for Kalahari, Cape Province for Bechuanaland Protectorate, Dutch-sounding names for African ones: Lobatsi, Mahalapye, Palapye.

The succulent green scrub of the Karoo was replaced by low trees with long white thorns – which were much more like Col. Patterson's 'interminable Nyika thornbush'. The sheep were replaced by scraggy cattle with great horns and humps on their shoulders, storing nourishment for the long, rainless days.

Skins of game animals, including lion and leopard, were on sale at the stations, and the height on the boards still remained around 4,000 feet. Small boys danced for pennies beside the track and reached up to the windows of the carriages offering toys and carved animals for sale. South African and Rhodesian coins appeared in the change, and they seemed to mistrust English money a little.

Rumour, still our only guide in these matters, now said that our destination was Southern Rhodesia but was uncertain whether it would be Bulawayo or Salisbury.

Three days after leaving Cape Town the train reached Bulawayo and pulled into the longest platform in the southern hemisphere. It was just under half a mile long and the altitude was 4,464 feet above sea level. We were almost surprised to hear that this was really the end of the journey. The authorities were equally surprised, and not a little embarrassed, by the arrival of 400 aircrew cadets when all the stations of the group were already full. We were marched off to a temporary army camp in the middle of the Bulawayo Agricultural Society's Showgrounds and billeted mainly in the cattle pens, with bed-boards and straw.[4]

How the next years of war were spent officially is unimportant. My eventual posting was to a station sixteen miles out in the bush.[5] There was very little transport into town and my interest in the surrounding country grew with the months.[6]

It was considered distinctly 'off' to like the country. Many new arrivals were heard to say, 'If I stay here twenty b----y years I shan't like this country.' Yet it is amazing how many of the 'Blue Plague' came back after the war to settle in Rhodesia. After three or four years of this climate and the feeling of freedom that the great spaces of plateau inspire, England may remain 'home' but a home to which one goes on holiday. Once Africa has admitted you, she has a way of bringing you back again.

[4] The first Bulawayo Agricultural Show had been held in May 1899. The history of the Agricultural Society is told by Robin Rudd, *Show Business* (Bulawayo: Bulawayo Agricultural Society, 1997).

[5] Air Vice-Marshall Sir Charles Meredith was Air Officer Commanding of the Rhodesian Air Training Group during the Second World War: Sir Charles Meredith, 'The Rhodesian Air Training Group 1940–1945', *Rhodesiana* (1973), 28: 16–29.

[6] An article about later recruit to Bulawayo, Norman Enders, has this to say:

'Life on and off camp contrasted with life in wartime Britain. The base had its own swimming pool, and fresh food (especially fruit) was un-rationed. The weekly highlight was the Friday night bus into the local town to visit the pubs and dance halls, where South African wines and beers were readily available. He recalls his billet mate, a Glasgow chap inevitably known as "Mac", had expertise in a useful combination of unarmed combat and dancing, often having to be dragged away from the mirror to ensure he got on the bus. No wonder the RAF was known to the locals as the "Blue Plague".' James Baillie, 'From Southern Rhodesia to Walthamstow: One RAF pilot's journey', *Waltham Forest Echo*, 24 March 2021, <https://walthamforestecho.co.uk/from-southern-rhodesia-to-walthamstow-one-raf-pilots-journey>.

During the war years I gained confidence in the bush. At first, its size, the heat, and general lack of water and landmarks had made me respect it. The fear of getting lost was always present from the beginning and has never left me completely. Until I know the general layout of a new piece of country, I am extremely cautious.

I gradually came to realize that the middle part of the Zambezi Valley was the part of the country where the things I wanted to find out could best be tackled. It seemed at that time completely inaccessible. The map showed only one road into a triangular piece of country – 200 miles along its longest side and about 150 miles from the base to apex. But it was to be several years before I could get into this intriguing piece of the country.

It was not only remote but very sparsely populated, and over the years I developed a high regard for the so-called 'primitive' Batonka of the Zambezi Valley. They are wise in the ways of nature, and I have depended on them day after day for water, directions and protection from wild animals. I have never been let down in matters of honesty or personal safety. In the bush they are men. I am sure that I was regarded by them as quite mad. They could not understand my meaningless traverses, and my Sindebele name, *Umantswane*, the honey badger, caught their feelings exactly. The honey badger hunts in such a meaningless pattern that his spoor cannot be followed; so it is with a geologist.

My wanderings in the Zambezi Valley really became possible only through my association with the National Museum in Bulawayo,[7] and my first contact with that institution and the people who worked there happened soon after the arrival of the 400 unexpected aircrew and while I was still living on my heap of straw in the cattle pens of the Showgrounds.

I was interested in museums and had done research in the Natural History Museum in South Kensington. The 1940 telephone directory told me that Bulawayo had a museum, and I lost no time in making telephonic contact. It was a woman's voice that answered and announced that she was Keeper

[7] Now the Natural History Museum of Zimbabwe. As the Rhodesia Museum, it came into being on 1 January 1902 and occupied several buildings as it grew. The government acquired the museum in 1936 and it was renamed the National Museum of Rhodesia. Bulawayo City Council later made available land in Centenary Park, and the present building that was begun in 1960 was the fifth museum. It was opened to the public on 20 March 1964. 'History of the Natural History Museum', *Natural History Museum of Zimbabwe*, <https://naturalhistorymuseumzimbabwe.com/history-of-the-museum>. See also C.K. Cooke, 'The History of the Museums in Bulawayo, 1902–1985', *Heritage of Zimbabwe* (1986), 6:61–9.

of Geology. This announcement contained three important ingredients: one that there was a department of Geology; two that the Keeper was female; and three that she used the familiar, but archaic, title of Keeper, which I was used to in the Natural History Museum. Evelyn Ferrar asked me to drop in to see her department and meet the rest of the staff. Naturally, no time was lost in the dropping.

Evelyn was tall, almost red-headed, a New Zealander, a graduate, and young. Her father had been geologist on one of Scott's Antarctic explorations, and the Ferrar Glacier in Antarctica is named after him.[8]

She had already collected round her a group of young RAF officers, but none of them was a geologist or even a scientist. In spite of being very nearly the lowest form of non-commissioned animal life in the Air Force (my rank was LAC),[9] I was added to the collection as a rather special specimen, perhaps because I could help with her work of identifying rocks and minerals for the local small-workers and prospectors.

The Bulawayo Museum was founded in 1901 at the instigation of Cecil Rhodes, who realized that the early prospectors needed expert help if their work were to succeed. They could all recognize gold in the prospecting pan and could estimate values in pennyweights per ton from the length of the 'tail' derived from crushed rock measured in a round fifty-cigarette tin. But they knew very little mineralogy. He insisted that the first appointment should be a trained geologist who could determine their finds and advise them on their economic significance, if any. F.P. Mennell, from the department of Geology at the Royal College of Science in London, became the first director and set the pattern followed for the next sixty years – that part of the Museum geologist's work, and an important part of it, was rock and mineral determination.[10] His register of this work still exists and contains some notable 'firsts' for this country.

Evelyn was struggling with this kind of work, and my second opinion was eagerly sought. Spare time from the Air Force was spent exclusively on this work. Soon a third and very valuable opinion became available in the form of another LAC from the camp, Bill Swift. Bill not only had a Bachelor's degree in Geology from Edinburgh but a Ph.D. as well, and he had by then had two years' experience in this country as a member of the staff of the

8 'Hartley Travers Ferrar (1879–1932) – Biographical notes', *Cool Antarctica*, <https://www.coolantarctica.com/Antarctica%20fact%20file/History/biography/Ferrar-Hartley-Travers.php>.

9 LAC: Leading Aircraftman. 'Leading Aircraftmen ... would have done everything from maintain aircraft to cooking and basic clerical tasks. Sometimes nicknamed 'erks', although the origins of the nickname is not known.' 'Armed Forces & World Wars', *Genes Reunited*, <https://www.genesreunited.co.uk/contents/armedforcesbritishroyalairforce>.

10 See F.P. Mennell, *The Miner's Guide* (London: Gerrards, 1909); F. P. Mennell, *The Rhodesian Miner's Handbook* (Bulawayo: Ellis Allen, 2nd edn, 1909.)

The Museum in Bulawayo

1901 – 1905
Upstairs in Bulawayo Public Library.
Fort Street & 8th Avenue

Photo: Collections of National Railways of Zimbabwe, Railway Museum, Bulawayo

1905 – 1910
The former Congregational church.
Main Street, between 10th & 11th Avenues

Photo: Scrapbook in the collections of Historical Collections, Bulawayo Public Library

1910 – 1922
On Fort Street at the rear of the stand where the next was built. Fort Street & 8th Avenue

Photo: Postcard in collection of Andrew Wilkie

1922 – 1960
Built in stages in front of the previous one,
centre and wings added in 1930 and 1936.
Incorporated the previous one into the back.
Fort Street & 8th Avenue

Photo: Collections of Natural History Museum of Zimbabwe, Bulawayo

1960 –
The Present Museum. Centenary Park

Photographs of its construction and opening:

<http://bulawayomemories.com/PARK/museum_history_construction.html>

official Geological Survey, based in Salisbury. He already knew a great deal about the country's rocks and minerals, and things that were strange to both Evelyn and myself were familiar to him. We both learned a lot in a short time from him. We lost him when he was posted, but by then our joint confidence had been established, and we did our mineral and rock determinations with a new air of assurance.

With Bill Swift. The Cattle Pens,
No. 2 training camp, Bulawayo, June 1941

In addition to her physical attributes, which were considerable and much admired by all her collection, Evelyn had two material assets in the form of a comfortable flat and a reliable motor car. Both were shamelessly over-utilized by her adopted nephews, and we all called her Auntie. She, in her turn, was the adopted niece of the only two other permanent scientific members of the Museum staff: her 'uncles', Neville Jones and George Arnold, both in their late fifties or early sixties, it was hard to know which. They were the ageless kind.

This happy family seemed for a time to be quite stable, but Evelyn was introduced to a young District Commissioner on leave in Bulawayo from Uganda. She met Forbes Mackenzie on a Saturday, announced her engagement to him on the next, and they were married exactly one week later. I was suddenly on my own, battling in my spare time, as Honorary Keeper of Geology, with the problems brought in by local prospectors and small-workers. This work and the complete revision of the catalogue and storage system of the entire collection of the Geology department in the Museum took me three years, and almost unwittingly paved the way for my permanent appointment to the Museum staff after the war. Some research was also possible during this time.

The old Museum building in Fort Street was by no means an ideal one by modern standards. It had grown piecemeal as funds became available, and these had never been anything but meagre. When I first worked there in my honorary part-time capacity, the total budget for the year, and this included the Director's and the one other scientist's salaries, was £3,000, so there was very little over for fieldwork expenses, even though petrol was only 2/6 per gallon.

The two colleagues were an oddly assorted couple. Dr George Arnold was the Director and had served in that capacity since 1912. In that time, he had seen a succession of geologists come and go and had a slightly contemptuous attitude towards them for lack of staying power. This was a little unfair. One had died in the flu pandemic of 1918, another had been killed in a car crash, one had left to join the Geological Survey. In desperation at one stage, the board had appointed a man close to retiring age. Evelyn, of course, had, inevitably, got married, and one had left to join the more lucrative mining industry as a consultant.

George Arnold

George Arnold was an entomologist and a world authority on bees, wasps and ants. During his lifetime he discovered, named, and described over 1,200 new species of insect.[11] His choice of insects, those which generally have a sting of formic acid, suited his temperament admirably. He could be acid and had a stinging tongue. He not only declined to suffer fools gladly, he suffered them not at all. To be accepted by George Arnold was reassuring. He delighted in the reputation of being the rudest man in Matabeleland, but under his crusty exterior was a very warm inner person, whom few managed to find. A confirmed bachelor - though it is rumoured that in his young days he had been married briefly to a glamorous redhead - and small and neat, George had systematically built up the Museum and his own international reputation. His field of research would now be considered old-fashioned, but proper naming of an animal is a prerequisite for any further study, and this is particularly true of insects, so many of which are agricultural pests. Wrong identification can be expensive.

George Arnold can be typified by two of my own experiences with him. I took him to Khami Ruins for an afternoon. He collected insects and I looked at rocks. Coming back after a while to the car I found him digging in the sandy path. 'Keep well away,' he said. 'I'm digging out the nest of the fiercest ant in Matabeleland, and it's named after me.'

The other concerns the Bulawayo Club, of which august body he was a very long-standing member and deeply immersed in its traditions. Such

[11] '... many years after his death [1981] most of the Hymenoptera collection was sent to the South African Museum in Cape Town in exchange for the stone Zimbabwe birds, 'Arnold, Dr George', *S2A3 Biographical Database of Southern African Science*, <https://www.s2a3.org.za/bio/Biograph_final.php?serial=97>.

clubs are exclusive, and their exclusiveness is jealously guarded even today. During World War II the Club opened its doors and its hospitality to the RAF, but only to commissioned officers, who used it as an extension of their own exclusive club, the Officers' Mess. George insisted on taking me into the Club as his guest on every possible occasion, even though I was in the uniform of a lowly order of the non-commissioned ranks. Because of his own seniority, no member would have dared question the presence of a mere Leading Aircraftman, and he seemed to take a malicious delight in watching the scandalized looks on the faces of young junior officers, lounging with their beers and girlfriends. I became very fond of my adopted uncle.[12]

After the meticulously curated collections of the Natural History Museum on which I had worked before the war, I was horrified by the chaotic state of the magnificent collection of minerals, rocks and fossils in the Geology department. The successive Keepers of Geology had created this chaos by initiating new systems of numbering specimens and storing them, but no new system had ever been completed. It was virtually impossible to locate any specimen from the card index, and a museum specimen that cannot be found is as good as lost. George Arnold approved of my resolve to disentangle this mess in my spare time, gave me the keys to his kingdom and *carte blanche* in the Geology department. After three years of this part-time work - a lot of it had been done at night and at weekends - the collection of about 15,000 items had been completely overhauled, and I could boast that any specimen could be located inside five minutes.

Arnold's relations with other scientists were peculiarly mixed. He could not stand the Museum's first Director, F.P. Mennell, who was an even more senior member of the Club than he was. Even among geologists, F.P. was not universally regarded as highly as his pioneering work in Africa demanded. As an entomologist, George Arnold had no right to take sides, but he did.

Neville Jones, the only other member of the scientific staff, was a complete contrast to George Arnold. He, too, was Evelyn's adopted uncle, but this was all they had in common, though they were firm friends. Neville was the mildest of men, though his African name could only be translated as 'the one who

[12] 'Dr George Arnold ... was short and crusty, physically and temperamentally. He arrived at lunch one day to find a stranger sitting in his customary seat. He paced back and forth up and down the dining room until the interloper had finished his meal; then he called a waiter and demanded, "Bring me a cold chair!"' Peter Gibbs, *The Bulawayo Club: A History* (Bulawayo: The Bulawayo Club, for private distribution on the occasion of the 75th anniversary of the founding of the Club, 1970), 86.

makes the sparks fly', and for good reason. He had begun life as a missionary, had served in Madagascar, where he acquired a moderate fluency in French, and had then been posted to Matabeleland, where he acquired an enviable fluency in Sindebele, becoming a noted scholar in this difficult language. On the rare occasions when I heard him speak French, it was obviously rusty, for when a word failed to come, he unconsciously slipped in the Sindebele equivalent, to the bewilderment of his listeners.

Neville Jones

On fieldwork, he dressed sensibly, which was especially necessary when excavating the floor deposits of caves, which consisted largely of ash. George, on the other hand, collected insects in jacket and waistcoat, collar and tie, with town shoes on his feet and a crease in his long trousers; but then, entomology is a nice clean discipline. His only concession to hot weather, and it had to be very hot, was to take off his jacket. Trilby hat and butterfly net completed the outfit.

Neville Jones's work on the Stone Age had earned him an international reputation and had laid a solid foundation for all subsequent research, but he was self-taught. His interest had begun when he was made Head of the Hope Fountain Mission, which included a boarding school for African girls. One of the country's classic Stone Age sites was on his doorstep at the Mission. His African name came from the ferocity he developed if anyone dared to molest any of his girls.[13]

Gradually, his interest in archaeology outstripped his missionary zeal, and in 1936 he joined the Museum staff as Keeper of Archaeology, built up the collections, and founded the Commission for the Preservation of Natural and Historical Monuments. He served as its Secretary for many years.[14]

Southern Rhodesia was known to be rich in Stone Age material. Jones set

[13] Hope Fountain Mission is the second oldest in the country. <https://zimfieldguide.com/matabeleland-south/hope-fountain-mission>. The name of the school seems to have recently been changed to Tennyson Hlabangane High School.

[14] Neville Jones, 'The Commission for the Preservation of Historical Monuments and Relics', in Rowland J. Fothergill (ed.), *The Monuments of Southern Rhodesia* (Bulawayo: The Commmission, 1953), 9–13. The Monuments and Relics Act, 1936, 'brought in the new concept of ranking sites by affording the status of National Monuments. The Act brought into being the Commission for the Preservation of Historical Monuments and Relics, also known as the Monuments Commission. It was believed that more was required than just a protective Act and that a regular form of inspection was required. The 1936 Act was repealed in 1972 by the National Museums and Monuments Act which amalgamated the Monuments Commission with various city museums.' 'Southern Rhodesia: 1936 Monuments and Relics Act', *The Rhodesian Study Circle*, <http://www.rhodesianstudycircle.org.uk/southern-rhodesia-1936-monuments-and-relics-act>.

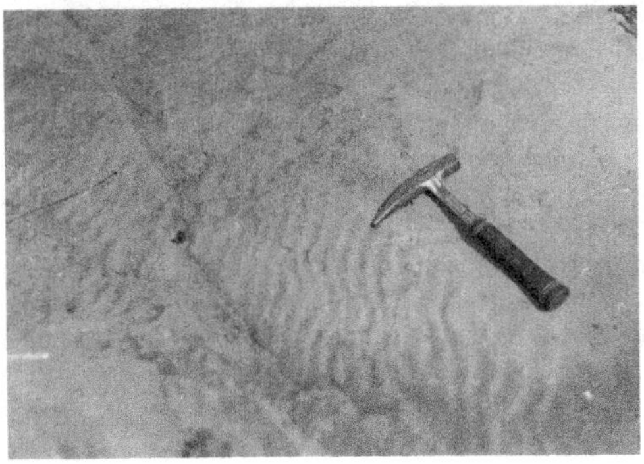

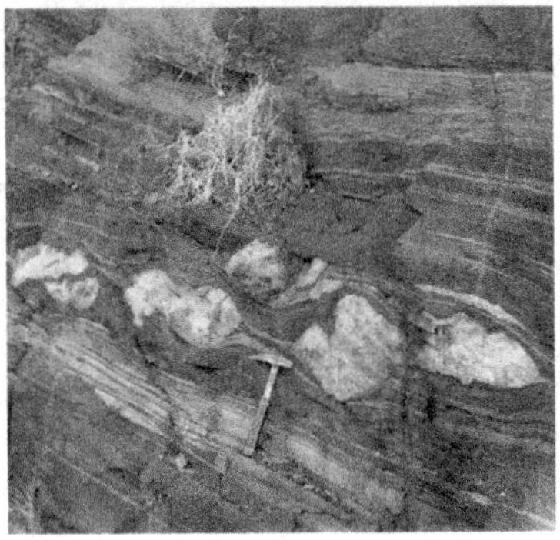

about establishing the relative time sequence of the various phases. To do this it was necessary to discover stratified sequences where the various cultures could be shown to be in order, the oldest at the bottom and the youngest at the top. A great number of sites were in the open, in or near rivers, but most of them had been naturally disturbed after they had been occupied by Stone Age Man. Caves in the granite country, often with Bushman paintings, were more satisfactory and he excavated a number of them, including such famous sites as Bambata and Nswatugi in the Matopo hills, south of Bulawayo.

The Stone Age can be divided into Old Stone Age, Middle Stone Age, and Late Stone Age, and none of the cave sites seemed to go back in time before about the middle part of the Middle Stone Age. For the unravelling sequences in river valleys, where Old Stone Age material was known to exist, he needed a geologist. I found myself getting more and more deeply involved in Stone Age archaeology in collaboration with him. I never expected to get interested in anything so geologically young as river terraces and beds of quite recent alluvium, but Uncle Neville's enthusiasm was infectious.

I travelled by goods train with him to the Sawmills on the Umguza river and together we worked out a stratigraphic sequence of the Old Stone Age to the Late Stone Age, which Neville wrote up and published with suitable acknowledgement to 'Sgt. Bond of the RAF'.

I was stationed at Heany, sixteen miles north of Bulawayo, at the time and, having tasted research in this field of geology, I walked miles at weekends and on leaves, investigating all the river valleys that I could reach that flowed towards the Zambezi.

Heany had a small satellite landing ground on Lochard ranch, about forty miles north of Bulawayo and a few miles west of the main Bulawayo–Salisbury railway line. This little airfield was fitted with radio equipment for teaching a blind-approach system for bad-weather flying and had to be manned continuously. The strength was one sergeant (officer commanding), one electronics corporal and one general purpose airman. They all did stints of a month at a time in the bush. Being so isolated and so far from the bright lights and fleshpots of civilization, it was an exceedingly unpopular posting, but I found it fascinating and a relief from the discipline and order of a big station. I volunteered frequently for duty at Lochard and, since flying stopped at lunchtime, afternoons were free for roaming around the bush. The river valleys were rather badly gullied, and it soon became plain that the area was rich in Stone Age material of all ages, and that it was possible to work out the relative time sequences in even more detail than was possible at Sawmills.

One area of gullies, and sheet erosion in particular, proved to contain the actual factory sites where Old Stone Age parties had made the tools they needed. There were hundreds of axes and other tools in varying stages of manufacture but very few finished products, presumably because the good ones had been taken away and used. There were three of these factory sites, each about ten yards in diameter and roughly circular, in one small section of the valley. It could be shown that these early hunters, who lived there perhaps 100,000 to 200,000 years ago, had used a variety of rock types, some of which they had imported from a minimum of forty miles away. By making certain assumptions that seemed reasonable, I calculated that the total human population of Rhodesia could, at that time, have been as little as 2,000. Man was rare, the environment hostile, but the human race survives, largely because of superior brain power and the ability to overcome environmental problems and hazards by technical inventions.

Neville Jones published the results of our joint research at Lochard, again with acknowledgement to Sgt. Bond of the RAF.[15] Lochard still remains an important archaeological site, and there must be many more of such sites in Matabeleland. We tried to work out the sequence of climatic changes from the field evidence at Sawmills, Lochard, and the other north-west-flowing river valleys. They all showed similar sequences of events, but it is very doubtful whether our interpretations of climatic changes are now acceptable. Our pattern was too simple. There have certainly been changes of climate, measurable in periods of thousands of years, but subsequent worldwide research has shown how complex the overall picture will eventually turn out to be. The simple pattern of Pluvials (more rain) and Interpluvials (less rain) that was fashionable then is extremely suspect now.

All this work, and a paper I wrote on some fossil shells that Dr A.M. Macgregor had found in the depths of the Zambezi Valley,[16] gave me a certain standing in scientific circles in this country and convinced me that my future lay here, but the war was still dragging on, and it was, in truth, an uncertain future. The Allies might be defeated, I might never have the chance to return to Imperial College to complete the fourth year of my first degree, and in time I was sure to be posted away from Southern Rhodesia. Inevitably the posting arrived, back to the UK, where, in January 1945, I found myself at Wick in the extreme north of Scotland in the middle of a harsh, freezing winter after nearly four years in the tropics. My prospects had taken a plunge, but another lucky chance took place.

[15] N. Jones, 'The climatic and cultural sequences at Sawmills, Southern Rhodesia', *Occasional Papers of the National Museum of Southern Rhodesia* (1944), 2(11): 39–46; N. Jones, 'The archaeology of Lochard', *Occasional Papers of the National Museum of Southern Rhodesia* (1946), 2(12): 116–27.

[16] See Appendix 3: 'Geoffrey Bond's Publications'.

On the way to Wick, I had spent a brief posting near Cambridge, and on a browsing expedition to the department of Geology at Cambridge, the famous Sedgwick Museum, I had run into Professor W.B.R. King, who casually asked me what I was doing in the RAF. When I told him, he muttered something about a waste of time and dropped the subject.

I forgot all about it, and was surprised to be summoned to the Orderly Room at Wick some weeks after my arrival and to be confronted by a puzzled adjutant waving a signal from the Admiralty, summarily posting me to something called ISTD in Oxford. Since the Navy is the Senior Service, there was nothing he could do about it. I reported to my new unit in Oxford, still wondering what ISTD stood for. I soon found out. It was Inter-Services Topographic Department, was both inter-service and international, and was concerned with pre-invasion planning in all its various aspects.[17]

I was attached to a small geology section of ten people, and my immediate Commanding Officer was Major J.L. Farrington, who was resident geologist at the Wanderer gold mine at Selukwe before the war.[18] Professor King was Head of Military Geology and had quietly rescued me from a routine RAF existence.[19] ISTD occupied Manchester Theological College and everyone wore civvies. I was promoted to Flight Sergeant, but rank counted not at all. I put on uniform only once a fortnight to attend a pay parade.

Military Geology is a very strange thing, just about as far removed from my previous experience of the subject as it was possible to get. Pre-invasion planning involves such weird subjects as beach composition in relation to landing tanks and other tracked vehicles as well as ordinary wheeled traffic, soil and vegetation colour for camouflage netting, the selection of suitable sites for rapid construction of forward airfields, underground water supplies,

[17] 'Inter-Services Topographic Department', *DBpedia*, <https://dbpedia.org/page/Inter-Services_Topographic_Department>

[18] 'From May 1931 to July 1936 [John Leonard Farrington] was in Southern Rhodesia ..., working for the Gold Fields Rhodesian Development Company, for valuation and development of prospects, and as the only geologist at the Wanderer Mine.' P.A. Farrington, 'Discussion of "Terrain evaluation for Allied military operations in Europe and the Far East during World War II: 'secret' British reports and specialist maps generated by the Geological Section, Inter-Service Topographical Department", by E.P.F. Rose & J.C. Clatworthy, *Quarterly Journal of Engineering Geology and Hydrogeology*, 41, 237–256', *Quarterly Journal of Engineering Geology and Hydrogeology* (2009), 42: 389–92.

[19] 'William Bernard Robinson King was awarded the Military Cross for bravery with the British Expeditionary Force before being evacuated from Dunkirk in 1940. He was a Cambridge graduate and World War I veteran who pioneered the use of geological expertise in the theatre of war. King went on to become the 11th Woodwardian Professor of Geology at Cambridge in 1944.' Douglas Palmer, 'W.B.R. King – the Cambridge geologist who went to war', 11 November 2018, *Earth Sciences, University of Cambridge*, <https://blog.esc.cam.ac.uk/?p=1275?>. See also 'The Papers of William Bernard Robinson King, 1909–1961', *Sedgwick Museum of Earth Sciences, University of Cambridge*, <https://archiveshub.jisc.ac.uk/data/gb590-wbrk>.

and sites for the easy extraction of concrete aggregate. By the time I arrived in Oxford, Europe had been successfully invaded and my first major job was to make a military geological map of Hainan island off the south coast of China. Ordinary maps of the island, which is about the size of Wales, were rudimentary, to say the least, and real geological information was scant, and mostly in Chinese.

The Admiralty has for decades run a photographic library in which only dated, orientated and accurately located photographs are kept, but these have a worldwide coverage. During the war it was situated in the Bodleian Library in Oxford and was freely available for staff of ISTD. I thought it was worth trying but expected to find perhaps half a dozen prints because Hainan was so remote and half-explored. They produced 234. Strange information came from them. There was one, taken by missionaries on a picnic, of a goose standing on a rutted cart-track. The goose gave scale, and the ruts were a good indication of the kind of soil in that area. My final printed map and report contained such guesswork. They were never used because the war ended without the Allies needing to invade Hainan. I still have a copy of that map.

The war ended after the biggest bang in history destroyed the city of Hiroshima. Within a fortnight, ISTD had copies of the Intelligence air photographs taken after the explosion. In 3D under the stereoscope, they were an awesome lesson on the destructive power of that comparatively small bomb. They are permanently etched on my memory.

– 5 –

The Bridge Completed

The war ended abruptly and by Christmas I was a civilian again, once more a poverty-stricken student under Professor Read at Imperial College, with two terms to go before, hopefully, achieving the graduate status I needed - nearly ten years after starting my full-time degree work.

By then I had the offer of the Keepership of Geology in the National Museum in Bulawayo, conditional only on my getting a satisfactory class in my Honours degree. By July 1946 I knew that I had the necessary qualification and by the beginning of September I was on a Union-Castle liner bound for Cape Town. What a contrast it was to travel as a civilian compared with the discomforts of troopships.

This time I was expected in Bulawayo, and I was soon installed in the familiar little office opening on to Fort Street, ready to pick up the threads where I had left them so many months before.

I quickly forgot my Professor's irritation that I had chosen, as he put it, 'to bury myself in the bush' instead of taking up his offer of a job on his staff at Imperial College. George Arnold was still Director and Neville Jones was still Archaeologist, but both were on the point of retiring and, with the coming of peace, there were plans to expand the scientific staff to the tune of two more appointments. There was still the irritation of petrol rationing, which limited the amount of fieldwork, but there was plenty to do in the Museum itself and I was soon immersed in it.

In spite of petrol rationing, prospecting was increasing, and base-metal mining in particular was getting a lot of attention. Monday morning saw queues of up to a dozen prospectors with bags of samples for determination. Mining was still dominated by small-workers, but the tempo was quickening and the big companies were moving in. My determinations register bulged, and up to 2,000 samples a year passed through my office. Much of this material was worthless, and the heap of rejects grew to vast proportions in the backyard. Some samples came through the office at least twice: small boys with an interest in minerals picked through the dump avidly and familiar bits often reappeared this way.

The Museum is full of inanimate objects, from stuffed animals to bits of prehistoric pottery, but it is a strangely alive place in which to work. In stark terms it justifies its existence at the taxpayer's expense by providing a number of services. There is the obvious one of exhibition, but only a small fraction of the total collection is on public display. There is the big task of curating the specimens, which are partly a reserve from which to make new displays and partly exist for research. Our museums have a most distinguished research record and have often attracted scientists from all over the world to work in them. This side of their activity is not seen by the public.

Then there is education and the use made of museums by school parties, which can take up quite a large slice of time. Finally, museum staff are in some demand as guest speakers by a range of organizations who run regular meetings. If any organization is let down by an invited speaker, particularly at short notice, it is a fair bet that they will turn in their hour of need to museum staff, and I hope we never let them down. Conducting school parties round galleries and going to schools to lecture can occasionally touch off an answering chord in a young mind, and some quite distinguished scientists of today seem to have got their first inspiration in this way. I can think of several, one of whom is even on the present scientific staff.

Museum life is very varied, and teatime, when all the staff get together, often with their visitors, is nearly always spent 'talking shop'. It is the time to learn about other disciplines and is time well spent. Members of the public bring in weird queries; they even ring up to ask the queerest of questions.

George Arnold got one from a new immigrant just after World War II that went something like this, on the telephone: 'Last night there was a commotion in the hen run. The gardener says it was a snake. Nobody saw it. Was it poisonous?' This tried George Arnold's patience beyond breaking point, and he replied: 'Madam, my initials are G.A. They stand for George Arnold, not God Almighty.' At which point he put the phone down.

My approach was more tolerant, and because my office was the most accessible to the public, perhaps this was just as well. I filtered enquiries off to the appropriate officer whenever possible, but occasionally I was landed with questions on all sorts of topics. Often a quick visit to a particular exhibit would clear the matter up, but not always.

A man came to my office one day and said that he had seen a strange bird but it happened that the ornithologist was away. I knew very little about birds, but there was a chance that this one might be common enough for even a geologist to recognize, so I suggested to him that a good description might help me. He said he would do better than that and would draw it for me. I began to fear the worst: museums attract cranks, and this seemed to be one, for it takes an artist to do more than make vague birdlike sketches. To draw a really specific bird was asking too much. He took pencil and paper and drew a long narrow horizontal rectangle, which he said were the wings. Then he drew a narrow vertical rectangle crossing the first at right angles and said that was the body. This was obviously a crank. But then he took his pencil, and from each wing drew a single wavy line, trailing backwards. There is only one bird this could be. It was a stylized picture of the male Pennant-winged Nightjar in full breeding plumage. A quick trip up into the bird gallery confirmed the identification and off went a satisfied customer.

In 1947 the search for a new Director to succeed Dr Arnold was on in earnest. While I had been a part-time member of the staff during World War II, it had been half hinted to me that I might be considered, but by the passage of time, and perhaps by a change of heart, the subject was never mentioned again. I began to feel that the inclination was towards a zoologist, and preferably an entomologist, to carry on where Arnold's research had led with such distinction. The choice finally fell on Reay Smithers, a zoologist, but not one directly concerned with insects. He was a spider man and had worked on the deadly button spider in the Cape Province. His work had helped materially in the development of a serum to be injected into victims of button-spider bite. This work had been very favourably received in scientific circles and the new serum had done much to improve the chances of survival of the victims.

This spider, which was later found to be identical with the infamous American black widow spider, was the cause of deaths annually in the Cape wheat belt among harvesters. It is an inoffensive spider but had the habit of nesting in corn stooks, and when these were handled by harvesters, bites were not uncommon. It was virtually unknown in Southern Rhodesia at that time, but Arnold hoped that Smithers would concentrate on the spider fauna that until then had been neglected, whereas true insects had been investigated in detail.

Reay Smithers showed soon after his arrival that button spiders were common enough in Matabeleland. In fact, they lived in some numbers under the overhanging damp course just outside the Fort Street door of the geologist's office, where I found them very useful. Visitors quite often asked about them, and it was easy to show them the ones just outside my door. The egg sacks look like cotton-wool balls, and with a long pair of forceps it was easy to find the female, pull her out for inspection by the visitor, and slip her quietly back in her nest. Like all spiders, the males are insignificant, and are eaten by the females after mating.

I have found them since in many strange places, including under a stone hand axe in a riverbed, under the handle of a PK (toilet) in a house, and outside the window of the Principal's office at the university in Salisbury. They are quite common but shy and retiring, only their characteristic egg sacks giving them away. Reay Smithers's appointment was duly confirmed and announced, and I waited to see what manner of man was destined to become my boss.

Late one afternoon there was a knock on my Fort Street door; I unlocked and opened it. There was a man about my own age but much taller. He was wearing a trilby hat at a jaunty angle, a tweed jacket, rather baggy grey flannels, and sported a small, rather gingery, moustache. 'I'm Reay Smithers,' he said, and so began an association which has lasted ever since, and which has survived strange changes. We began as colleagues of about equal status on the staff; he then became Director and I worked under him. Many years later, I became, in turn, a Trustee and then Chairman of the Board, and Reay was my Executive Director. None of these changes made the slightest difference to a solid friendship between us and were accepted by both of us as quite natural.

In due course George Arnold officially retired and Reay took over, but he never worked on spiders. His interests soon turned to birds, and later to mammals, on both of which he became an international figure.[1] Very early on in his time at the Museum in Bulawayo he developed a passing interest in bats, and he asked me if I knew of any old abandoned mine workings near the city where he could collect some bats. I thought an old adit[2] on Hope Fountain farm might contain a colony, and an inspection soon proved that it was literally hanging with them, and so we arranged a collecting trip. Bat collecting has now become a highly sophisticated operation and, looking back on our efforts, our singular lack of success could easily have been foretold.

Reay Smithers, Kariba, 1959

[1] A review of his *Check List of the Birds of Southern Rhodesia* (Bulawayo: Rhodesian Ornithological Society, 1957) said that it 'presents for the first time a thoroughly authentic list of the birds of Southern Rhodesia. ... After rejecting 32 species previously listed on the basis of vague or unsubstantiated evidence, the authors present an annotated list of 565 species which have been collected within the political boundaries of the country.' John T. Emlen, *The Auk* (1957), 74(4): 513.

Perhaps his two best-known books on mammals are: *The Mammals of the Southern African Subregion* (Pretoria: University of Pretoria, 1983), a third edition of which, revised by J. Skinner & C. Chimimba, was published in 2005 (Cambridge: Cambridge University Press); and *Land Mammals of Southern Africa: A Field Guide* (Halfway House: Southern Book Publishers, 1986; 2nd edn, 1992), a fourth edition of which, revised and updated by Peter Apps, was published as *Smithers' Mammals of Southern Africa: A Field Guide* (Cape Town: Struik Nature, 2012).

[2] A horizontal passage leading into a mine.

Our simple plan was that I should go into the adit, stir up the bats, and Reay would stand at the entrance and hit them with a tennis racquet. To our innocent minds it was simple. Stirring up the bats was easy, and neither of us gave a thought to the possibility of getting cave disease - histoplasmosis, a fungal infection of the lungs. The spores of the fungus accumulate on the guano-covered floor of caves inhabited by large colonies of bats, and it is a real hazard. The bats flew towards Reay but quite easily avoided his flailing racquet, broke back over my head each time, and retreated into the depths of the old adit. Bats may be blind, but their sonar is so remarkable that they could dodge the tennis racquet every time and not one was hit. The only live thing that Reay hit was me. I stopped a stinging blow on the side of the head, and we gave up the unequal battle.

Perhaps not surprisingly, he showed no further interest in bats but rapidly developed his interest in birds, which can be collected much more easily, either by bird nets or by shotgun. The Museum's collection of bird skins was small. All exhibited specimens had been prepared overseas, and there was no one on the staff who could be called a skinner, let alone a taxidermist. Reay set about correcting both deficiencies. He imported an African skinner from Cape Town. This man, Ali, was running a herbalist practice in the Cape but had previously been personal skinner to a world-famous ornithologist, Admiral Lynes.[3]

Ali had travelled widely with his Admiral and had been as far as Siberia, where he claimed to have eaten roast mammoth, the only man I have ever met who claimed this distinction. And distinction it is, for the last mammoth died about 15,000 years ago. Rare carcases have been found in frozen bogs in Siberia in a state of perfect preservation; nature invented the deep freezer before man. The Tsar put roast mammoth on the menu of a state banquet in 1912, and there is at least one stuffed specimen in a Russian museum.[4]

These frozen mammoths had finally cleared up an apparent paradox. The extinct mammoth was obviously an elephant. Its bones were well known by the mid-nineteenth century from all over Northern Europe, including the Thames valley, Siberia, and even north America. Mammoth ivory was a

[3] Rear Admiral Hubert Lynes (1874–1942). 'Throughout his service life and during retirement, Lynes was a noted and experienced ornithologist who contributed to numerous books on the subject and was in his lifetime considered the leading expert on African birds.' 'Hubert Lynes', *Wikipedia*, <https://en.wikipedia.org/wiki/Hubert_Lynes>.

[4] Apparently, '... real mammoth tastes just dreadful, according to Russian scientist Alexei Tikhonov. Once, after a few shots of vodka, Tikhonov took a bite of a woolly beast that had been dug out of the Siberian tundra – as Richard Stone recounted in his 2002 book *Mammoth*.' Angus Chen, 'Two Grad Students Use Science to Bust the Dinner Hoax of the Century', 5 February 2016, *NPR: The Salt*, <https://www.npr.org/sections/thesalt/2016/02/05/465309625/two-grad-students-use-science-to-bust-the-dinner-hoax-of-the-century>.

flourishing export from Siberia and was used for, among other things, making billiard balls. Yet the deposits in which the remains were found indicated Arctic conditions, and all modern elephants are tropical animals. How could a tropical animal live in the arctic wastes without freezing to death?

The first clue came from prehistoric paintings, some 20,000 years old, found in Spain and southern France. These showed elephants with what looked like a thick covering of long brown hair. These pictures were painted by Stone Age men who hunted the mammoth and may have hastened its ultimate extinction, though the retreat of the ice at the end of the Great Ice Age no doubt ruined its habitat. The discovery of deep-frozen whole carcases finally cleared up the puzzle: the mammoth had a thick coat of long brown hair to keep out the cold.

Ali set about skinning the birds for Reay in the proper expert way so that the result was scientifically acceptable for study skins and also available to the taxidermist for exhibition. He trained other skinners in the art and established a new tradition of excellence not only for birds but for small and even medium-sized mammals as well. The tradition lasts to this day.

A further recruit was an amateur taxidermist, Terence Coffin-Grey, who joined the Museum staff as a technical officer. He founded a tradition that has flourished ever since, and the excellence of display in zoological exhibits in our Museums stems from his brilliant work as a taxidermist and trainer of taxidermists.[5] With these two new appointments, the Bulawayo Museum (more correctly the National Museum of Southern Rhodesia) moved into a new era under Reay Smithers's direction.

His appointment was indeed a stroke of genius by the Board of Trustees. He had the vision to see how the Museum service should be expanded to cater properly for the needs of this country. The Smithers era was truly the golden age for our museums. For the first time there was adequate administrative and technical backing to make modern display techniques possible here. Artists were brought in so that the proper backgrounds could be provided for dioramas. The team for a new exhibit in any department could at last be more than one. In my first years I did my own labels laboriously by stencil, drew my own diagrams, did my own photography, and even had to be my own electrician for internal lighting of new displays. Under Reay Smithers all that changed. The Keeper of a department became the captain of a team of experts, and the translation of his scientific brainchild into a public display was done

[5] 'On July 24, 2021, Terrence Coffin-Grey was officially inducted into the Taxidermy Hall of Fame by THOF Board Member Ken Edwards of Georgia during the 2021 National Taxidermists Association's awards banquet.' 'Terence Coffin-Grey', *Taxidermy Hall of Fame*, <http://taxidermyhalloffame.org/terence-coffin-grey>. And see Appendix 1.

under his direction by people properly skilled in the various techniques to be employed. The results are there for all to see, and our museums rank highly - not only in southern Africa.

Reay also envisaged a museum in each centre and set about finding money to rebuild existing museums on better sites and with custom-built architecture. From the single unit, founded in 1901 in Bulawayo, he oversaw the expansion to Salisbury, Umtali, Gwelo and the Zimbabwe Ruins near Fort Victoria, all of which now have good museums in good new buildings.

One facet of the expansion policy may sound a bit brutal but it has proved its worth. If a new museum is started by local enthusiasm, it is carefully nurtured under its local committee of citizens - but it is taken on, and becomes part of the main organization, only once it has demonstrated its viability in terms of local support.

Reay also encouraged his scientific and technical staff to do long collecting trips to broaden their research, for he was a believer in research being published so that the Museum's standing should be enhanced in the eyes of the world scientific community. To this end he built up the Museum's transport fleet so that journeys to very remote areas, even beyond our borders, became possible. All these policies were entirely in line with my own ambitions and I made full use of them. George Arnold and Neville Jones were at last able to get around and extend their own research in a way never possible before.

Little George Arnold was never much of a driver or camper in the bush, and it fell to my lot to take him on collecting trips. The first of these was to one of his favourite places, Mount Selinda Forest in the Eastern Districts. I was understandably apprehensive. George was fastidious about food, comfort, cleanliness, as well as scientific accuracy. I wondered how he would tolerate my cooking and camping techniques, which were both a bit crude. Additionally, it was the beginning of the rains, for few bugs are out in the dry season. There was little I could do on the geology in the forest, so I was provided with a butterfly net and a killing bottle and told to collect. The trip was for a week to ten days, depending on how good the bug hunting was; to begin with it was terrible. There had been little rain before we arrived and for a day or two George caught practically nothing that interested him.

As a complete novice I collected everything I could find, only to have 99 per cent rejected with the caustic and characteristic comment from George, 'Pah - common as dirt'. Then it rained, accompanied by a truly magnificent thunderstorm. Flying termites (never call them flying ants in George's hearing) came swarming into our small tent, and after them all the frogs and toads in the forest. George was in his element. He watched the frogs and toads gorging themselves and even fed them by hand, speculating why one termite was accepted greedily while another, seemingly identical to us, would be rejected. We were sure the next day's collecting would be good but it was still poor – except for a rather beautiful beetle that I collected accidentally. While walking along a path with my net over my shoulder, something fell into it, and it turned out to be a newly emerged beetle that George was sure was new to science. So it was, but I do not think it was named after either of us.[6]

Another journey with George was to the Robins end of Wankie National Park, where we collected moths and other insects at night with a pressure lamp hanging in front of a large white sheet to which all manner of night-flying insects were attracted. Just for reassurance I kept a loaded rifle handy, but George was unimpressed. On the return journey we were late, and by sunset we were still 150 miles north of Bulawayo. I stopped and got out the whisky and water-sack: George was a stickler for ceremony at the end of the day. I noticed some limestone nodules in the roadside drain and broke one open. It disclosed a fossil snail shell, which was eventually shown by a specialist at the Natural History Museum in London to be a new species and the first fossil snail from Gondwanaland. Scientific progress is not all a matter of careful planning.

Another newcomer at this time was Roger Summers, who was eventually to replace Neville Jones as Archaeologist.[7] He had begun life as an actuary, served as a gunner in World War II, and turned archaeologist after the war – largely, I think, to avoid being an actuary again. He had qualified at the London Institute in Regents Park, but he was new to Africa, whereas Reay had grown up in the southern most part of the continent and was a good bush man before he joined the Museum.

Roger's first major field trip was to the Bumboosie ruins just outside the north-western boundary of Wankie National Park, in very wild sandstone

[6] See the reminiscences of Susan Warner (now Stocklmayer) in Appendix 1.

[7] In his obituary of Summers, Robert Soper wrote: 'In addition to this academic approach to archaeology, [Roger Summers] was also concerned to disseminate the information to the general public in clear non-technical language. The title of his *Zimbabwe: A Rhodesian Mystery* [Johannesburg: Nelson, 1963] was perhaps intended to attract the attentions of a white population brought up on romantic ideas of the Queen of Sheba and King Solomon's Mines, but it is a scrupulous and clear account of the archaeology and economy of the site based on the 1958 excavations, detailing the history of research, as well as oral traditions.' Robert Soper, 'Roger Summers, 1907–2003', *Azania: Archaeological Research in Africa* (2003), 38(1): 217–18.

country.[8] I went with him and we had Ted Davison as our guide.[9] The ruins are among sandstone hills with much rocky outcrop, and close to them is a rock shelter with spoor carvings, a great rarity in this country. Roger was to excavate the cave floor and see if any link could be found between the rock carvings and any Stone Age culture in the cave earth. Fossil wood is quite common in this area, and any link there may be was through bits of this very hard silicious rock, which breaks easily into sharp points. The sieving from the cave earth produced a good deal of Late Stone Age cultural material, including pencils of fossil wood with worn points, which could have been used to make the spoor engravings. This rather tenuous connection would make the carvings as old as the Late Stone Age, which is now known to go back several thousand years.

The spoor carvings in the cave were all identified easily by Ted Davison, and included many antelope species, big cats and even otters. In the cave they are quite fresh, but there are some in the open that are deeply weathered and can be seen only when the sun is low and the light is oblique. There are also remains of at least one *tsoro* board for playing a game still common in this part of Africa. The board has four rows of eight pits and is played nowadays with crown corks. If it goes back to the late Stone Age, as it appears to do at Bumboosie, it may be the oldest game known in the world, but I have never learnt to play it.[10]

These, then, were my scientific colleagues just after World War II: Reay Smithers, Roger Summers and, of course, George Arnold and Neville Jones, both of whom actively carried on their research in honorary capacities right up to the time of their respective deaths. This, too, was the stimulating environment that encouraged me to get on with research in my own field.

[8] Now known as Bumbusi National Monument, it was declared a National Monument in 1946. 'The National Museums and Monuments of Zimbabwe ... owns the Bumbusi National Monument, but does not have sufficient funds to protect, maintain, or restore the site.' '2008 World Monuments Watch', *World Monuments Fund: Bumbusi National Monument*, <https://www.wmf.org/project/bumbusi-national-monument>.

[9] Ted Davison was the first warden of what is now Hwange National Park. He took up the post in 1928 at the age of 22. In his book about the game reserve, he acknowledges 'with thanks the advice of Dr G. Bond, Mr R. Smithers and Dr W.A. Wragg.' Ted Davison, *Wankie: The Story of a Great Game Reserve* (Salisbury: Regal Publishers, 2nd edn, 1977).

[10] See J.B. Matthews, 'Notes on Some African Games', *NADA, the Rhodesian Ministry of Internal Affairs Annual* (1964), 9(1): 64–6.

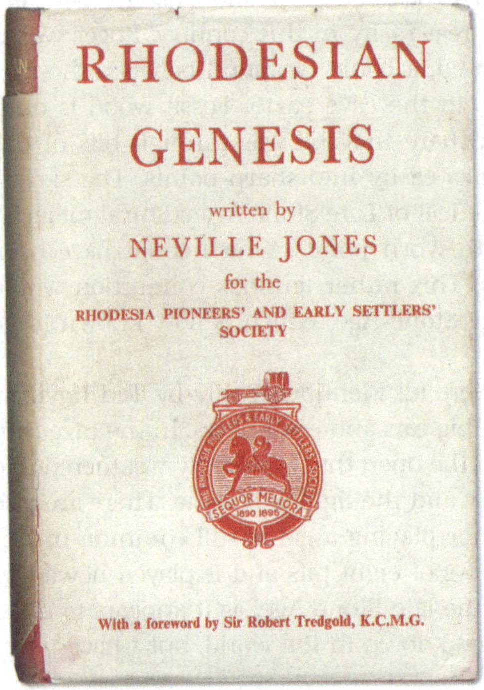
RHODESIAN
GENESIS
written by
NEVILLE JONES
for the
RHODESIA PIONEERS' AND EARLY SETTLERS' SOCIETY
SEQUOR MELIORA
1890 1896
With a foreword by Sir Robert Tredgold, K.C.M.G.

Guide to the Waterfowl of Rhodesia
REAY H. N. SMITHERS
and M. J. S. MacKENZIE
illustrated by M. J. S. MacKENZIE
NATIONAL MUSEUMS & MONUMENTS OF RHODESIA

WANKIE
The Story of a Great Game Reserve
Ted Davison

ZIMBABWE
A
RHODESIAN
MYSTERY
Roger Summers

Not all the fieldwork was research. Geologists were rare in Matabeleland and there was growing pressure on me not only to do mining consulting but also to get involved in underground water and foundation engineering,[11] for neither of which did my rather scientifically pure upbringing particularly fit me. I had to tread warily and learn as best I could. For foundation engineering, the best advice I ever had, came very early on from a consulting engineer who told me bluntly that civil engineers were not interested in fancy rock names but only in rock defects.

My first venture in this new field came from the City Engineer when a new water pipeline was to be installed from Ncema Dam to the city, a distance of thirty-seven miles. He suspected that the contractor had estimated that the pipe trench would need blasting every inch of the way and had tendered accordingly. He asked me to examine the full length and make an estimate of the percentage of the distance that would need blasting to make a six-foot trench. My estimate was less than one third. Perhaps this saved the ratepayers some money.

The next job was less tiring and involved the planners' choice of a new cemetery site – ideal on paper, but no one can have looked at it except on a map. The City Engineer suggested that it should be vetted geologically. At a quick glance it looked like a fossil cemetery already, for it was on steeply dipping hard schists that stuck out of the ground in long lines like miniature tombstones. All graves would have needed explosives to get them down to the regulation six feet.

Matabeleland is dry country, and the Rhodesia Railways' motive power was entirely steam. Locomotives are thirsty machines: for instance, a Garratt articulated engine needs about five thousand gallons of water every twenty-five miles. The 1946/7 wet season was a disaster, and water was so scarce on the Bulawayo–Victoria Falls line that, at the peak of the drought, traffic had to be cut by fifty per cent simply for lack of water.

The Railways had two sources of water – dams and boreholes – and their boreholes had always been sited by water-diviners. The engineer in charge of water was a somewhat crusty individual. By chance he had been taught as a naval cadet at Dartmouth by my father, but illness had ended his career in the navy and he had turned to civil engineering. Reluctantly, I think, he had appointed a young geophysicist, Don Macdonald, trained in the Western Desert during the war. He joined the Railways on demob and set about

[11] 'Foundation engineering is the application of soil mechanics and rock mechanics in the design of foundation elements of structures.', 'An overview of foundation engineering', 14 August 2020, *Constro Facilitator*, <https://www.constrofacilitator.com/an-overview-of-foundation-engineering>.

revolutionizing railway water supply over the entire network, including the drought-stricken Bulawayo–Victoria Falls section. Heavy on geophysics, he was light on geology, and I soon found myself working with him on borehole sites.

The water engineer obviously mistrusted these new-fangled ideas. For a full year, he solemnly sent a diviner to check Don's scientifically chosen site. Don laughed it off and his rate of success soon convinced High Authority that he really could find water with his mysterious box of instruments.[12]

[12] This acknowledgement appears in P. Eldon Dennis and L.L. Hjndson, *Ground-Water Provinces of Southern Rhodesia* (Washington DC: United States Geological Survey, Geological Survey Water-Supply Paper 1757-D, 1964): 'Maps, reports, data, and discussions were freely given ... Dr Geoffrey Bond of the National Museum at Bulawayo; and Mr D. Macdonald, Water-Supply Engineer for Rhodesia Railways.' <https://pubs.usgs.gov/wsp/1757d/report.pdf>. See also Appendix 3: 'Geoffrey Bond's Publications'.

– 6 –

Divination

I have always been interested in the mysteries of divining, dowsing, or whatever you like to call it. Traditionally, in England, a hazel twig is used. There were plenty of hazel bushes growing near my home village, but I had no idea how to use one, though I knew it was done with a forked stick; it just wasn't needed in my part of England, where surface water was almost too plentiful and springs were quite common. Only in exceptional summers, like that in 1926, did our rivers and springs show any stress.

A famous case of the success of a dowser, in rather unusual circumstances, helped to sharpen my interest but gave me no further practical help. The river Wharfe near Bolton Abbey, some twenty miles downstream from Arncliffe, goes through a narrow cleft in the rocks before entering a large deep pool that is surrounded by rocks with many overhanging ledges. The narrow cleft is called the Strid, presumably because it looks so narrow that one can stride across, and so one almost can. It is an easy jump for the young, but only in one direction. It is said that no one has managed the return leap and that failure inevitably means drowning. Stories of such tragedies go deep in local history.[1]

I think it was in the early 1930s that a well-known local landscape painter disappeared. The tools of his trade – easel, paintbox, stool and so on – were found along with his jacket on one side of the Wharfe near the Strid, and the police concluded that he had tried his luck and jumped one way successfully. His gear was on the bank from which the leap is easy, but that he had missed the return jump, fallen into the chasm and drowned. However, his body could not be found, and after a week of dragging the pool the mystery was no nearer solution. After all, plenty of people have faked disappearance, and without a body there is always doubt.

At that stage, the local police received an offer of help from a water diviner, I think a woman, living in Harrogate. I can imagine a certain scepticism

[1] 'The Strid gets its name from the Anglo Saxon "Stryth" meaning Turmoil or Tumolt; corrupted to Strid, from the possibility of striding across the channel.' 'Strid Wood and the Strid', *Bolton Abbey*, <https://boltonabbey.com/what-to-see/strid-wood-and-the-strid>.

among the authorities, but, since the offer had been made, there was nothing to lose by its failure and everything to be gained if it succeeded. So the diviner was duly turned loose on the problem and, after a period of going round the banks on both sides (without jumping the Strid presumably), she pointed to a spot and said that the body was lodged under an overhang below. And it was.[2]

This is quite baffling. It's hard enough to believe that underground water can influence a hazel twig, or anything else, when it is properly held in the human hand. Many diviners claim that there is a mysterious force of attraction between certain metals and their ores, though science seems unable to find it. But what attraction can a human corpse exert below rock cover and many feet below water? But she was right. Did the twig really help, or did she subconsciously weigh up the shape of the pool and the direction of the currents, and so pick the most likely place? I don't know, and it was not until I came to this country that I began to make direct contact with diviners.

Even before I took up full-time work at the National Museum in 1946, I had learnt that divining for water was a common practice in the rather dry environment of Matabeleland, and that geologists were fair game for water diviners. Since very few geologists had, at that time, any training in the mysteries of underground water, myself included, it was hardly surprising that water diviners considered themselves, and were considered by the farming public, to be the real experts. I was constantly told of their successes but never seemed to hear of their failures. There were several very famous diviners in Matabeleland in those days, and they were famous, quite logically, because of their successes, especially as at that time geophysics was in its infancy in this country. It had no more prestige than the geologists, and I heard much more of its failures than of its successes. Being scientists, they keep proper records of success and failure; diviners, being mystics, do not seem to be so systematic.

It was soon obvious to me that water was not the only commodity that diviners claimed to be able to find: metals and metal ores were included in their repertoire, though I never seemed to find a rich one. An early experience involved a man who claimed that he could divine gold in quartz reefs, but he was not a practising small-worker. I found this rather surprising, but he

[2] The water-colour artist was Arthur Reginald Smith, and the incident occurred in September 1934. Andrew Robinson, 'The innocent looking stretch of Yorkshire river that's one of the deadliest places in the world', *ExaminerLive*, 27 February 2021, <https://www.examinerlive.co.uk/news/local-news/innocent-looking-stretch-yorkshire-river-18615167>.

offered me a demonstration and this was the first time that I actually saw a diviner at work.

He took me to a spot near Bulawayo where he said he had found two quartz reefs with gold, which intersected at right angles. It was late in the dry season, so the grass was down. There was no sign of outcrop, not even sparse quartz rubble, but it was on greenstones of the Bulawayo gold belt, and quartz reefs are not always obvious. The ground was not pegged by the diviner or anyone else, and I wondered why not.

He cut a fresh mopane twig of the appropriate shape and I persuaded him to let me mark the point of intersection with a small stake that I drove in with my geological hammer. He then set to work about twenty feet from the point of intersection and walked around in a square. The rod went down sharply at four points, which I marked with sticks pushed into the ground. They very neatly marked out the four arms of the crossed reefs. Intersecting reefs are not unknown, but since they often follow structural lines such as joints and faults, right-angled intersections are not the commonest pattern.

But I was not prepared to discount his demonstration simply on theoretical grounds, so I suggested blindfolding him to see if the rod still dipped on the lines he defined. To my surprise he agreed to the test, was duly blindfolded, and I led him to one of the pegs I had put down, turned him round a few times to destroy any sense of orientation, and off I set at a slow pace holding him by one elbow. The rod went down after about ten paces and the spot was duly marked. I took him round the exact circuit that he did in his first demonstration and marked each point where the rod dipped. There were quite a number, and they were more or less evenly spaced, but none of them coincided with the four arms of the cross.

He was quite cheerful about it, but he had to admit that he had not been very successful. We parted at my office and I have never seen him since. I was impressed by the amount of twist he put into the twig, and by the regularity of its dips after a regular number of paces, and fell to wondering if this had something to do with muscle fatigue, but, from subsequent tests on myself, I rather tended to discount this idea. The effort he put into the performance was considerable, and he was visibly tired by his work that day, a point made many times in later contacts with diviners. On hot days they really perspire with physical and mental effort. The twig works for me, but I do not exert the same intensity of effort, and I don't believe my results anyway!

Beryl was much sought-after about the mid-1950s, especially when it was realized that a lot of our beryl is much like quartz vein material, not always being green or having a nice hexagonal crystal form. Colour is no guide. Hardness and specific gravity have to be determined very carefully and confirmed by optical tests under the microscope, so it was not surprising that I had a lot of customers bringing material in for determination and to get advice on how to determine it by eye in the field. Lustre, a property hard to define, is a fair guide. It really means the look of a freshly broken surface in strong light and is quite a good guide in experienced hands. Quartz has a glassy (vitreous) lustre; beryl looks distinctly greasy or waxy. I was not surprised when a diviner who was also a prospector, or would-be prospector, turned up who said he could divine beryl and wanted me to see him do it.

He had three rods, of different metals, bent into a conventional forked-stick shape. One was thick fencing wire, the second was thick aluminium wire, and he claimed that the third was thick magnesium wire, though I can't imagine where he got it from. The theory was that beryl is a light metal (it is the fourth element up in the scale of atomic weights and is the lightest metal except for lithium) and that it would attract a light metal 'twig'. Hence the magnesium rod would work but not the other two, even aluminium being too heavy.

The demonstration was done in my office, and the first specimen was a large lump of beryl weighing several pounds. He walked backwards and forwards over this specimen with each rod in turn. Only the magnesium rod 'reacted' and he turned to me in triumph. I was politely interested but suggested trying further specimens. He agreed, but I played a bit dirty. Without telling him, I used topaz, feldspar and quartz in lumps of approximately the same size. In all cases, only the magnesium rod 'reacted'. He was a bit put out when I told him that I had deliberately used other minerals, but as a test it would have been useless if he had known. I have never seen him again either.

There was a great deal of publicity on radioactive materials when there was an office of the British Atomic Energy Authority in this country, and much interest was generated among prospectors. The office was under the direction of the late Ken Branscombe, a geologist well trained in uranium prospecting.[3] They brought with them supplies of the latest solid-state detecting devices, Geiger counters, scintillometers, and continuous recorders, including probes

[3] See Chapter 13 for more detail about Ken Branscombe and the Atomic Energy Authority.

for logging boreholes. Naturally, this sort of equipment was beyond the means of the average weekend prospector, who could do no better than crude gold-leaf electroscopes, fluorescent beads and photographic plates.

I was not surprised when a well-known water diviner came to my office and suggested using his rods. However, he would not be able to experiment unless he had something that contained uranium and was, therefore, strongly radioactive. Could I lend him something with which to work? Of course, I agreed, and I selected a specimen from the famous Shinkolobwe uranium mine in the Belgian Congo.[4] I knew him well and trusted him. The specimen was about the size and shape of half a cricket ball, and consisted of shiny black pitchblende in the middle, with concentric shells of weathering products, the so-called uranium ochres, on the outside. These are orange, yellow and pale green, and quite soft and powdery.

I did not see him again for a fortnight, but he appeared and was obviously pleased with his experiments. So, we piled into his car and went to a big open space in the northern end of Bulawayo for a demonstration. This time I felt it better to do it outside my office, which at that time had an assortment of radioactive specimens temporarily housed there and must have been rather full of radiation in consequence. Looking back on it now, I realize that I must have been exposed to rather too much radiation, particularly first thing in the morning after the windows and door had been closed for the night. The atmosphere must have had radon gas in it, though I do not seem to have suffered any ill effects even twenty-five years later.

The diviner parked his car on a broad grassless path, pulled out the specimen, and planted it in full view on the bare earth. He then produced a variety of rods from the boot of his car. They were all fencing wire, but some were 'twigs' and some were 'Ls'. Some had black tips, others were tipped yellow or orange. The demonstration began.

There was the now familiar slow walk, the usual dreamy far away look in the eyes, and the expected reaction of each kind of rod as he passed slowly over the specimen. There was also the expectable air of triumph at the successful conclusion of the demonstration. As usual, too, I protested that I was unconvinced, on this occasion because the specimen was too obviously visible on the bare earth. I argued that it would be much more convincing if I could hide it and he could find it. I was surprised how readily he agreed to such a

[4] 'The Shinkolobwe mine [in Katanga province] – named after a kind of boiled apple that would leave a burn if squeezed – was the source for nearly all of the uranium used in the Manhattan Project, culminating with the construction of the atomic bombs dropped on Japan in 1945.' Frank Swain, 'The forgotten mine that built the atomic bomb', *BBC Future*, 4 August 2020, <https://www.bbc.com/future/article/20200803-the-forgotten-mine-that-built-the-atomic-bomb>.

stringent test of his skill. He sat in the car, hid his face in his hands, and even bent down below the dashboard, while I wandered around in a random way and, without stopping, slipped the specimen under a tuft of grass.

For nearly two hours he persevered, first with one rod and then another, quartering the ground quite systematically. At one stage a rod reacted, and he pounced, only to come up with a rusty tin can. In the end he was exhausted. I was sceptical but realized with horror, that I did not remember where my precious museum specimen was. I could not find it. So we made a quick trip back to the office, collected my portable scintillometer, checked the batteries and returned. By equally systematic traversing, it was very quickly found. By then, my prospector had recovered somewhat physically, but he was undismayed by his failure. He claimed that he had gone within twenty feet of it. Such is the faith of the dowser.

I tested his rod tips: they were all radioactive. Scrapings from my specimen had been mixed with glue and applied to the tips so that 'like would attract like', in defiance of the general observation of physicists that like repels like, except in the case of gravitational attraction between masses.

So far, as a result of my rather unkind and not very scientific tests, I remain unconvinced by metal diviners. Perhaps the most damning evidence, as mentioned above, is that I have never met a rich one. This, however, does not dispose of the age-old use of divining in finding underground water. Incidentally, was Moses a dowser?

Some of the jargon they use disturbs me. There is always talk of underground streams, and even lakes, and of strong flows. There is plenty of opportunity in mines to see underground water, often in quite uncomfortable quantities. Some small gold mines have made so much water that they have produced nothing else for weeks or even months, until the water level could be reduced enough to get at the reef. Others, of course, do not make enough water to run the mill. Seldom does a mine make just the right amount.

Underground water certainly occurs in fairly well-defined fissures, but they are very narrow and water moves fast in them only when they are artificially intersected. The yield then depends on the pressure head and the permeability of the fissure filling. Except in limestone country, where the rock is to some degree soluble in water containing dissolved carbon dioxide, I have never seen anything that could warrant the description of an underground lake, river or stream. Fissures certainly have direction, and groups of parallel fissures are common, all having more or less the same orientation. They are generally steep and often nearly vertical.

A friend of mine built a house just after the war in the Bulawayo suburb of Hillside. This is on an interesting little intrusion of syenitic rock, which physically resembles granite, and this area is notoriously difficult for wells and boreholes. He needed a well and somehow (by divination?) found a site, put on a gang of well-sinkers and went down 130 feet in dry, or nearly dry, rock. At that depth, the well produced a few gallons of water which was easily baled out by the gang in the morning.

When I was called in to inspect this hole in the ground, there were six small stones neatly laid out near the collar of the well, indicating that six well buckets of water had been baled out that morning when the gang started work. The prospects for success were low.

I took my lamp, a carbide/acetylene one in those days, put one foot in the bucket, one hand on the rope, and started the descent as two of the gang lowered me by windlass. This method of going underground is not ideal. The bucket tends to swing and also to rotate from the twist in the cable. There is no braking mechanism, and it depends on the two operators on the handles. It tends to be jerky. It is not a recommended mode of transport. The theory is that swing and rotation are controlled by the victim in the bucket by using the spare arm and leg. Speed is supposed to be slow and controlled, partly by shouting instructions up the shaft to the operators. I got wet once by mixing up *phansi* (down) with *phezulu* (up) and instead of stepping off into a dry horizontal drive, I was lowered into a sump full of water.

There was no problem in my friend's well. It was practically dry at the bottom. I stepped out after the bucket made a relatively smooth landing and proceeded to examine the rocks. They were fresh, unweathered and very little fractured, but there were some hairline fractures that oozed a little water. I noted their direction and climbed back into the bucket for the return journey. It was uneventful. In neither direction did the bucket get caught on a projecting rock with the opportunity of being tipped out.

I told my friend that going deeper seemed useless. He should try following the fissures to see if they opened up at all. I left and returned two weeks later to find literally dozens of stones in neat lines near the collar, and a thoroughly satisfied customer who had all the water he wanted in his well. I felt I had done a good job for an old friend, until he told me that there had been a mix-up in the directions he gave to the gang. They had indeed driven horizontally but, by mistake, at right angles to the direction I had suggested. Within a few feet they had intersected a strong water-bearing fissure and had to get out fairly quickly as the bottom of the well began to fill. I left with mixed feelings.

A borehole at Mpofu siding near Wankie Main Camp gave me a chance to study water divining more closely because the driller was a keen and experienced diviner. He always checked the sites he drilled by divining; one wonders what the geophysicist who sited them thought.

This diviner had elaborate methods. He first used the conventional forked stick to find the best site. He then used L-shaped lengths of fencing wire held loosely in his hand so that the long legs of the L pointed horizontally forward and lay parallel, about four inches apart. As he walked over the site, the Ls suddenly swung apart until they were at 180 degrees, and this gave the direction of the underground stream. He then took a very long forked stick, about 2 foot 6 inches long, knelt down, twisted it up, and put his wrists on the ground. The height of the point of the stick was measured by his assistant with a tape measure, in inches; this was then called feet and multiplied by ten. For instance, if the assistant measured the point at twenty inches above ground, this became two hundred feet, and this was the depth to water.

The logic escapes me, but he firmly believed it. Why should the short stick, held at about hip height, dip down on water, whereas the long stick, held at ground level, turn up over the same water? Why measure in inches, call them feet, and multiply by ten. This particular hole was in horizontal permeable saturate sandstone. In theory, the site was not critical. Anywhere on that sandstone should have been satisfactory, yet the precise spot had to be defined and the direction of the stream determined. By the same reasoning, why had the geophysicist chosen that precise spot?

After I learnt these methods from that driller, I had to do the journey back to Bulawayo, and rather than drive on well into the night on an old strip road, I decided to make a one-night stand and go on the next morning. I stopped at a spot I had used before, an uninhabited area in rather hilly country. After setting up camp, I took out the L-shaped rods (the driller had given me his) and walked down a game track. The rods reacted strongly next to a small bush on the side of the path. I walked backwards and forwards past it, and the rods reacted each time, but only at that spot. I sat down to think and smoke.

After some time, an African youth came along the path. I stopped him, showed him how to hold the rods, and got him to walk past the bush where I kept getting a reaction. Exactly the same thing happened when he reached the exact spot. He left me, looking thoroughly mystified. So was I, but it makes you think. What activates the stick? Why can some people use it, and others cannot? Does it really find water? I do not know the answers.

I have more than a little faith in experienced water diviners working in areas that they know. So many different kinds of implements are used that one wonders whether any of them are really necessary. Perhaps subconsciously the diviners observe subtle signs on the surface, such as trees with better canopies and perhaps some sort of alignment. Termite-heap distribution could also be an indicator, as it is known that these creatures go at least 200 feet down fissures for water, which is essential for their fungus gardens. There could be alignment or a vague pattern of termite mounds.

Dolerite dykes can act as underground dams and their outcrop is generally marked by rounded boulders and red soil. Whatever diviners do, many successful boreholes have been sited by them, but I have yet to meet one who keeps a strict record of success and failure.

I have given many talks on divining to many groups of people. I can always find a site, even in the middle of the lounge in a private house, but I do not really believe in what I am doing, and I have never had it examined by the acid test of actually drilling a borehole. At X dollars a foot, I am not prepared to risk it!

I did a *Science Forum* radio programme in one series, inviting a water diviner and a geophysicist, and looked forward to a good half-hour of hot argument. It was a damp squib. They spent far more time agreeing than disagreeing, but one thing emerged: the geophysicist did not believe that any unknown kind of force was involved.

One of the first books I ordered for the main University Library when I set up the department of Geology, was a new investigation of water divining carried out in Holland. It eventually arrived and was said to have been catalogued under 'witchcraft', perhaps appropriately. But I have never been able to find it!

– 7 –

Hull and Nigerian Interlude

The year 1948 was a disturbed one in Bulawayo. There were strikes, and conditions seemed to be getting progressively less settled.[1] There had been petrol rationing during World War II, as might have been expected. The regulations were taken off too-optimistically soon and had to be put on again, which was a severe handicap for the Museum staff and their need to get back into the field to collect the basic data for their research. The same situation was to recur during the fifteen years or so that Rhodesia was in the international wilderness.

Rationing happened twice more, making four times in all that I have had to endure it. For a field scientist, there is no more basic requirement than mobility, but it was not for these reasons that I seriously considered an offer from the then University College Hull to join their staff and found a new department of Geology.[2] Such an opportunity is rare: few geologists have the chance to found either a Geological Survey or a university department.

I wondered, too, if I was dropping out of the mainstream in my profession by 'burying myself in the bush' in Matabeleland. Certainly, great reputations had been acquired in similarly isolated situations. Sydney Haughton and Alex du Toit, both well known to me personally, had done it. Both were world famous.[3] Dr Alex Macgregor, nearer home in Salisbury, had certainly

[1] The first general strike by African workers in Salisbury and Bulawayo began on 10 December 1948.

[2] Now the University of Hull, it was founded as University College Hull in 1927. The foundation stone was laid the following year by the Duke of York (later King George VI). It was granted university status in 1954.

[3] 'In more than seventy years' service, [Haughton] had become a leading exponent of the concept of the Gondwana supercontinent; he had collected the fossil remains of the animals that roamed it and the fishes that inhabited the ephemeral seas associated with it; and he had traced the evolution of these creatures.' Kingsley Dunham, 'Sidney Henry Haughton. 7 May 1888 – 24 May 1982', *Biographical Memoirs of Fellows of the Royal Society* (1983), 29: 245–67.

Du Toit was the first South African-born geologist to have been elected a Fellow of the Royal Society, '... a man who will, for many years to come, be thought of, and rightly so, as South Africa's greatest and most eminent geologist.' S.H. Haughton, 'Alexander Logie du Toit, 1878–1948', *Obituary Notices of Fellows of the Royal Society* (1949), 6(18): 385–95.

done it.[4] Jim Wayland had done it, first in Ceylon and then in Uganda.[5] The opportunities were definitely open for a mass of new work in many parts of Southern Rhodesia in many branches of geology. The pioneers of geology in any country can hardly help carving a niche for themselves, but I wondered if the pioneering days were already over in this country.

The offer from Hull came, too, at a time when my doctorate was at a critical stage. The fieldwork, in spite of petrol rationing, was more or less complete, but access to better library facilities would be a great asset. Furthermore, the offer held out the hope that, if the new department was a success, a chair – that piece of mythical furniture that every academic covets for its title and prestige but certainly not for its comfort – would be forthcoming. I would stand a good chance of getting it. In spite of the attractions of the Museum Keepership, I weakened and accepted, albeit with regret at leaving Africa and with some misgivings about the future; but I was ambitious.

History does not relate how many young optimists were in the running for that job in Hull. Probably those in Britain knew enough about the place not to apply. Older and wiser men would not have been tempted by the salary anyway. It was to be as a Lecturer Grade I, at the princely salary of £600p.a., which even in those days was not a large sum of money and certainly not enough for a family man to survive on.

The Geology departments in Britain were small in these days and were not turning out many graduates. Most of them went into the Colonial Geological Surveys.[6] Jobs were scarce, and Survey jobs offered free travel to a variety of overseas territories and escape from war-torn Britain, which was taking a remarkably long time to make the effort necessary to recover from the ravages of bombing. A little determination to do some solid work, on the German pattern, would have made all the difference, but lethargy was the order of the day. Germany lost the war but won the ensuing peace.

So the offer was accepted and preparations made to leave the Museum of Southern Africa with all its unsolved geological mysteries and its vast areas that were still almost unexplored and geologically unmapped. The newspapers dubbed me 'the small-workers' friend' and splashed my departure quite hand-

[4] See Chapters 4 and 9.

[5] Edward James Wayland CBE was awarded the Bigsby Medal of the Geological Society of London in 1933 and the Victoria Medal of the Royal Geographical Society in 1935. *S2A3 Biographical Database of Southern African Science*, <https://www.s2a3.org.za/bio/Biograph_final.php?serial=250>.

[6] 'Colonial Geological Surveys', *Nature* (13 May 1944), 153: 582.

somely. I flew off to Johannesburg to catch a cheap charter flight to London. University College Hull was too mean to pay my fare.

With a flight originating in Durban, there should have been only an hour or two to make the connection. Time passed, and it was said to be delayed. More time passed, and it was said to be further delayed. After still more time had passed, it began to leak out that all was not well; in fact, something serious must have happened. Indeed, it had. The flight had taken off on schedule, but had flown into a hill, called, I think, Witkop, and was a total wreck with no survivors.[7] In the resulting confusion, the wrong passenger list seems to have been handed out to the press. I am told that my name was on it but, like Mark Twain, who was once similarly treated by the newspapers, 'the report of my death was somewhat exaggerated'.[8] Unlike Mark Twain, I was not famous, and the report seems to have passed off unnoticed.

The accident put the charter company in a spot; they had no spare aircraft available, and we were all scattered around Johannesburg hotels with instructions to report daily to the airline to find out when we could expect to travel. It took ten days to find a serviceable aeroplane and crew, and although the airline paid basic hotel bills, they produced no spending money, of which I was desperately short. However, my hotel had a snooker room in which a crowd of rather pleasant young men congregated nightly and saw to it that I got drinks and plenty of games of snooker. It was the most concentrated spell of snooker coaching that I have ever had, but the basic skills on the table, instilled into me by my mother at an early age, stood me in good stead and the improvement was spectacular. The evenings were good, but the days were awful. The geography of Johannesburg has always baffled me, and it is no place in which to be young and virtually penniless. It was a long ten days.

Just when it seemed that it would never end, the airlines announced that we would take off the following day, flying by way of Entebbe and Khartoum.

[7] 'The aircraft left Durban in the early morning [15 May 1948] on a flight to London via Johannesburg. While cruising in poor weather conditions, the aircraft hit the slope of Mt Spitzkop located 32 km south of Vrede, in the Witkoppens Mountain Range. All 13 occupants were killed. At the time of the accident, the aircraft was off track by eight miles, probably due to navigational error on part of the crew, caused by the lack of visibility due to poor weather conditions.' 'Crash of a Douglas C-49K in Mt Spitzkop: 13 Killed', *Bureau of Aircraft Accidents Archives*, <https://www.baaa-acro.com/crash/crash-douglas-c-49k-mt-spitzkop-13-killed>.

[8] Mark Twain had actually written 'The report of my death was an exaggeration' in response to an English journalist from the *New York Journal* who had contacted him to enquire whether the rumours that he was gravely ill or already dead were indeed true. 'Apparently, many of the misquoted versions stem from a Mark Twain biography by Albert Bigelow Paine published in 1912, two years after Twain's death. According to Paine's embellished version, Twain had told the reporter, 'Just say the report of my death has been grossly exaggerated.' Emily Petsko, 'Reports of Mark Twain's Quote About His Own Death Are Greatly Exaggerated', *Mental Floss*, 2 November 2018, <https://www.mentalfloss.com/article/562400/reports-mark-twains-quote-about-mark-twains-death-are-greatly-exaggerated>.

The journey would take about thirty-six hours. We made London without further mishap and I retired to Arncliffe for a few days with my mother to recover before facing up to the ordeal of a new job in a strange place.

One look at Hull and I began to regret leaving the Museum. The port and the city had been badly knocked about by bombs during the war and there was little sign of any effort to repair them. The climate seemed cold and windy but not particularly wet. Hull has, if anything, less annual rainfall than Bulawayo, though it comes at any time, usually as a steady, depressing drizzle.

I duly reported to the Principal, a rather colourless little man in a dingy office, and began to learn the worst. Hull, which is always coupled in Yorkshire with Hell and Halifax,[9] was, as a potential place to live, bad enough, but the University College was quite dreadful. I knew a little about Hull from my singularly unsuccessful sojourn at the prep school in Hessle, where I had met Fred Bagley, and been fired by his enthusiasm for geology, more than twelve years before.

The University College, was, in 1948, still a College of London University and worked for London degrees. This put a damper on me right away, as I would be entirely constrained in my teaching by the London syllabus, which I thought was narrow and archaic, with no scope for using my African experience to embroider my lectures. Much worse than this was the discovery of a crucial situation that had been carefully concealed from me during the negotiations: Geology was to start as a sub-department of Geography. This was a relationship that no young geologist could stomach. Geography has an overlap with geology in the field of Physical Geology or Geomorphology, the scientific study of the landscape, but even here the rocks are paramount because they are the raw material on which the processes of landscape sculpture operate. So it is really more the province of the geologist than the geographer.

The situation looked grim. It looked worse when I met my 'boss', the head of the department of Geography. Herbert King was more than middle-aged, rather short of inches, ample in weight, and embittered by not having been

[9] From 'The Beggars' Litany' by John Taylor (1578–1653), an English poet who dubbed himself 'The Water Poet':

> There is a Proverb, and a Prayer withal,
> That we may not to three strange places fall:
> From Hull, from Halifax, from Hell, 'tis thus,
> From all these three, Good Lord, deliver us.

The wooden huts housing Geology at University College Hull

given a chair. Neither did he have a doctorate. I took an instant dislike to him, which blossomed into mutual antipathy as time went by, without ever erupting into open hatred.[10]

The more I probed, the worse the situation seemed. Obviously, the College had no interest in its embryo department of Geology. The sum of money

[10] 'The death was announced in *The Times* of March 6 of Herbert King, Professor of Geography at Hull University. Professor King took his degree at Liverpool with First Class Honours in geography in 1922. He was one of the first members of the staff of the University College of Hull when it was founded in 1928, and played a principal part in building up the flourishing school of geography there of which he was the first head. He was the first Professor of Geography when a Chair was established at Hull in 1954.' 'Obituary: Professor Herbert King', *The Geographical Journal* (1958), 124(2): 288.

available with which to set it up had already been allocated in the budget and was not negotiable. It was £3,000 for equipment and the necessary modifications to a wooden army hut that was set aside for it. Academic staff was to remain at one (me) for the time being, and the chance of any students enrolling for Geology Part One, I was blandly told, was nil. For technical assistance, I was given a young man, Bernard Conway, who had been thrown out of nearly every science department as insubordinate, lazy, and generally useless. I explored him with some caution in consequence, but what I found was totally unexpected.

Bernard was a keen amateur geologist and already a member of the venerable Yorkshire Geological Society, of which I was also a member.[11] He was well versed in the stratigraphy of the complex Jurassic rocks of East Yorkshire and knew their fossils. His own interests had led him to explore the region in detail and he knew every quarry and cliff of significance for miles around. For financial reasons, he had not had a chance to take any formal qualifications, even as a junior technician, and I resolved to make the most of him. He soon became a firm friend, a loyal and hard-working helper, and a part-time student who attended all lectures, practicals and excursions during my year at Hull.

We agreed that the only reason they had hired a lecturer in Geology was that, with more subjects on the calendar, the College stood a better chance of becoming a fully independent university with its own Charter. He shared my dislike of the place, which was perhaps not surprising in view of his own background, but together we resolved to make the best of a bad job. He left the place soon after I did, got qualified and joined the Geological Survey in London. He has never looked back and must now be near an honourable retirement.[12] Not only did he go to all my lectures and practicals, he gave the lectures on the stratigraphy and palaeontology of the Yorkshire Jurassic, and I sat in on them.

[11] Geoffrey Bond had been elected as an ordinary member in 1937. *Proceedings of the Yorkshire Geological Society* (1 January 1943), 25: v–ix.

[12] Bernard Conway was described in 1985 as 'recently retired from the British Geological Survey'. As the on-site geologist during excavations at the Swanscombe Skull Site National Nature Reserve, near Dartford in Kent, between 1968 and 1972, he wrote a summary here: Bernard W. Conway, 'The Geology and Archaeology of Barnfield Pit', *Earth Science Conservation* (1985), 22 (May), 15–20. See also: Bernard Conway, John McNabb and Nick Ashton (eds.), *Excavations at Barnfield Pit, Swanscombe, 1968–72* (London: British Museum, Department of Prehistoric and Romano-British Antiquities, Occasional Paper 94, 1996).

This was, in short, the situation that I faced in Hull, and the prospect appalled me. Within a week I had decided this was no job for me. I cabled the Director of Museums and asked if I could withdraw my resignation and have my precious job as Keeper of Geology back again. His reply earned my everlasting gratitude. I had to give six months' notice to the College, though I was inclined to walk out of the poke in which I had bought such an unsavoury pig,[13] without warning, and hang the consequences. The cable in reply said that the Trustees would give me a year, if necessary, in which to extricate myself from this predicament and get back to Bulawayo.

Armed with this assurance, I told Herbert King that I was highly dissatisfied with the whole scene, that I was affronted by being a sub-department of Geography, and that if no students were to come my way, I proposed to leave at once. The whole interview gave me immense satisfaction and to some extent restored my self-respect, which had been almost totally destroyed. He was astounded that anyone should react as I had done to his precious college, in which he had lurked with minimal distinction for so long.

My next call was on the Principal, to whom I delivered a similar broadside and who seemed equally astonished. It was then discovered that there were actually four students who had opted for Geology One in the next session: two Kenyans and two Englishmen. Having secured my line of retreat, I resolved to make the best of it all and at least finish my doctorate as I had intended.

That year was in many ways a nightmare experience, but all four students were successful at the end of it, and I successfully acquired my much desired Ph.D., though not without four hours of agony at the oral examination, at the end of which I was totally convinced that I had failed, and failed utterly.

Since becoming a real Professor, and having a real department of Geology, with genuine research being done by well-grounded postgraduates, I have taken part in many Ph.D. orals. None of them has ever lasted four hours, but I expect most of the candidates felt, as I did, that they had failed. Looking back on my own performance, I am surprised that I passed. I am sure that standards have risen significantly since 1948 in regard to the volume of research, its depth and its final expression as a thesis.

My own Ph.D. experience was typical of the young enthusiastic innocent abroad. I had no supervisor, and I doubt if I had even read a Ph.D. thesis

[13] A Pig in a Poke: 'A blind bargain. The French say *Acheter chat en poche*. The reference is to a common trick in days gone by of substituting a cat for a sucking-pig, and trying to palm it off on greenhorns. If anyone heedlessly bought the article without examination he bought a "cat" for a "pig;" but if he opened the sack he "let the cat out of the bag" and the trick was disclosed. The French *chat en poche* refers to the fact, while our proverb regards the trick. Pocket is diminutive of poke."' (E. Cobham Brewer, *Dictionary of Phrase and Fable* (London: Cassell, 1898).

when I compiled mine and blithely submitted it, full of hope and confidence, based on the fact that I had never failed a Geology examination and was building on the foundation of a First Class Honours degree from the Imperial College of Science and Technology, with its great tradition of excellence.

I had discovered that one of my two examiners was Dr Kenneth Sandford from Oxford, who was known for his research on the Pleistocene in the Sudan.[14] I could not discover who the other was until I met Dr Kenneth Oakley at the Senate House in London.[15] This Kenneth I knew well, and I had spent a month with him in Kenya and Tanganyika on the first Pan-African Congress on the Pleistocene period, in January 1947, based in Nairobi. It had been organized by Mary and Louis Leakey.[16]

The Pleistocene period is roughly the last part of geological time and more or less coincides with the span of time during which man and man-like creatures have existed. My thesis dealt with this period of geological time in Matabeleland. Very little previous geological work had been done, though the succession of Stone Age cultures had been explored by Dr Neville Jones of the National Museum of Southern Rhodesia in Bulawayo, with whom I had worked for several years.[17]

The two Kenneths set about me with great enthusiasm and took my work apart with the proverbial toothcomb. All I could say about it was that it was virtually all new information, but Kenneth Sandford's final question – was the second part of my thesis really necessary? – was direct, frightening and full of menace. Ever since, I have appreciated that peculiar academic expression, 'the candidate defended his thesis'. I must have put up a reasonable defence, for, in the fullness of time, I was officially told that the Senate of London University were pleased to inform me that the Degree of Doctor of Philosophy in the Faculty of Science had been conferred upon me.

[14] Kenneth Sandford was Vice-President of the Royal Geographical Society and died while still in office. 'He was a distinguished geologist and during his many years at Oxford he earned the gratitude of successive generations of undergraduates through his work for the Oxford University Exploration Club. ... He was a valuable member of our Expeditions Committee and an able Chairman of our Publications Committee. No one could have more wholeheartedly devoted himself to the interests of the Society.' Lord Shackleton, 'The Presidential Address at the Annual General Meeting of the Society on 5 June 1972', *The Geographical Journal* (1972), 138(3): 281–6.

[15] Kenneth Oakley was a 'physical anthropologist, geologist and palaeontologist, best known for his work in the relative dating of fossils by fluorine content. ... He won renown in 1953 when he played a decisive role in the exposure of the Piltdown hoax. The famous Piltdown man, remains of a skull found in a ditch in Sussex, England, had been unchallenged as the missing link between man and ape since its discovery in 1912.' 'Kenneth Oakley', *Britannica*, <https://www.britannica.com/biography/Kenneth-Page-Oakley>.

[16] See Wendell Phillips, 'The First Pan-African Congress on Prehistory', *Science* (13 June 1947), 105 (2737): 611–13. A photograph of the participants, including Geoffrey Bond, appears in Peter Robertshaw (ed.) *A History of African Archaeology* (Cape Town: James Currey, 1990).

[17] See Chapters 4 and 5.

Herbert King was not in the least impressed when I told him that I had been 'doctored'. He did not suffer, as Hilaire Belloc's Henry King did, from having the defect of eating little bits of string, but he was knotted up internally in other ways.[18] I am sure he relished my departure at the end of the academic year quite as much as I did. He got in exchange a much less difficult customer, in the shape of a brand-new graduate with no experience at all, which merely shows what Hull thought of its new geological department.

That year in Hull did, however, have one redeeming interlude, which greatly broadened my experience, gave me a two-month respite from the rigours of the East Yorkshire winter, and refreshed my spirit of adventure. The International Geological Congress – the mecca of all in the profession, which seems to meet only in leap years – met in London in 1948.[19] I was able to go to some of the sessions, at which I met Bernard Fagg, at that time the Official Government Archaeologist in Nigeria and about my age.[20]

Bernard had not been able to attend the Nairobi conference in January 1947 for some reason, perhaps tied up with Colonial Office red tape. He should have been there because he had made some quite startling archaeological discoveries in the plateau tin fields of Northern Nigeria that had attracted worldwide attention. Like so many new and unexpected things, these discoveries had first been made accidentally. The tin in Northern Nigeria is alluvial, and the deposits in which it is found are geologically very young. They are late in the Pleistocene period in valley-fill deposits on the Northern plateau, which is centred around Jos, where Bernard was based. The tin-miners began to turn up very remarkable terracotta figurines, mainly of human heads. They were real prehistoric works of art, generally six to twelve inches high, with slightly stylized features but of an advanced craftsmanship of totally unknown antiquity.

Bernard Fagg had himself excavated and preserved some of these figurines, the first of which had been turned up by the miners, who very wisely referred

[18] Hilaire Belloc, 'Henry King, Who chewed bits of String, and was early cut off in Dreadful Agonies', in his *Cautionary Tales for Children* (London: Duckworth, 1918):

The Chief Defect of Henry King
Was chewing little bits of String.
At last he swallowed some which tied
Itself in ugly Knots inside.

[19] S.E. Hollingworth, 'International Geological Congress, London, 1948', *Journal of Glaciology* (1949), 1(5): 252–3.

[20] Bernard Evelyn Buller Fagg (1915–1987). 'Bernard Fagg', *Wikipedia*, <https://en.wikipedia.org/wiki/Bernard_Fagg>.

them to him for safe keeping and investigation. He was aware of my work on the Pleistocene in Matabeleland and, of course, of Louis Leakey's work on the Pleistocene in East Africa. This had become the standard against which all work on this late period of geological time was measured. Its cornerstone was the theory that there had been oscillations of climate during this period, in some way connected with periods of glaciation and deglaciation in higher latitudes. In Africa these were expressed as periods of wetter and drier climate compared with the present day, the so-called Pluvials and Interpluvials. Since they were thought to have occurred simultaneously all over Africa, the East African succession of climatic changes was accepted as a yardstick.

The Stone Age deposits on which I had worked on the plateau around Bulawayo were all in the headwaters of the main rivers draining towards the Zambezi to the north-west. The deposits associated with the various Stone Age cultures were in beds of alluvium, marking periods of deposition in these shallow valleys, separated by periods of erosion. They can briefly be called a succession of cut-and-fill episodes, with stages of chemical alteration, leading to lateritic ironstones or ferricrates, and calcretes due to deposition of lime. The most logical explanation of the observed facts, the same succession being repeated in all the valleys, seemed to lie in regional shifts of climate about the present mean, in other words Pluvial and Interpluvial periods. The range of change was estimated to lie between 75 and 150 per cent of present annual rainfalls.

I had already published papers on this succession of events and compared them with Leakey's East African sequences by the time that Bernard met me in London.[21] To my delight, he had read my papers. To my even greater delight, he explained that the alluvial tin deposits of the Jos Plateau, which contained his figurine culture somewhere in the succession, were also cut-and-fill deposits in shallow valleys, and invited me on behalf of the Nigerian government to go out, all expenses paid, for two months over Christmas and New Year. He thought my methods could be applied to his region, and that some idea of the antiquity of his figurine culture might be forthcoming.

The chance to get back to Africa was too good to miss, and my experience of the Pleistocene would be broadened to include a key area of West Africa, to add to a nodding acquaintance with East Africa gained during the 1947 Nairobi Congress, and my extensive knowledge of Matabeleland, where I had more or less invented the Pleistocene. Hard-rock geologists tend to look on the Pleistocene deposits with amused benevolence because they are so young, tinged with irritation at the way they obscure the older and more economically interesting rocks.

[21] See Appendix 3: 'Geoffrey Bond's Publications'.

I wondered if Bernard's invitation was real or whether his enthusiasm was beyond his means of persuading his government, but, in the fullness of time, the official written invitation came from the Colonial Office in London. It was complete with air ticket and instructions.

The flight was from London to Kano and the plane a converted four-engine wartime bomber, known in its civilian form as a York.[22] It was an un-pressurised high-wing monoplane, which flew at moderate speed and low altitude with wonderful visibility from its ample cabin windows. The route was over central and southern France, with excellent views of the Alps, the Mediterranean and the Western Sahara. My old field notebook records my impressions quite vividly until the landing at Kano, where troubles began.

Kano airport is almost on the southern fringes of the Sahara. Even in 1948 they had black Immigration and Customs men. They were not in the least impressed by my official invitation and made life as difficult as possible for me. Returning civil servants were waved through with a minimum of formality, but they gave me the full treatment. Everything was examined and re-examined in detail, but no fault could be found with my travel documents and I was eventually cleared and allowed out of the airport. There was no Bernard Fagg to meet me as arranged, so I spent some time examining my strange surroundings, hoping he would appear.

In those days Kano airport sported one of those multi-armed signposts saying X thousand miles to Mombasa, Cape Town, London and so on, utterly useless but good tourist material. Even better was the method of announcing the arrival of rather infrequent aeroplanes. A local gentleman in very picturesque costume armed with a long, straight hunting horn blew blasts on it to the four points of the compass as soon as an incoming plane was on the circuit. The ritual seemed to serve no useful purpose, but was again a good tourist effort, which would have been more productive if there had been any tourists. I seemed to be the one and only on that day.

Time hangs heavily in an unknown land when travel arrangements have gone awry. The plane had arrived in the early morning and by noon there was still no sign of Bernard Fagg and his motor car. Jos is a long way from Kano. There is a railway line from Kano to Jos via Zaria, Kaduna and Kafanchan, a total distance of nearly 300 miles of single-line uncertainty. Rather than try to find somewhere to stay overnight in Kano, it seemed better to explore the possibility of a train journey.

[22] 'The Avro York was a British transport aircraft developed by Avro during the Second World War. The design was derived from the Avro Lancaster heavy bomber, several sections of the York and Lancaster being identical.' 'Avro York', *Wikipedia*, <https://en.wikipedia.org/wiki/Avro_York>.

Subsequent exploration of Kano on the return journey two months later bore out the wisdom of my choice. In those days Kano was no place for staying in. There was still no sign of Bernard, so I made enquiries about trains. Surprisingly, there was one that day, and a taxi could be found to get me to Kano station. Since the train came from Nguru, 150 miles north-east of Kano, its time of arrival could be problematic, but, if it was on time, it would fit in with the predicament I was in and would land me in Jos after breakfast the next day.

There seemed to be no alternative, and I was assured that a telephone call to Jos was out of the question. I took a taxi, bought a ticket and waited. Vultures wandered around on the track like chickens. I discovered that they were royal game since they were the only sanitary arrangements. There was a £5 fine for killing one, and they were so tame that they must have been aware of their immunity. The train came in on time, and I found a 2nd class compartment with only two or three people in and climbed aboard. Climbing was necessary. There was no platform. I was used to this from various journeys to and from sidings in Matabeleland such as Igusi and Redbank.

It was hot, and I dozed a bit between frequent stops. After leaving one of these small country stops, the train pulled up with a great outcry of brakes and came to a sudden halt, apparently in the middle of nowhere and for no obvious reason. My companions were baffled. One was a Greek, whose English was remarkably good. He volunteered to find out the reason for this unscheduled stop.

The guard and the driver were sitting beside the engine and he came back from his consultation with them in fits of laughter, assuring me that all was quite normal. What had happened was briefly this: At the previous siding, a local worthy had boarded the train but forgotten his luggage. He realized this fact about one mile out and promptly pulled the communication cord. The brakes were then automatically fully applied and hence the sudden stop. The passenger, apparently completely unabashed by his unorthodox behaviour, had told the guard of his predicament and had been assured that the train would wait while he walked back, collected his luggage, and re-joined the waiting train. This took well over an hour, after which the train resumed its journey. According to Bernard, to whom I told the story, this was by no means an isolated example.

The afternoon wore on and stop followed stop, but at one of them an agitated Bernard appeared. He had had mechanical trouble, which had caused a long delay and missed my arrival. He guessed what had happened and set off to overhaul my train, not a very difficult undertaking on Nigerian Railways in those days. We drove on through the night and made Jos in the early hours. I was sorry to miss so much new scenery in the dark. After a couple of days rest and preparation we were ready to begin fieldwork on the Jos Plateau tin fields in earnest.

Jos is nearly ten degrees north of the equator and the Jos Plateau is 4,000-5,000 feet above sea level. It was the middle of winter and the middle of the dry season. The Harmattan wind was blowing off the Sahara and is the driest wind I have ever met. The rainfall in Jos is 40–50 inches per annum in a short heavy season, hot and humid. This is followed by the intensely dry winter dominated by the Harmattan, dry and often dusty. The effect of these contrasts on anything made of wood is devastating. All door and window frames in Jos and the whole surrounding region had to be metal to avoid the swelling, shrinking and twisting that would otherwise have taken place. But the temperature reminded me of Matabeleland in September, just right for fieldwork.

I now tasted the luxury of fieldwork under Colonial Office conditions. It was a great contrast to what I was accustomed to working from the Museum in Bulawayo, which operated on a shoe-string budget (our Museums still do). This meant, if I was lucky, that I had one general-purpose attendant to look after camp, a vehicle, and the absolute minimum of equipment, which in the dry season did not even include a tent. Camp consisted of a shady tree as conveniently near water as possible, doing a large share of the domestic work, as well as doing a reasonable day's fieldwork, which might involve traverses up to twenty miles or even more.

Not so with Bernard Fagg, as I soon discovered. His vehicle was a five-ton truck, complete with driver. He had mapped out very carefully the route we were to follow. At each projected night stop, an advance party had built at least one hut, cleared the site, erected camp tables and collected firewood. In addition to the driver, we had several camp attendants, who unpacked, set up camp, got beds and tables and chairs ready, and made sure that the *two* small paraffin refrigerators were working properly.

Baths were prepared and meals expertly cooked and served. We did not exactly dress for dinner in the traditional manner, but we at least changed out of our bush clothes. It was all very decorous, very elaborate, and must also have been very expensive. The contrast between this and my own simple style of camping, both before and after the Hull episode, was marked. I must say that, on the whole, I preferred my own way of life, but at least this was a complete novelty.

Practically all our exposures were in tin mines either active or dormant. All tin operations on the Jos Plateau ultimately came under the wing of the giant organization Amalgamated Tin Mines of Nigeria, ATMN for short.[23] They either owned and operated the mines themselves or tributed them out to individuals or small local syndicates. The main economic deposits were alluvial gravels of the Pleistocene age. The primary source of the tin mineral, cassiterite (tin oxide), was in certain granites of peculiar chemistry known locally simply as the Younger Granites. Cassiterite is a heavy almost unweatherable mineral. Weathering of the primary host releases it, and water transport does a first concentration of the mineral, generally in gravelly alluvium. There were quite well-defined beds in the valley filling between the hills of hard ancient granites and gneisses, and rather resembled the dambos of the Copper Belt or the vleis in the country round Salisbury.

The tin mines were opencast and referred to locally as paddocks. There were often two or even three horizons of tin gravel in the alluvial profile and the deepest excavations were 30–40 feet deep, ending on weathered bed rock. In some areas there were even older tin washes, dating from some time in the geological era called the Tertiary, which were several million years old and lay under basalt lavas as much as 130 feet thick. Under less tropical conditions of weathering, these could have been mined by horizontal adits, but the originally hard-rock basalts had been completely altered to soft pastel-coloured clays that could be entirely stripped by mechanical shovels until the tin wash was exposed. One mine we saw spent ten months of each year stripping overburden in this way to reach a tin gravel only two feet thick but very rich in cassiterite.

Once overburden had been stripped, in most cases by hand labour using head pans, all that was needed to concentrate the tin was abundant water for use in a variety of simple gravity separators. Small tributors merely panned the wash to get a heavy concentrate. This consisted of cassiterite, columbite, ilmenite, zircon, garnet and a few odds and ends. These simple operators winnowed the dry concentrate in flat baskets, rather like separating grain from chaff, and produced a tolerably clean residue which was mainly cassiterite with some columbite. This was sold to ATMN, who cleaned it up magnetically along with other parcels of concentrate.

These operations had been going on for decades on the Plateau tin fields. For most of this time, ATMN had stockpiled columbite, for which there was no known market, but after the war, with the rise of jet engines in the aircraft industry, this mineral suddenly became very valuable for alloying with steels for turbine blades. In some way it prevents the carbon in the steel from coming out of solution along the crystal boundaries in these special steels.

[23] See Bill Freund, 'Nigerian Tin Mining and Imperialism: From the Niger Company to ATMN', (Johannesburg: University of the Witwatersrand, African Studies Institute, Seminar paper, 23 Aug. 1982).

This leads to brittleness and failure of the blades. ATMN admitted that they were making a killing out of what had been a useless mineral, which had to be separated from cassiterite. Fashions in minerals change as the back-room boys develop new products, and fortunes change with fashion even in the mineral industry.

Bernard had been in contact with all mines, however small, before our visits, and everywhere we were free to examine, collect and record. There were indeed similarities between the valley deposits of the Jos Plateau and my Matabele valleys, but there were also differences, as might have been expected. The Nigerian rivers had also gone through successive phases of erosion and deposition, and the only possible process to account for their behaviour seemed to be changes of climate during the Pleistocene, for the general successions were similar in all the deposits we examined, whatever direction the rivers flowed in.

The one major difference occurred round a residual plateau formed by the Lirue hills, which stood several hundred feet above the general level of the Plateau, forming a remnant of an even older land surface. Since these hills included tin-bearing Younger Granites, the rivers that flowed radially off this plateau were particularly rich in tin. They all fell steeply on to the main Plateau surface, generally in spectacular smooth slides at the very steep angle, which were referred to, not quite accurately, as waterfalls. The Zenebi Falls and Gaya Falls were two we paid much attention to, as active tin mining was going on in large well-ordered paddocks below them.

Such rivers with steep upper reaches, falling several hundred feet on to the plains, had carried large quantities of sediment with them, which, because of the sudden change to a relatively flat profile, had been deposited as great cones of outwash debris, in some cases stretching a few miles on to the plains below. These contained two or three rich tin washes. As the climatic cycles changed, the rivers lost much of their energy and deposited progressively finer gravels, sands and, finally, clays, until the next active cycle began and started the process all over again. If we were to work out climatic cycles to compare with the East Africa succession, such a locality seemed to hold out the best hope. We concentrated much time and effort on Zenebi, making rough sketch maps of the great alluvial cones at the foot of the falls. In those days there were no air photos, and no topographic maps, and time did not allow doing pace and compass traverses or plane tabling.

I thought my notes and sketches were pretty good at the time, but thirty years later I had cause to wonder. Nigerian archaeology had moved well beyond the pioneer stage and a new generation of archaeologists, well versed in modern techniques, has arisen. One of them, Dr Philip Allsworth-Jones, wrote to me in 1979, announcing that he was doing more work on our Zenebi site, and asking for information and, if possible, photographs of it as it was in 1948. The mine must have closed soon after our visit, and nature had done its best to obliterate man's handiwork, which it does quickly under Jos Plateau conditions. Dr Allsworth-Jones was therefore faced with the expensive and laborious task of making his own excavations, whereas tin mining had provided us with its beautiful great pits. Luckily, my photographic records were still in good order, even to the negatives. So he got, much to his surprise, good prints, well documented, plus measured sections of the main walls of the paddocks.[24]

In the Zenebi succession we were able to find good evidence of the relationships between deposits and at least one period of Stone Age activity. There were abundant flake tools unrolled - and therefore contemporary - with the gravel containing them, which was one of the main tin washes in the Zenebi paddocks. They belonged to what was known as the early part of the complex called the Middle Stone Age in East Africa and southern Africa, tools like those of the early Middle Palaeolithic of Western Europe. Cultural similarity does not necessarily prove correlation in time, but it strongly suggests it, and this Zenebi material was very important to us.

The stone-implement horizon was covered by a few feet of sticky clay, representing the waning activity of this cycle in the alluvial cone below the falls. This contained a good deal of wood in the form of branches, twigs, and even occasional trunks, washed over the falls and caught in the swamp below. It was hard and heavy, and very like the bog oak of Ireland. Here was a chance to convert at least one horizon of our stratigraphic relative timescale into an absolute date. The Carbon-14 absolute dating method had just been evolved, by which it was possible to date pieces of wood from archaeological contexts if they were less than about 50,000 years old. In Rhodesia, the same method has been used on wood from a drain under the main enclosure wall at Great Zimbabwe and to date the centre of a fifteen-foot diameter baobab felled in clearing fishing grounds in the Kariba lake basin. This came out at about 1,000 years. We really had no idea of the antiquity of the Middle Stone Age in Nigeria, and so here was the golden opportunity to make history.

By chance, the stripping of overburden had exposed part of a prone tree trunk about a foot in diameter, and labourers were immediately put on to

[24] Philip Allsworth-Jones's publications are listed on ResearchGate: <https://www.researchgate.net/profile/Philip-Allsworth-Jones>.

excavate it. It proved to be more than twenty feet long, and it took twenty labourers to lift it out and put it diagonally on the back of our five-tonner. It was indeed a prize specimen. Only a few grams would be needed for C-14 dating, but Bernard was sure that he could dry it out slowly under controlled conditions in his garage in Jos. It could then be cut into planks, if it did not split while being dried, and he could have a piece of furniture or some occasional tables made out of it. This would be unique, and the whole idea appealed strongly to his archaeological soul, so it was looked after with great care until we got back to Jos for a Christmas break with his family from fieldwork.

It reached Jos safely and was put carefully in his garage and covered with damp sacks to protect it from the ravages of the Harmattan. It was still there when I returned to Britain, but tragedy befell it later. Like all Colonial officials, Bernard had one foot in Jos and one in England, to which he went at the end of each 'tour' of three years or so, and one of these visits was due soon after I left. It duly took place and he returned 'home' for his six months' long leave from his 'White man's grave' in West Africa. On his return to Jos he went to see how his precious log was faring. It had gone. While he had been away, it had dried out, and his staff had chopped it up and used it for firewood. I think other bits have been dated and came out at about 40,000 years; an occasional table of that age would indeed have been something to boast about. No doubt Dr Allsworth-Jones will get more wood from Zenebi and have the date re-run. The technique has been greatly refined in the last few years, and it would be interesting to get a new determination.[25]

The Christmas break gave me a chance to meet up again with my old Commanding Officer, Major J.L. Farrington, from my time at Oxford in the last few months of the war when I was in that strange job of Military Geology in that equally strange unit, Inter-Services Topographic Department. By 1948 he was Chief Geologist of ATMN. My field notebook shows that we took only Christmas Day and Boxing Day off before going back to the bush.

The second half of our work was to include the tin mine at Nok, where the first terracotta figurines had been found, and which was thought to be rich in

[25] Dr Allsworth-Jones was a staff member at the University of Ibadan from 1975 to 1991, and it was in that capacity that he carried out his own research in northern Nigeria. He reports that, in fact, the only date obtained on the material submitted by Bernard Fagg was no more than 5,440 ±100 years old. He was not able to get more wood from Zenebi and re-run the date. He has recently published a book to make known the Palaeolithic sequence in West Africa in so far as it concerns the Middle Stone Age: Philip Allsworth-Jones, *The Middle Stone Age of Nigeria in Its West African Context* (Oxford: Archaeopress, 2019). He said that Geoffrey Bond's account 'had and has a continuing utility, and I was grateful to him for replying to me'.

such remains. The mine had already given its name to the hazy period of their manufacture, the Nok Culture, and my primary task was to try and date it.

By now we knew that tin washes were of several ages: Zenebi was mainly Middle Stone Age, but at least one deposit elsewhere had yielded hand axes and cleavers of the Old Stone Age. At the mine at Nok, they were working at least two Middle Stone Age tin washes but had not gone below the hard lateritic layer at the bottom of the existing paddock, which they believed was the bottom of the succession. Our archaeological evidence strongly suggested that it was worth going through this laterite to see if there was a tin wash below. We suggested this to them, and it was subsequently found that there was a tin wash down below. In at least one other mine we made the same suggestion on the same archaeological evidence, and again we were proved right. It is not often that Stone Age archaeology can be put to economic use.

Nok lived up to our expectations and hopes. Bernard laid out a 'dig' in an undisturbed part of the deposit and turned up a figurine in a matter of hours. As less than twenty such items were then known, this was a rare success. It proved beyond doubt the geological horizon in which the Nok Culture occurred, and from our observations on that site we could place this in the stratigraphic ladder to a time before the very last phase of erosion and deposition. This meant an age rather greater than expected, and on comparing our climatic sequence with that in East Africa we concluded that the figurines must have been made 1,500–2,000 years ago.

When the results were published, they were for a time received with much scepticism.[26] It was said to be far too old. It was impossible for anything so advanced to have been made in West Africa so long ago. Our geological evidence seemed to us reasonable enough, and confirmation came much later by the C-14 method, which gave just that age. However, I am now not nearly as happy with the geological dating method we used. As more and more local sequences have been worked out, it has become clear that the Pluvials and Interpluvials are by no means the same all over Africa, and the modern picture is so much more complex than was then thought that I would now have to admit that we got the right answer but by quite unreliable means! Nevertheless, an advanced figurine culture did exist in West Africa far earlier than was credited thirty years ago.

[26] The results were published as contributions to the *Proceedings of the III International West African Conference held at Ibadan, Nigeria, 12th to 21st December 1949* (Lagos: The Nigerian Museum, 1956): B. Fagg, 'An Outline of the Stone Age of the Plateau Minesfield', 203–22; G. Bond: 'A Preliminary Account of the Pleistocene Geology of the Plateau Tin Fields Region of Northern Nigeria', 187–202. Their work is also acknowledged in Peter Breunig (ed.), *Nok: African Sculpture in Archaeological Context* (Frankfurt: Africa Magna Verlag, 2014) and in Gabriele Franke, 'A Chronology of the Central Nigerian Nok Culture – 1500 BC to the Beginning of the Common Era', *Journal of African Archaeology* (2016), 14(3): 257–89.

The rest of the fieldwork was notable for only one thing. Bernard had a slightly lame leg, the result of polio. Walking was difficult and tiring, and he did most of his work on horseback and enjoyed it. He was constantly at me to try it, but I had not been on a horse since I fell off a Shetland pony at the age of five, and I resisted all his efforts until just before the end of our journey. Then I allowed him to persuade me to try it the next day, which involved a fairly long walk for me and an easy ride for him.

At breakfast next day, two horses duly arrived, one each. Mine was a full-blooded Arab stallion, which I viewed with the greatest suspicion. It was decked out in full Hausa saddle, stirrups and bridle and was in the charge of a very small boy. I told him that before I would set foot on such a fiery-looking steed, he must gallop it around for at least half an hour to take the steam out of it. He did so, and on I got, full of foreboding. It was the tamest and dullest horse imaginable. It must have had every pest and disease in the local catalogue. It plodded slowly along behind Bernard's horse, stopped willingly each time I needed to get off to look at something and stood with its head drooping until I climbed on again. It would have been just as easy to walk without having to bother about this creature, but I persevered and have a photograph to prove it.

Bernard laughed at me most of the day, and I have never been on a horse since. They are mechanically unreliable, their steering is erratic, their brakes eccentric, to say the least, and any conscientious vehicle inspector would instantly put them off the road as unsafe in modern conditions. My mother was a great horse rider, side-saddle of course, in her youth.

The time came to pack up and go back to Hull. Africa had claimed another two months of my life, and it was two months nearer my escape from Hull to the haven of Bulawayo. We packed several figurines for me to take to the British Museum. This was to be the bulk of my hold luggage and was very precious. We used rice as the packing material. It was excellent for the purpose and plentiful in Nigeria but very scarce in England at the time. The box was sealed and stamped by every high-sounding official we could lay hands on in the hope that the English Customs would be persuaded to let it through without opening it and so not disclose the precious rice cargo. It worked. There was some hesitation by a junior official, but I called for a senior man, who let me through. It is uncertain whether the figurines or the rice were more welcome at the BM but I was given a warm welcome there, perhaps on both accounts.

Hull seemed more dreary than ever in February after the joys of the Jos Plateau, but it had to be endured till the end of June. Hull seems to have nothing to recommend it. There is often a smell of fish from the docks when the wind is from the east. I lived in a dreary boarding house in Beverley Road and walked to College. Bernard Conway was a great support, and together we made the most of our wooden army hut and did collecting excursions at weekends. The £3,000 was stretched to its limit but was woefully inadequate, even though prices then were quite reasonable. My Nigerian notebook records that a single ticket from Hull to London then cost £1 15*s*. 5*d*. My ticket on Nigerian Railways from Kano to Jos cost £5 1*s*. 9*d* and was worth it, even though I did only part of the journey, for the strange experience.

The whole exercise of the Jos Plateau had been most rewarding. I had collected my prized Ph.D. but, apart from that, Hull was a disaster, just as disastrous as my previous year in Hessle just up the Humber. I cannot get a warm feeling for any city that has, in the wisdom of its City Fathers, built a mock Roman Ruin in its only public park.[27]

One last memory of the Humber is perhaps worth recording. Fred Bagley and I built a wooden framed canvas canoe in the school workshop in Hessle, with a view to disporting ourselves on the Humber in the summer. We finished it in mid-winter and tried it out on a flooded abandoned brick pit, one of a series near Hessle that accounted for the plague of mosquitoes, including Anophelines, which happened every summer. It floated and we launched it on the Humber on a cold February day in 1934. We were told we were quite mad, as the currents were dangerous, but we paddled upriver until we were frozen stiff. We landed on a mud bank, abandoned the canoe to its fate, and got home tired, wet through and aching cold. Many years later, local historians found the scanty remains of a Viking ship just about where we landed. I wonder?

In July 1949, I left Hull's University College. You don't shake the dust of Hull off your feet even in July. It is too damp for that. I had learnt my lesson. Life in England was not for me. Henceforth I would stay where I belonged, in Africa. And so it has been, for richer for poorer, in sickness and in health, etc., as it says in another context.

[27] This seems to be what is now East Park, opened to the public in 1860. Zachariah Pearson, the mayor, had donated land for Hull's first public park, which became known as Pearson Park. There was a small lake and a folly called 'The Ruins', along with a cast-iron drinking fountain and a statue of Queen Victoria and Ceres. 'Fact 12. Hull's first public park was The People's Park', 29 March 2014, *150 Facts about Hull* <http://factsabouthull.blogspot.com/2014/03/fact-12-hulls-first-public-park-was.html>.

– 8 –

Back Where I Belong

The academic year in Britain ends in June, and by July I was once more travelling by train from London to Southampton to catch the Union-Castle Royal Mail Ship for Cape Town. Those ocean journeys are past history now, and in many ways this is regrettable. Modern jetliners do the journey in roughly the same number of hours as the mail boats took days, but there is no comparison between the idle luxury of a liner and the air trip. Those journeys were fun, in spite of the Bay of Biscay at one end and the 'Cape rollers' at the other. The enforced idleness on board was a holiday in itself, with a stop at either Las Palmas or Madeira. Cape Town docks felt just as good on this landing as they had eight years before, and the journey up by train just as fascinating, and this time also had the prospect of a warm welcome at the end, the return to familiar places, and all the interest of Rhodesian geology. It was an immense relief to be back where I was convinced I belonged. The threads were soon gathered up, and work began in earnest.

Don Macdonald had continued his successful search for underground water, and now had sufficient grasp of the local geology to dispense with my help most of the time, but Sawmills on the Bulawayo–Victoria Falls line and Dett further up the same line were still problems, and we teamed up again to tackle them. Sawmills was our first target. The dam had silted up and there was a severe shortage of water for engines.

Under Sawmills there is a small thickness of loose Kalahari Sand, which makes up the surface of so much of western Zimbabwe. It covers nearly all the Wankie National Park and stretches westwards into the Kalahari country of Botswana. It carries the very valuable teak and mukwa indigenous forest, but it is so permeable that surface water is rare, apart from a few pans that manage to survive in the dry season. Below this sand layer there is a thick succession of hard basalt lavas, like those of the Victoria Falls, and under these again there are soft porous sandstones, which, around Bulawayo, lie on the hard rocks of the greenstones that make up the gold belts.

To hold water, underground rocks must be permeable: soft sandstone can hold as much as thirty per cent of its own volume of water. Basalt, on the other hand, is impermeable, though weathered fissures may hold water. The thickness of the various layers could only be guessed at, since no boreholes had penetrated deeply enough to reveal the succession. This succession of bedded rocks could be shown to dip to the north-west at a gentle angle but a bit more steeply than the fall of the surface in the same direction. We inferred from this rather scanty evidence that there was a chance that the sandstones under the basalts might contain water confined by the impermeable lavas, so that a deep borehole might find water which would rise under artesian pressure. With luck it might even be fully artesian and flow from the hole without pumping.

The proposition was put, rather hesitantly, to the water engineer, remembering that he was only just out of the divining era. His immediate question was how deep the exploratory hole would need to be. Our reply, also rather tentative, was a minimum of 800 feet (243.8 metres). We were curtly told that the railways had never drilled deeper than 350 feet for water but that, as a concession, he would allow us to go to 500. This was ridiculous and would prove nothing either way. Don took a chance and fitted the drilling machine with 1000 feet of cable, and the experiment began.

The site was at one end of the station area at Sawmills, and we kept a very close watch on progress. Drilling began after the rains had started, and we usually travelled to Sawmills in an engineer's trolley. This was driven by a Ford V8 engine and had all the controls of a car except the steering wheel. It had to be treated like a full-blown train, complete with crossing orders for meeting oncoming traffic, but we could do the journey in about two hours, unless we had to wait for a train coming south.

At quite shallow depth we went through the Kalahari Sand layer and into the hard basalt, in which drilling was slow. As expected, the hole was dry at 100 feet, 200 feet, 300 feet, 400 feet and 500 feet, but we pressed on, hoping. We eventually got through the basalt, but the hole was still dry at 600 feet, 700 feet and 800 feet, close to our minimum guess for finding water. One morning, we got the message that the hole had reached 814 feet (248 metres) and that water was suddenly flowing out of the top. It had become truly artesian, but the flow was only about 200 gallons (909 litres) per hour. The important thing was that we had correctly predicted the artesian conditions and had even been pretty close to the depth at which it would be found.

The analysis of the water was so alarmingly close to rainwater that we feared the driller had sent in the wrong bottle, but samples we took ourselves confirmed its purity. It was a bit too alkaline, and the engineers muttered awful warnings that it might give the locomotives a mysterious disease called 'caustic embrittlement of the boiler tubes', but alkaline water can easily be corrected with a little acid.[1]

The water engineer was generous in his praise for our effort, overlooked the fact that we had acted quite against his instructions about depth, and gave carte blanche to drill to any depth we needed in future. He also authorized two more holes at Sawmills, which we sited at a lower altitude and proceeded to drill to about 1,300 feet (396 metres). This allowed us to bottom the full sedimentary succession and make full use of it for water extraction. Both holes were fully artesian, and each flowed at 2,500 gallons (11,365 litres) per hour without pumping. Don tested each by fourteen days' continuous pumping at 15,000 gallons (68,191 litres) per hour, and the Sawmills problem was solved. The water withdrawn by test pumping ran to waste and into the Umguza river, which confused a water inspector somewhat when he found no flow in the river at one point but a considerable flow suddenly appearing a short distance downstream.

Our success at Sawmills led to the development of large reserves of underground water, in the Nyamandhlovu farming area, some of which was also artesian. Large volumes are extracted from this aquifer for intensive irrigation. We were pleased with ourselves and set about the problem of finding water for Dett, where half a million gallons (over two and a half million litres) were needed daily for engine-watering and for the local inhabitants of this small but important railway village.

Dett station is just about on the line where the Kalahari Sand feathers out on to the ancient gneisses. Many boreholes had been drilled into those old hard rocks with no success at all. They were all virtually dry. The floor of old rocks slopes southwards from Dett and the Kalahari Sand thickens in the same direction, but our first try was not based on this because we could trace a little rift valley filled with rocks, rather like our aquifer at Sawmills, across the railway line, though it was hidden by the Kalahari Sand cover. We could pick it up by geophysics in spite of the cover of sand. We sited a borehole at Mpofu siding, right at the centre of this miniature rift valley. Drilling began and we were full of hope.

At about the predicted depth, the driller reported abundant water; we were elated. A sample was sent in and analysed. Our elation turned to immediate gloom. The analysis showed that the water was five times saltier than sea water and was quite useless to man, beast and locomotive. The hole was

[1] 'Caustic embrittlement is a phenomenon that occurs in boilers where caustic substances accumulate in boiler materials. It also can be described as the cracking of riveted mild steel boiler plates. This occurs at temperatures of 200°–250°C as a result of local deposition of concentrated hydroxide. ... Caustic embrittlement is also known as stress corrosion cracking.' 'Caustic embrittlement', *Corrosionpedia*, 14 February 2020, <https://www.corrosionpedia.com/definition/240/caustic-embrittlement>.

plugged and abandoned. Dett, and nearby Main Camp, were still going to be short of water. We had to think again.

From what little underground information we could gather, there seemed to be a depression in the floor of old rocks on which the Kalahari Sand rested, centred about twenty miles south of Dett. There could be a large volume of saturated sands in this old hollow, which was several miles in width. It was worth drilling to find out, and an exploratory hole was begun to test the idea. This time we were right, and the hole gave a good yield of very pure water. Don laid out a circle of holes round the pilot one, and all gave good results. They were connected to a central point, and a six-foot main was laid along the railway track to Dett. It still supplies Dett's needs, and careful records have been kept now for more than twenty-five years, which show no loss of capacity. My university training had taught me nothing about underground water, but I had learnt a great deal from working with Don, who became by far the greatest authority on underground water in this country. And, incidentally, his only son became my godson.

At about this time, a wealthy but eccentric Bulawayo industrialist became interested in an abandoned copper mine on the Gwai river, the rough country east of the Bulawayo–Victoria Falls road. On the map it was rather confusingly shown as the Mercury copper mine. It had produced hand-cobbed malachite ore round about 1910, which had been taken by ox wagon to Dett station, but it had soon closed down. We went up to look at it together.

It had been pegged by an old European, badly crippled by arthritis and so poor that he was living in a mud hut and eating practically nothing but mealie meal. He hobbled around his kingdom leaning on a long stick, dressed in rags, and looking like a worn-out Old Testament prophet. This abandoned mine was his last hope. His decline to this level of poverty puzzled me, for he had an honours degree in metallurgy from Sheffield University, but whatever the cause, the end result was pathetic.[2]

[2] This seems to be one George Shaw Ludlam, who 're-pegged the mine in 1945 and again in 1949, when the Messina Company had its first look, but without enthusiasm. ... Some work started in October 1955, when George Ludlam was living in reduced circumstances on a small hill overlooking the Gwai River which became known as Ludlam's Hill. He was already 80 years of age and had spent his life prospecting and mining in Tanganyika and Northern Rhodesia, interspersed with visits to Paris to spend the money he made! Unusually for a bush prospector, he read classical Greek with ease and had an extensive knowledge of Greek literature. He formulated a theory that the Carthaginian general, Hannibal, had financed his 17-year war against Rome from silver mining, probably on the Guadalquivir river, in the south of Spain.' *Gwai River: Valley of Copper*, published by George Hindley on the occasion of the official opening of the Gwai River mine on 8 August 1970 by the Prime Minister of Rhodesia, Ian Douglas Smith.

The rich industrialist for whom I was consulting seemed unmoved, but some of his own habits startled me. He had beer with every meal, including breakfast, and even had a beer when I had my first cup of tea. He refused to go underground and warned me seriously about the danger of snakes in old workings, though he had never been down one to see for himself.

The workings were hazardous in the extreme. The old rough-timber ladders were still in place but had rotted in some sections. I explored very carefully. The place was full of bats, which flew wildly about, and my carbide lamp was periodically blown out by the beat of countless wings. There were also several snakes, and I had to report them to my boss. Naturally, he was elated.

I pronounced that the mine had some promise but would need an expensive exploratory drilling programme. My rich industrialist lost interest and the ancient owner got nothing out of it, but many years later it became the Gwai River copper mine and produced for a few years.[3]

The 1946/47 drought had not only posed problems for the railways but emphasized the potentially precarious state of the City of Bulawayo's water supply. Bulawayo straddles the main watershed of this country. To the north, the rivers flow north or north-west to one of the two main drains, the Zambezi river. To the south, the rivers flow south-east to the main drain, 'the great grey-green, greasy Limpopo River, all set about with fever-trees'.[4] The Zambezi tributaries have gentle gradients and offer few dam sites, but those heading for the Limpopo have much steeper profiles and have a number of potential sites on offer. A growing city on a watershed poses problems of supply that offset the main advantages. In the early days, these were its comparatively cool climate and its consequentially greater healthiness. It also allowed the main railway line to pass through the area with a minimum of major bridges.

The City Fathers of Bulawayo have always been far-sighted over its water supply. Just before World War II they had built Ncema Dam and waterworks and put in a pipeline, with pumps to bring water to the city from about thirty-seven miles away, and with an effective pumping head of nearly 2,000 feet (609.6 metres). This problem of distance and pumping has always made Bulawayo water comparatively expensive.

Not long after the end of World War II, it became plain that Bulawayo's

[3] The Gwai River mine, 1916–1975, is discussed by Harald Solberg in his chapter, 'Copper', in M. Prendergast and J. Hollaway (eds.), *Mining in Zimbabwe: From the 6th to the 21st Centuries* (Harare: Chamber of Mines of Zimbabwe, 2019), 350–3.

[4] Rudyard Kipling, 'The Elephant's Child', in his *Just So Stories*.

growth could easily outstrip its water supply. The aim was to have about three years' supply in the dams in case a succession of drought years resulted in no flow into them. London keeps about three weeks' supply, but three weeks without rain in the Thames valley is a severe drought. The Bulawayo City Engineer's eyes turned to the Umzingwane river, which runs along the north edge of the Matopo granite hills. I was invited to do the geology of the foundations for the dam on this river. It was my first major consulting job and, again, I felt the lack of proper training in both foundation geology and in the proper relationship between geologist and consulting engineer. There were books available on engineering geology, but nothing seemed to have been written on the relation between engineer and geologist. I had to think it out for myself.

I concluded that it had to be the engineer who made the decisions but that it was the geologist's job to provide him with the right kind of information on which to base them. This should include all the underground data that could be gathered from a programme of surface mapping - and as many boreholes and trenches as the geologist could persuade the engineer to pay for to obtain it. The aim is to allow the engineer to design the foundations so that when they are excavated no unforeseen snags appear to hold up construction while re-designing is carried out. If this can be done, it avoids breaks in construction, which exasperate the contractors, and delay in the completion of a scheme, which infuriates the customer.

The geologist's job is not finished when he hands the engineer his report. One of the joys of foundation geology is seeing how right (or wrong) one's forecasts have been when excavation takes place, so the progress of the excavation must be carefully monitored. A second satisfaction comes from seeing the scheme finally completed - and, hopefully, even being invited to the official opening, but, by then, the geologist's work is deeply buried and often forgotten when the invitations go out for the opening ceremony. In this respect the Umzingwane Dam was no exception.

Because there was no convenient narrow gorge with rocky walls that could accommodate a concrete arch, the Umzingwane Dam was to be an earth wall; at the time of its completion, it was the longest earth dam in southern Africa. It was a major undertaking for all concerned.

I camped at the site for two weeks in the rains, did my fieldwork, and caught tick fever. The geology dismayed me a great deal; it was a bad site from my point of view. The rocks were old sediments and lavas of very varied kinds, but all of them were dipping vertically and all had varying degrees of a slatey structure. The really bad feature was that these slatey partings were up and down the river valley, so there was an almost infinite number of potential leakage paths under the dam unless they were sealed down to the base of the zone of weathered rock. The only way to do this is to drill holes down which liquid cement is pumped under pressure. When it sets, it hopefully seals up all the cracks. This process is called grouting, and the resulting barrier is a grout curtain. A useful maxim is, 'When in doubt, grout.' Prevention of leaks under a dam can be expensive, but it is much better than trying to cure them after the dam is filled. Umzingwane needed a very extensive - and hence expensive - grout curtain.

The first contractor on the site was concerned mainly with bush-clearing and other preliminary chores. I happened to go there soon after he arrived, and we talked. He was full of complaints. Never had he worked on a site that had been so badly prepared, and he blamed it all on the consulting geologist. I confessed, and the full battery of fury was turned on me. Somehow, I got the hint that this angry man was a licensed radio amateur, for which ticket I was actively striving. The anger evaporated, and the site was pronounced to be not so badly investigated after all. I have often talked to Tom Blackburn on amateur radio since I got my licence, but we only talk radio shop and dam sites are never mentioned.

The dam began to take shape, but it was the usual race against time to make it safe before the rains came. Then came crisis. The grouting contract fell behind the construction contract, and the south bank was still ungrouted. The decision had to be made whether to grout the south section, which I forecast would leak if not grouted, or to abandon the grouting, finish the wall, and hope that my forecast was pessimistic. The consulting engineer opted for completing the wall on ungrouted foundations. It leaked.

There is some satisfaction in saying, 'I told you so.' The Umzingwane holds back a lake about nine miles long and provides quite a beautiful addition to the scenery. Even the dam wall, being earth and well grassed, is unobtrusive. As an American engineer once said, there is merit in making permanent earth dams, rather than temporary concrete ones. They become part of the scenery and being natural materials, they settle and compact with age better than brittle concrete can do. But there are plenty of records of failures, some of them disastrous.[5]

[5] The Umzimgwane Dam was decommissioned on 3 June 2019 after its water levels dropped to 3.95 per cent. 'Umzingwane Dam decommissioned', *The Chronicle*, 4 Dec. 2019. The Ncema Dam, known now as the Lower Ncema dam since the Upper Ncema dam was commissioned in 1973, was decommissioned on 15 May 2020. Tanaka Mrewa, 'Byo water woes: Lower Ncema Dam decommissioned', 15 May 2020, <https://cite.org.zw/byo-water-woes-lower-ncema-dam-decommissioned>.

However, three of the city's supply dams, Umzingwane, Upper Ncema and Lower Ncema, were recommissioned at the end of 2020 after good rains: Mashudu Netsianda, 'Bulawayo prepares to recommission three dams', *The Chronicle*, 22 Dec. 2020. By February 2021, 'the second highest inflows in dams in 50 years [had] been recorded countrywide, following incessant rains', Nqobile Tshili, 'Dams record second highest inflows in 50 years', *The Chronicle*, 17 Feb. 2021.

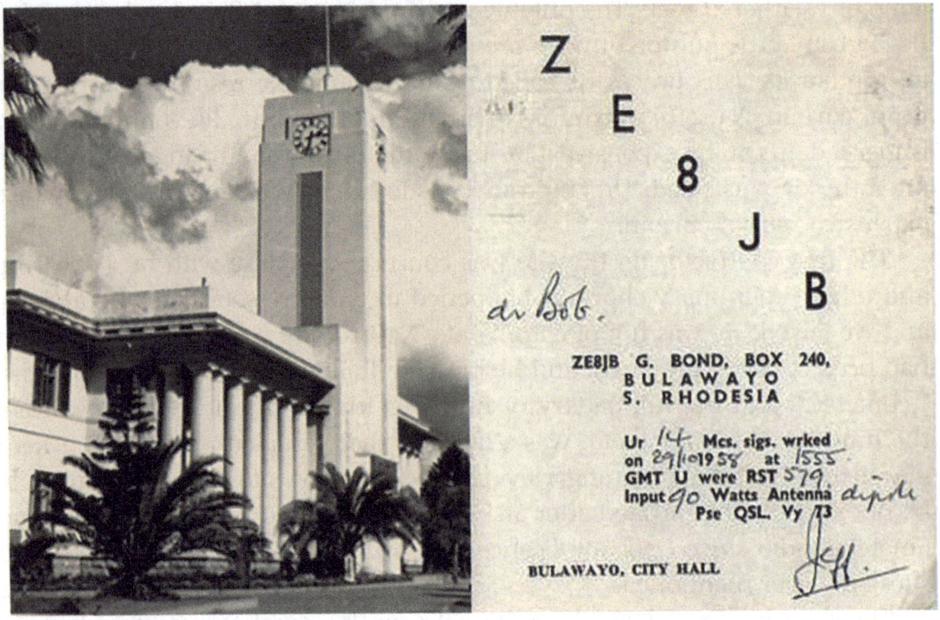

Geoffrey Bond's QSL card on the Rhodesian Study Circle's website.

A QSL card is a written confirmation of either a two-way radio communication between two amateur radio stations or a one-way reception of a signal from an AM radio, FM radio, television or short-wave broadcasting station. It can also confirm the reception of a two-way radio communication by a third-party listener.

QSL card derived its name from the Q code "QSL". A Q code message can stand for a statement or a question, when the code is followed by a question mark. In this case, 'QSL?' means "Do you confirm receipt of my transmission?" while 'QSL' means "I confirm receipt of your transmission."

A typical QSL card is the same size as – and is made from the same material as – a typical postcard, and most are sent through the mail as such.

QSL Cards,
<http://www.rhodesianstudycircle.org.uk/qsl-cards>

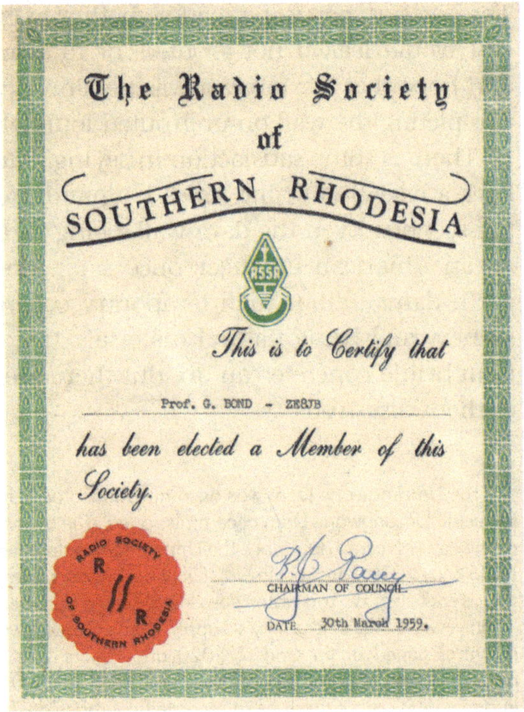

– 9 –

The Valley

Even as early as 1950, there were rumours that plans were afoot to build a hydroelectric scheme in the Kariba gorge, by damming the mighty Zambezi itself. The resulting lake would be about 200 miles long and would flood out the people living on its banks. These were the Valley Tonga, a little-known people who had been described by Livingstone in some detail in his book published in 1865.[1] Dr Desmond Clark, who was Director of the Livingstone Museum,[2] was primarily a Stone Age archaeologist, but he decided that a combined archaeological and ethnographical exploration of the section of the Valley threatened with inundation should be done without delay. It was totally unknown archaeologically, and the traditional material culture of the Valley Tonga would inevitably be lost in the resulting resettlement if the dam were built. We planned a joint venture from the northern side, into a part of the Valley called the Gwembe, which distinguished it from the southern side, where the equivalent country was called Sebungwe. Both sides were little known, but access looked easier from the north.

My part in our joint venture would be to try to work out the later erosional and depositional history of this part of the Zambezi Valley, and to link it up with the work that had been done in the vicinity of the Victoria Falls. Desmond Clark had already published this work, in which he had linked the succession of Stone Age cultures with the terrace gravels above the Falls, and had shown how the earliest men had lived in the area when the Falls had been some distance downstream from their present position. The zigzag gorges below the Falls mark the positions of eight or nine old fall lines, and quite a number of them were active in turn during human times.

[1] David and Charles Livingstone, *Narrative of an Expedition to the Zambesi and Its Tributaries* (London: John Murray, 1865).

[2] 'The Livingstone Museum is the largest and the oldest museum in Zambia, established in the 1934 as the David Livingstone Memorial Museum and was located in the old Magistrate's Court building. In 1939, The Rhodes-Livingstone Institute was started and incorporated with the David Livingstone Memorial Museum. In 1939 the collection was expanded to include relics of Cecil Rhodes and the British South African Company. In 1946, the Museum and the Institute were separated and the Museum's name changed to the 'Rhodes-Livingstone Museum'. 'Livingstone Museum', *Rhodesian Study Circle*, <http://www.rhodesianstudycircle.org.uk/livingstone-museum>. The name was changed to the Livingstone Museum in 1966 after Zambia's independence.

NARRATIVE

OF AN

EXPEDITION TO THE ZAMBESI

AND ITS TRIBUTARIES;

AND OF THE DISCOVERY OF THE LAKES SHIRWA AND NYASSA

1858—1864.

BY DAVID AND CHARLES LIVINGSTONE.

WITH MAP AND ILLUSTRATIONS.

LONDON:
JOHN MURRAY, ALBEMARLE STREET.
1865.

Many years later, I was severely castigated for referring to these gorges as a 'dirty great donga' eroded by the Zambezi. The Victoria Falls publicity people regarded this as poor propaganda for their beautiful tourist attraction. But in a geological sense it is a valid description. Livingstone explained the formation of these gorges as a mighty split in the crust, but modern geological thinking attributes the whole length of the gorge, something like sixty river miles, entirely to the relentless erosive power of the river itself. While we were in the Valley, I would have chance to study all the geology, including the coal-bearing rocks, and I even speculated on why there was a major gorge with a waterfall in it. We arranged to meet in Livingstone in May 1950 and spend up to two months in the Gwembe section of the valley.

Desmond arranged with the government of Northern Rhodesia to have with us a young professional photographer from the Information Department, Nigel Watt, whose job would be to collect a good series of pictures of the Valley Tonga.[3] Perhaps even more usefully, he had arranged with the Gwembe District Commissioner (DC) to borrow one of his Boma messengers, complete with uniform, as a mark of our authority with the local people. Boma messengers were the outstretched arms of the all-powerful DC and, because they did frequent patrols, they knew all the local chiefs and headmen.[4]

It was easy to get to the Gwembe Boma from the Great North Road, and a road of sorts led from there to the Zambezi at Kanchindu Mission, which was on the north bank, opposite where the village of Binga now stands.[5] Just upstream from Kanchindu was Mwemba's village, which is shown on Livingstone's map of the area and was visited by him on his travels through the valley 100 years before us. Livingstone's spelling was Moemba.

At this point in my field notebook there is the significant entry: 'stopped

[3] Some of his photographs appear in the digital libraries collection of 'Vernon McKay Photographs' of Northwestern University: <https://dc.library.northwestern.edu/collections/1d849df1-eb32-43f4-b7b9-e435cff18f7c>.

[4] Called District Commissioners (DCs) in Northern Rhodesia and Nyasaland but Native Commissioners (NCs) in Southern Rhodesia. In 1962 the Native Affairs Department in Southern Rhodesia was renamed the Ministry of Internal Affairs, and the NCs were renamed DCs. Literally meaning an enclosure, the word Boma was used in Northern Rhodesia for the place where their offices were situated and from where a district was administered.

[5] 'The Primitive Methodist Missionary Society (PMMS) expanded its reach in Northern Rhodesia (Zambia) to the BaTonga (Tonga) peoples with the establishment of Sijoba Mission in 1901. Soon after arriving at Sijoba in 1907, Rev. John R. Fell moved the mission down river, leaving Sijoba as an outstation. The new mission station became known as Kanchindu. ... In 1958, as part of the forced resettlement of the BaTonga peoples in connection with construction of Kariba Dam, the mission was moved inland to its present location near Maamba village in the Sinazongwe district of Zambia. The United Church of Zambia assumed responsibility for the school in 1965. The school was taken over by the government in 1975, but handed back to the church in 2011 and now runs as a 'grant-aided mission school'. 'The Primitive Methodist Church: Kanchindu Mission', *Rhodesian Study Circle*, <http://www.rhodesianstudycircle.org.uk/the-primitive-methodist-church-kanchindu-mission>.

shaving'. I have generally made it my rule to shave every day on fieldwork, but Desmond and I had made a pact that on this one we would both grow beards. Neither of us had ever tried it, and here was an opportunity to get over the scruffy stage in the privacy of the Valley, but Desmond omitted to warn his wife before leaving. We got back to Livingstone weeks later with beards of which we were both very proud, but we arrived long after dark and unexpectedly. Betty Clark came out to meet us, heard Desmond's voice, bounced up to him, flung her arms round his neck and bounced off again, obviously thinking she had hugged the wrong man. His beard was accepted and has survived to this day, trimmed down to an elegant Van Dyke.[6]

I took my full-bottomed naval beard back to the Museum in Bulawayo, where I was told I looked like something washed up on a raft. I suppose it was rather horrible, as there was a lot of grey in it, mixed with hair that was almost gingery. I shaved it off, which ended an experiment never to be repeated. It is something every man should try once in a lifetime.[7]

The DC at Gwembe honoured his promise, and we squeezed a fairly small Boma messenger in among our belongings and set off for Kanchindu. The track down the northern escarpment was steep and rough, which left us wondering whether we could persuade a vanette with two-wheel drive ever

[6] 'A style of facial hair named after the 17th-century Flemish painter Anthony van Dyck (1599–1641). A Van Dyke specifically consists of any growth of both a moustache and goatee with all hair on the cheeks shaved.' 'Van Dyke beard', *Wikipedia*, <https://en.wikipedia.org/wiki/Van_Dyke_beard>.

[7] Desmond Clark spoke about this trip:

> When I went the first time into the Gwembe Valley I went with Geoffrey Bond, the geologist from Bulawayo. And our cook was called Cabbage. He'd been the cook for the policemen down in Southern Rhodesia. Sounded all right – they said he could do things. Typically for breakfast we would have oatmeal, something like that. Jam, butter. For lunch there would be thickish slices of bread with butter on them, and if we had any meat, we'd put it on it, something in between the two sheets.
>
> Well, I remember the first time we went out, no butter on the bread. So I said to Cabbage, 'We would like butter on our bread.' Cabbage said, 'Yes, of course' It came back, we had butter on one piece but not on the other piece. So we said, 'Cabbage, we would like butter on both sides.' So what we got was indeed butter on both sides of one piece of bread. You could see, it was great fun. He was okay, but we were quite glad to see him go.

J. Desmond Clark, *An Archaeologist at Work in African Prehistory and Early Human Studies*, an oral history conducted in 2000–2001 by Timothy Troy (Berkeley: University of California, Regional Oral History Office, The Bancroft Library, 2002), 96.

Desmond Clark died on 14 February 2002. An e-memorial website was set up: 'This e-memorial was established in February of 2002 to help honor the memory of the late John Desmond Clark, a pioneer in studies of world prehistory. Desmond was a legend in his own time, leaving footprints on every continent. His lifetime of achievement will continue to impress us. His relentless pursuit of knowledge about the human past will continue to inspire us.' 'Tributes to Betty Baume Clark, J. Desmond Clark', *Human Evolution Research Center*, <http://herc.berkeley.edu/jdesmond_clark_memorial/index.html>.

to climb it again. Below the escarpment, the valley floor was undulating and the track much easier; there were no forks, so there was only one way to go. There were no real maps, but Kanchindu Mission was the end of the track and we duly arrived.

The Valley has its own beauty and fascination for its addicts. Where the soils are clays, there are great stretches of stately mopane (*Colophospermum mopane*) woodland. This is the geologist's ideal country, for there is only sparse grass under the trees and one can walk for miles without any problem. On the sandier soils there is a mixed woodland with much denser undergrowth. By the river itself there is discontinuous riverine forest, with the great evergreen musikili/mutsikiri trees (*Trichilia emetica*) with dense shade that make good campsites.[8] And there was the river itself, several hundred yards wide on average, with a current running at perhaps five knots between sand islands on which dense growths of the prickly *Phragmites* reeds abounded, and on which crocodiles basked on warm days: in the Valley, most days are warm, and generally very warm. Hippo were plentiful, and there was constant war between them and the Batonka, who struggled to grow crops on the terraces of fertile alluvial soil.

The river had an annual rise and fall of about thirty feet, and the Batonka cultivated the steep bank as the river was falling, leaving behind a skim of fertile silt. The lowest terrace was swampy at certain times and provided a crop at a different time of year, but the main crop was on a higher terrace and grown during the rains. The villages were on a higher terrace still, which was older, less fertile, and cultivated only when pressure of population forced it on them. Altogether, the Batonka in the Valley were very closely tied to the local geology and the limitless water of the river. Soon after our travels, Thayer (Ted) Scudder lived among them for about fifteen months and wrote it all up for his Ph.D. at Harvard. He was just in time, for the filling of the lake destroyed their highly evolved ecological links forever.[9]

The Valley Tonga were then 'primitive' in the best sense of the term. Reading Livingstone's account of them showed that they had changed very little in a hundred years.[10] In their own environment they were highly educated. There was a name for every tree, shrub and grass, and it was applied consistently throughout the Valley. Food and medicine were carefully

[8] Also known as Banket mahogany and Natal mahogany.

[9] He wrote: 'I am indebted to Dr Geoffrey Bond for explaining to me the classification and distribution of Zambezi alluvial.' Thayer Scudder, *The Ecology of the Gwembe Tonga* (Manchester: Manchester University Press for Rhodes-Livingstone Institute Northern Rhodesia, 1962), 4, fn. 1.

[10] Livingstone, *Narrative of an Expedition to the Zambesi*. A more recent book about the people of the Gwembe Valley before the Kariba dam was built is Pamela Reynolds and Colleen Crawford Cousins, *Lwaano Lwanyika* (Harare: Colleen Crawford Cousins in association with Save the Children Fund (UK), 1991).

collected, and the best woods were used for each cultural purpose – be it spear shafts, hut-building or fencing.

Each man carried two spears and an axe. One spear was the conventional blade, the other a fishing spear, square in section and barbed with innumerable spikes cut with a cold chisel from the square edges. The axe, which could be used either as a conventional chopper or as an adze, completed the unusual armoury. All the metal work was done locally: in Livingstone's day, by smelting local iron ores; in our day, by collecting old motor-car springs and reworking them. Smiths were skilled craftsmen, but all Batonka males could make fire quickly and easily with fire drills.

Although the Tonga had been in the Zambezi Valley for perhaps two to three centuries, they were not a river people and had a healthy fear of it. Fishing was done only in dying tributaries with basket fish traps, except at one place where we saw a unique method of fishing. Behind a low hill near Kanchindu, there was a hot spring, which issued at near boiling point and formed a considerable right-bank tributary of the sluggish Zimu river. Fish drives were organized from upstream towards this confluence, and the fish died instantly and floated when they met the hot water.

Nigel Watt had a real harvest photographing these people, particularly the women and girls who were topless long before this fashion was invented in more sophisticated circles. They also invented the micro-miniskirt quite independently. This was made of countless short lengths of baobab-fibre string, each with a knot at the end and attached to a belt around the waist.

Desmond Clark and Geoffrey Bond at a hot spring in the Gwembe Valley

Photo: ZANIS

Beadwork was everywhere, and the young girls wore little circles of cowrie shells in their hair. These must have come all the way from the coast, since they are the shells of a marine snail. Traditionally, the four upper incisor teeth were knocked out and the nose septum pierced so that a stick could be worn through it. Copper or brass wire was wound around both ankles, and the whole outfit was set off by a complete coating of a red ochre mixed with a natural oil. Livingstone was shocked by their lack of clothes, but we found the overall result very impressive.

As we moved among these people, we realized that we were seeing them very much as Livingstone had seen them. We even became convinced that some of the paths connecting the villages, all of which were close to the Zambezi, were the same paths that he had walked so long ago.

Although the riverine fringe is well populated, inland water is scarce, and very little habitation existed before dams and boreholes allowed the people to spread into that semi-arid country. This became necessary as the population grew, and imperative when Kariba Dam was built and the entire riverine habitat was flooded.

Their general fear of the river did not mean that the Zambezi in any way divided the population into northern and southern parts. There were traditional dugout canoes that frequently crossed from one bank to the other, and marriages took place across the boundary. Crossing by this means was something of an experience for the stranger. The canoes seemed heavy, unmanageable and unstable, but their owners were skilful and knew their section of the river. The drill was to pole upstream in the quiet water near the bank for up to half a mile, then launch out across the stream, poling and paddling to the far bank and landing plumb opposite the starting point. The main danger lay in disturbing a submerged hippo.

One of our objectives was to collect the traditional material-culture objects of these people for preservation in the Livingstone Museum, and the Boma messenger was sent round to make it known that on a certain day we would be prepared to buy what was on offer. The time came, and with it a motley collection of old women, old men and young children. The sale began when an old woman stepped forward and offered a very small, blackened pot, a 'relish bowl', for which she demanded ten shillings. None of us understood ciTonga and the Boma messenger in his immaculate uniform acted as interpreter. He looked scornfully at the little fire-blackened pot and said scathingly, 'Ten shillings! This is government, not business. Sixpence.' After this, prices were very reasonable, though it is doubtful if the sellers had much appreciation of money as there was no store in which to spend it for several tens of miles. Raw salt and unstrung beads, provided they were locally fashionable colours and sizes, were far more useful as currency than real money.

The Sebungwe, June 1954

It took only a few days of traverses around our camp at Kanchindu to get an idea of the local geology, and archaeologically it was disappointing. Desmond began to get restless. We did manage to get his vehicle as far as Mwemba's village, perhaps the first motor car to get there. We even persuaded Chief Mwemba, a very old man, to take his first ride, of a few hundred yards. He was obviously reluctant, but he did it, and emerged looking extremely thoughtful.

To our intense surprise we came across a grave in a small clearing, marked by a neat iron cross. It was close to Mwemba's village. The inscription read:

A TERÖRDE
12 May 1844 - 7 Sept. 1880

We got more details from the Gwembe Boma diary on our return journey. Terörde was a Dutch Jesuit missionary who was stationed at Zankie's (Wanki's) kraal, then on the north bank of the Zambezi, not far from the Deka confluence. With a colleague he had walked down the north bank of

Photo courtesy of Fr Edward Murphy SJ, Harare (original: ZANIS)

Photo courtesy of Ron Hidaka, Jesuit mission, Chikuni

Left: Nigel Watt's photograph of Fr Terörde's grave. Right: The cross, taken to the Fumbo Catholic mission in the Gwembe Valley, where it still stands today. Some Irish Jesuits at Chikuni mission in Zambia tried to exhume the remains at the time but failed to find anything.

the Zambezi as far as Chief Mwemba's village. The chief might have been our friend's great uncle but was unlikely to have been his father. The chief seems to have laid on some kind of festivity, but both Jesuits became ill and Terörde died. His colleague recovered, presumably buried his friend near Mwemba, and managed the return journey alone.

The cross we saw was erected by the Jesuits on 23 February 1916. The original site of this very early grave, the earliest known to me in the valley, would now be under water and I hope Father Terörde's remains were moved to higher ground before the flood overtook him.[11]

Desmond decided that we should do a walking tour up the Valley to a point about forty miles upstream of Kanchindu, exploring the archaeology and geology as we went. He aimed at Sijoba's village. He sent our Boma messenger to collect about half a dozen carriers, and selected the minimum of food, equipment and bedding. Nigel Watt and I prepared for our first taste of his traditional way of travelling. Desmond had already done miles of it in his archaeological travels and knew all the answers. On reflection, I am not so sure that he did.

His golden rule was to start very early on the first day, and cover at least twenty miles at the fastest pace that the party could manage. This meant that everyone was tired out - and that if any vital thing had been left at base camp, it was too far to go and fetch it. Desmond's pace was gruelling on that first day. His stride was too long for me to keep in step with him, and I struggled along behind. He was ten years my junior, and I silently cursed his back view most of the day, as he strode along with his ashplant walking stick, his neat long woollen stockings with tabs on the garters, a double Tiree on his head,[12] and shorts of the regulation Colonial Office length down to the kneecap. By the time he called a halt I was blistered and footsore, and a thorn had gone through the sole of one shoe into the ball of my left foot. Nigel, the youngest of us but totally unused to the bush, was also in poor shape. The next day we did another twenty miles and reached Sijoba's village, where we meant to spend a couple of days.

[11] The full story of Fr Antoine (Anton) Terörde's journey to the Batonga was told in Fr H. Depelchins's diaries and letters that were translated by Véronique Wakerley, edited by R.S. Roberts, and published as 'Part Two: Among the Batonga, 17 May - 19 December 1880', in *Journeys Beyond Gubuluwayo to the Gaza, Tonga and Lozi: Letters of the Jesuits' Zambesi Mission, 1880–1883* (Harare: Weaver Press, 2009), 77–177. It was believed that Fr Terörde and his colleague, Br Arnold Vervenne, who survived, were poisoned (p. 169).

[12] Similar to a beanie or bobble hat, the Tiree is named after the Isle of Tiree in Scotland, the western-most island in the Inner Hebrides.

Desmond was by now very disappointed by our lack of success in finding Stone Age material. This part of the Valley seemed almost barren by comparison with the riches around the Victoria Falls. We had found some Late Stone Age sites, and a certain amount of Middle Stone Age material, but none of the beautiful biface hand axes and cleavers of the Older Stone Age, so common at the Falls. He decided to make some, as sympathetic magic. He was good at it, though no Stone Age man would have said he was an expert. The ones he made were so good that they could have been misleading to archaeologists in the future, so they were ruthlessly broken into small pieces.

After a couple of days at Sijoba, it was time for the return walk, but my left foot was by now swollen and very painful where the thorn had gone in. I had red streaks up my thigh and a lump in the groin. Walking was difficult and painful, but it had to be done. The first mile was agony, but then it seemed to ease off. Desmond frequently asked anxiously how I felt, and the phrase 'numbing nicely' became the standard answer. It was moderately true so long as we did not stop. Then it had to be numbed all over again.

Two days saw us back at the base and the M&B tablets cleared up the infection quite quickly.[13] We toiled our way back to the Gwembe Boma and prepared for the second leg. This was to take us to Chipepo's village, seventy miles or so downstream of Kanchindu, and just west of what is now the Kota Kota narrows of Lake Kariba. The DC warned us that the track ended twenty-one miles beyond the Boma, leaving about thirty miles of footpath from a village called Munyumbwe to Chipepo. It would mean walking with carriers again.

We left the vehicle at Munyumbwe's village, collected carriers and set off as Desmond still insisted, as early as possible. Our first day was to take us to Musulumba, an unknown but obviously considerable distance away. At about two o'clock we met a Batonka man going in the opposite direction. On being asked how far the village was ahead of us, he said thoughtfully, '*Dusi bechana*' (a little bit near). This sounded quite hopeful and we pressed on. An hour or so later we met another, and his verdict was, '*Dusi*' (near). This sounded encouraging, so on we went. An hour or so later, we asked the same question, and the answer this time was, '*Dusi sterek*' (very near). We arrived half an hour after this. To the Batonka, so used to walking vast distances, 'near' is elastic and has odd shades of meaning.

[13] M&B 693, 2-(p-aminobenzenesulphonamido) pyridine, was one of the first generation of sulphonamide antibiotics. 'The most famous patient saved by the drug was British Prime Minister Winston Churchill, who contracted pneumonia at the height of the Second World War. The drug most likely saved his life and the success of the treatment was widely reported in the press.' 'Pre Antibiotics', *History of Science Museum*, <https://www.mhs.ox.ac.uk/backfromthedead/exhibition/pre-antibiotics>.

Once more, Desmond's first day had taken us about twenty miles. By now we were all fit, and the next day's walk to Chipepo, a mere fifteen miles or so, seemed like a gentle stroll. We made camp in the firm belief that we were the only white men in the entire Valley. Just before sunset, we were startled by loud blasts on a trumpet, which then went on to well-known hymn tunes. This was found to come from an American missionary, the Revd Blann, calling the faithful to prayer, equally sure that he was the only white man in the Gwembe Valley. He had only arrived a few days before us.[14]

At first, the Chipepo area seemed as poor in the Older Stone Age as the previous section had been, and Desmond set about making sympathetic magic again. Once more they were broken up, but this time the magic worked and we found quite abundant hand axes and cleavers in a proper geological context. The comparison of the middle Zambezi with the Victoria Falls area now seemed reasonably established, and the two main aims of our journeys were satisfactorily accomplished, at least on a reconnaissance level. There were plenty of details still to be filled in, and I resolved to mount much more detailed work on the Stone Age from the southern bank at a later date.

A minor puzzle near Kota Kota hill was a great heap of small stones beside a footpath. This heap contained several tons of pebbles, all of local rocks, but there was no obvious geological reason for its existence. We resorted to enquiry among the local people through the Boma messenger. There was some reluctance to talk, but gradually the story came out. The place was haunted. No one should pass it without tossing at least one small stone on to the heap, which in some way counteracted the spell. How the superstition had arisen, no one knew - or, if they did, would not say. Neither would they explain what would happen if the ceremonial stone was not added to the pile. For safety, we added quite a few to cover the times we had passed before we knew the significance of the place. I have not met this custom anywhere else in the Valley.

We packed up and left the Valley reluctantly. The journey to Livingstone was completely without incident, and nothing happened to me on the way to Bulawayo. We must have added an adequate number of pebbles to the heap at Nsengasenga.

[14] This is likely to have been Henry John Blann, who was a missionary of the Pilgrim Holiness Church and was based at Jembo Mission in Pemba from 1947. 'When more missionaries arrived from the United States to work in Jembo Mission, John and Eva [his wife] volunteered for the work of establishing a new mission in a malaria area on the Zambezi River, seventy miles from the main station in territory where most of the natives had never had any contact with white people. Their first temporary shelter was made of grass with no roof and a door of corn stalks. Before the permanent buildings could be made they had to make and burn the bricks, drawing the water for the bricks from the Zambezi River by oxen. With native help they made several buildings including their own home, one for the nurse, a dispensary, and a school for boys.' Caroline Brooks Adair, *Our Heritage* (York, PA: privately printed, 1954), Part II, 17.

Nigel Watt's photographs were excellent and provided a very good record of the Valley Tonga as they were in 1950. It is a pity that Livingstone was there before photography. A comparison would have been interesting, for his descriptions indicated little change in nearly a hundred years.

Major events in one place can produce minor ripples at great distances, and my next lucky chance came this way. The coal mines in Britain were nationalized, and William Baird & Co., who had owned and operated coal mines there, found themselves in 1950 with a lump sum in compensation for their loss.[15] This was the major event; the minor ripple was that they turned their attention to the coal potential of the eastern Sebungwe district, roughly the area between the Sanyati and Sengwa rivers. They employed Dr A.M. Macgregor, who had not long retired from the Directorship of the Geological Survey, to be their consultant, and 'Mac' asked me to be his assistant. This was just one more of my lucky chances, but a very significant one.

Coal in Southern Rhodesia had been known since the early 1890s, and Giese had pegged the Wankie Colliery concession in 1896 after trekking from near Francistown to follow up rumours of 'black stones that burn'.[16] That indomitable pioneer of Rhodesian geology, A.J.C. Molyneux, had explored some of the Limpopo coals near Beitbridge at about the same time, and had then begun his remarkable reconnaissance of the Sebungwe area, which at that time was almost unknown country, even geographically.[17] He had found coal outcrops in many places between the Sanyati and the Gwai rivers, and had kept meticulous photographic records, no mean feat around 1900. Cameras then used large glass plates and were huge boxes mounted on clumsy tripods.

It was largely the pioneer efforts of Molyneux that interested William Baird & Co., and Dr Macgregor was the right man to follow them up. He had

[15] William Baird & Company was established in 1830 by William and James Baird. 'Between 1946 and 1951, the whole of William Baird's coal, iron and steel interests were nationalized and the company began to diversify into other areas of business, including the textile industry.' 'Bairds and Scottish Steel', *Grace's Guide to British Industrial History*, <https://www.gracesguide.co.uk/Bairds_and_Scottish_Steel>.

[16] Albert Giese, a German geologist, had first arrived in Wankie (Hwange) in 1892 and registered the first claims in 1893, but he could not exploit them as the British South Africa Company owned all the mineral rights in the country. The Mashonaland Agency Ltd. acquired concession rights in 1894, and Giese returned in 1895 and pegged about 1040 km^2. See O.J. Maponga, 'Coal', in Martin Prendergast and John Holloway (eds.), *Mining in Zimbabwe: From the 6th to the 21st Centuries* (Harare: Chamber of Mines of Zimbabwe, 2019), 207–34.

[17] See 'Molyneux, Mr Arthur John Charles', *S2A3 Biographical Database of Southern African Science*, <https://www.s2a3.org.za/bio/Biograph_final.php?serial=1943>.

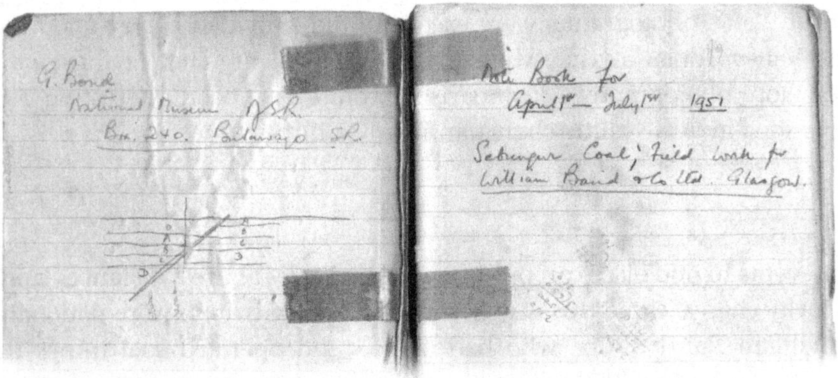

already done some traverses through the region just before World War II and, in January 1947, I had gone with him to rediscover some of Molyneux's fossil localities around Lusulu in the central Sebungwe. The chance to work with him again was too good to miss, and I was seconded by the Museum for the necessary time – about two weeks late in 1950 and for the whole dry season in 1951.

From his knowledge of the area, Mac marked out an Exclusive Prospecting Order covering 500 square miles. This was an early order and was in fact Number 14. There have been more than 500 since, but few as big. EPO 14 was an odd shape. The areas of interest were widely scattered, but the boundaries had to enclose a continuous piece of country, with no gaps, so it was centred more or less on the Bari salt pan near the Ume river and had narrow arms stretching out in various directions to include our target areas.

In theory, every single corner had to be properly beaconed, and line beacons should be placed where boundaries intersected roads. Since there were virtually no roads, this would be easy, but I am sure Mac never intended to identify and beacon all the corners, which were in totally inaccessible places, where Mines Department officials would never go to check up. We did put up a line beacon beside the only motorable track, and when later in the season we had a visit from the Chief Government Mining Engineer, he solemnly raised his hat to it, remarking that it deserved special recognition because it was probably the only one in the entire area. He was right.

Our first task, in December 1950, was to do a preliminary visit to our area in Mac's 30cwt Morris truck. This vehicle had huge wheels but only two-wheel drive, and we were well into the rains. We met up in Gatooma and set off. We followed old tracks that Mac knew, leading first to Robb's Drift on the Sanyati, and then, by devious means and many false starts, we somehow got to Gokwe and the Native Commissioner's office. Mac announced that we were going down into the Valley at least as far as Bari salt pan and would be back in the NC's office on 21 December. The reaction was utter disbelief. The NC obviously thought we were mad, and virtually said so. All the drifts had been taken up on the only track, the rivers were liable to be impassable, with such a vehicle we would get so bogged down that it would be impossible to extricate it, and we would end up with a 90–100-mile walk back to Gokwe. The whole scheme was preposterous, but we were, of course, welcome to try it.

Ivor Cockcroft knew his district and his opinions were well founded, but he underestimated Mac. At precisely 11.00 a.m. on 21 December, we reported back in his office to book out of his district. Furthermore, we had convoyed one of his BSAP members back, whom we had found struggling with a broken-down Land-Rover. He made no attempt to hide his astonishment at seeing us exactly at our Estimated Time of Arrival. After this, the district was ours. We could go where we liked and when we liked, a blanket invitation of which I made full use for many years.[18]

I doubt if Ivor Cockcroft ever really knew how nearly right his forecast had been, for it was an exceedingly tough few days. The main drifts were passable only when we re-laid the corduroy of mopane poles that had been carefully stacked out of reach of the floods. We bogged down innumerable times, chopped bush, and dug for hours on end to get going again. We sweltered in the heat, cursed the mopane flies, and camped rough when we could get no further. One day, after a heavy shower in the night, we managed no more than 300 yards. Mac could drive that truck through places where I got stuck immediately, so he did most of the driving. I did most of the chopping and digging. Considering our age difference, this was a fair division of labour. We had no maps and no air photos, but Mac's uncanny bush sense never faltered, and we always managed to get to our targets and back again whether it was on foot or going across country in his truck. I was sometimes totally lost, but Mac always knew what he was doing.

From Gokwe, we went down the escarpment using the track that Molyneux had used with ox-wagons pulling a steam-driven diamond drill almost exactly fifty years before. From his carefully located and oriented photograph, it was obvious that it had not changed much in that time.

From the traverses we made on this reconnaissance, the programme of exploration for the next dry season (1951) took shape. There were to be three diamond drills and crews, the whole area was to be mapped, and we hoped

[18] Ivor Cockcroft was posted to Gokwe, the administrative centre of the Sebungwe district, in 1949. 'The district was some 13,000 square miles and included some 150 miles of the Zambezi river frontage – on which most of the Batonka people lived. At that time the people lived in peaceful isolation – there was but one so-called track from Gokwe to the Zambezi.' I. G. Cockcroft, 'Kariba', *NADA, the Rhodesian Ministry of Internal Affairs Annual* (1967), 11 (4): 25.

'The Batonka were fortunate to be ruled by such a man during their years of trial,' wrote Frank Clements, *Kariba: The Struggle with the River God* (London: Methuen, 1959), 94. The Cockcroft Bridge across the Sebungwe river is named for him.

to drill at least ten holes in strategic places to test the coal potential of the area. The sedimentary rocks of the Karoo System had already become my prime interest for research, and this season's fieldwork with Mac and a drilling programme was to set me up for years of future work and give my geological work an entirely new dimension - and a positive direction.

Dr Macgregor looms large in my geological development, and was the last person to have a major influence on me, so some account of the man himself as I got to know him is essential, and this is the proper place to discuss him.

Alexander Miers Macgregor must have been expected to become a geologist, for his parents gave him the name Miers, who was a great mineralogist.[19] I have never found out what the connection was, but Mac certainly lived up to it, for he became known in the geological world for his pioneer work in this country, and for his remarkable ability to see far beyond the local field evidence and perceive things on the grand scale. His reputation in the world of science was very great. He was one of the select band of mortals who became something of a legend in their own lifetime, in his case not only internationally for his scientific wisdom but locally for his remarkable absentmindedness.

In appearance he was somewhat gaunt. He was very short-sighted and one of his most characteristic poses was to push his topi back on his head,[20] shove his glasses up on to his forehead, and then examine a rock specimen at very close quarters with the pocket lens he always carried slung around his neck on a piece of string.

He hated being photographed, except at long range near some outcrop of rock where he could be used as a scale, but getting a close-up of him was nearly impossible. He would never pose. Mrs Macgregor claims that the best close-up of him she has ever seen was one that I took. It was no use asking him if I could take his picture, but out in the field with him I took advantage of his habit of examining specimens without his glasses. I handed him one with some comment or other, but my camera was set up beforehand for a close-up shot, which was taken before he realized he had been trapped. The result was pure Mac, at his most natural.

The stories about his absentmindedness are many, some true and vouched for by his wife or by his associates in the office of the Geological Survey, some even happened while I was with him, and some are probably apocryphal.

[19] See Thomas Henry Holland and Leonard James Spencer, *Obituary: Henry Alexander Miers, 1858–1942*, <https://royalsocietypublishing.org/doi/10.1098/rsbm.1943.0009>.

[20] A sola topi: a sun hat made from the pith of the stems of sola plants, formerly worn in India.

'Mac' Macgregor

Two, vouched for by his wife, are worth telling. Mac was a shy man, particularly among women. Mrs Mac was short, outgoing, social and much addicted to bridge. On one particular wet afternoon, Mac walked home in the rain under his umbrella, and forgot to put it down as he walked through the French doors from the veranda into the lounge. To his horror, he found himself in a bridge party of Mrs Mac's lady friends. He stood under his dripping umbrella and murmured to his wife, 'Really, dear, we must get the PWD to mend this roof.'[21]

Mac came to this country as a new Cambridge graduate in 1912 to join the National Museum staff as Keeper of Geology in Bulawayo. The war soon

[21] The Public Works Department (PWD) was responsible for the construction and maintenance of government buildings.

intervened, but he survived and returned in 1918 to restart his work, but this time as a member of the staff of the Geological Survey in Salisbury. By the early 1930s he had married, a daughter had arrived, and the Geological Survey was experimenting with motor cars for transport. While his daughter, Rowena, was still very small, he had to go off to some spot near Gatooma for a day's fieldwork and took mother and baby along for the day. He made them comfortable under a tree with the picnic things and went off in the car. What he found was obviously very exciting geologically, whereupon he got into the car and drove back to the Geological Survey office in Salisbury to discuss it with his colleague. One of them pointed out that he had gone off in the morning with his family, and where were they now?

In the fullness of time, Mac became Director of the Geological Survey and received great international recognition for his work. He was honoured particularly by the Geological Society of South Africa, of which he became President, one of very few non-South Africans to hold this position. The Society also bestowed on him their highest award, the Draper Memorial Medal. The University of the Witwatersrand gave him an Honorary degree of Doctor of Science. This was the man with whom I was to spend a full six months in the Sebungwe district.

For me this was rare good fortune. His colleagues in the Survey rather cynically remarked that I was mad, and that he would probably walk me off my feet and involve me in all sorts of alarming encounters with wild animals. I was fully aware of his complete indifference to the larger mammals, having been charged by two rhinos simultaneously while walking with him during a short trip to the central Sebungwe in 1949. I knew, too, that, in spite of his age (he was already on 60), his stamina was immense, and that he was oblivious to heat, hunger, thirst and physical discomfort of all kinds. But the chance to work with him in the Sebungwe was too good to miss.

He belonged to the old school of geologists who could start a season with virtually a blank sheet of paper and finish it with not only a topographic map but also the geology plotted on it. The topography was done by rough methods, plane table and alidade, pace and compass traverses, bicycle wheel and compass, and motor-car speedometer (with tenths) and compass. Heights were measured by the controlled aneroid barometer. Mac was a consummate artist at this rough survey technique. His plane-table sheets were a messy spider's web of lines, smudges and rough notes of names and heights. Only Mac could decipher them, but when he made a fair copy, and added his traverses done by those rough methods, a very good map resulted.

After a day's work of many miles on foot, he would spend hours under the pressure lamp, plotting it all up before he went to bed. All the early geological maps of Southern Rhodesia were made in this way, and their accuracy has stood up to the test of modern air photos and contoured maps plotted from them. The modern generation of field geologists have missed something by having ready-made contoured maps and air photos in stereoscopic pairs, from which position can often be plotted to an individual tree. Mac's generation spent perhaps three fifths of their field time making topographic maps and only two fifths on the geology, but their results were remarkable.

Mac's dedication to his plane table caused me some hours of acute anxiety for his safety one night during our 1951 field season together. I had mapped the Sengwa river down through the Sidaga flats to the entrance of the gorge where the river cuts through the Sijarira (Chisarira) range. Mac came to join me in my temporary camp on the flats so that he could climb the scarp on the south side of the range and do some long shots with his plane table. After breakfast he set of with one Batonka man and his gear to walk the three or four miles to the foot of the scarp, climb it and work from some suitable point. I went off in the opposite direction, did my day's work, and got back to camp well before dark because it was rather difficult to find. It was perhaps no surprise that Mac was still out. He was still missing an hour after dark. I lit a large fire to guide him and waited. By 9 o'clock he was still missing, and I fired occasional shots to help him, but there was still no sign of him at 10 o'clock. This was serious. There were plenty of elephant and buffalo, quite a few rhino, and almost certainly lion and leopard, and no local inhabitants in the whole area.

At 11 o'clock, Mac strolled into the camp from a totally unexpected direction on a dark night with no moon to speak of. And he was quite angry to find me worried about his safety. The thought that he might have come to grief was totally foreign to him, but he condescended to give me a brief explanation. He had found such a good plane-table site, though it had taken a long time, that he had stayed at it until it was too dark to get down the scarp by the way he had climbed it. So, he traversed along it for about four miles, came down an easier slope, and then, by his almost uncanny instinct, headed straight across country back to camp. What his Batonka casual labourer thought is not recorded.

Alex Macgregor was a great field geologist. He had an 'eye for country', which enabled him to explore previously unmapped country, make the minimum number of traverses through it, literally tear the guts out of it, and come out with a remarkable grasp of the major elements of its geological make-up. This was my mentor for the field season of 1951.[22]

[22] See also Peter Fey, 'Alexander Miers Macgregor, OBE, MA, DSC, FGS, the Geological Survey's Third Director, 1946–1948', *Heritage of Zimbabwe* (2018) 37: 69–78. An 'Overview of the Life and Accomplishments of A.M. Macgregor' appears in the Macgregor Memorial Lecture 2019 by Hielke Jelsma, *Geological Society of Zimbabwe*, <http://www.geologicalsociety.org.zw/sites/default/files/news-attachments/Macgregor%20Memorial%20Lecture%20Jelsma%202019_c.pdf>.

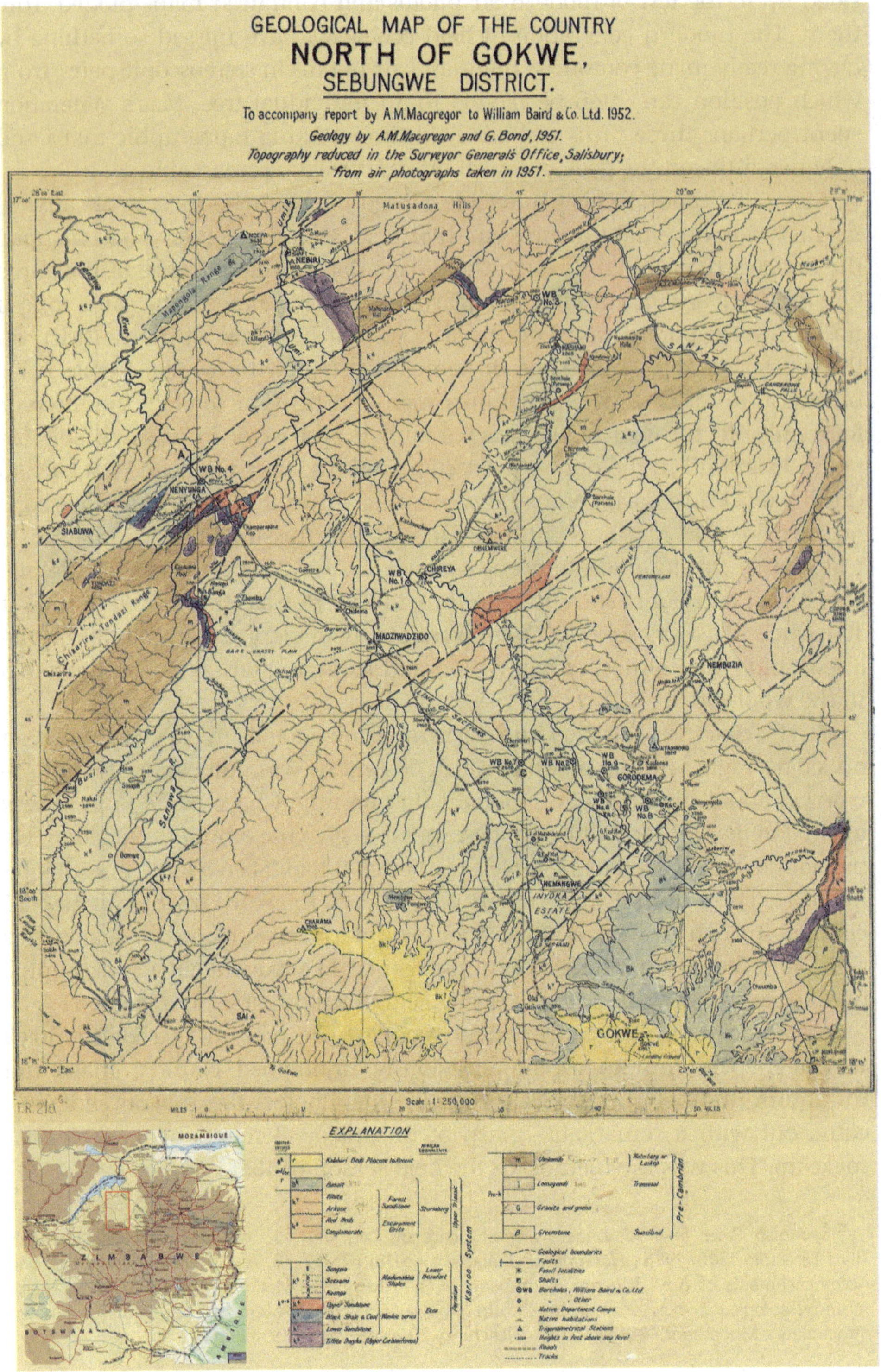
GEOLOGICAL MAP OF THE COUNTRY
NORTH OF GOKWE,
SEBUNGWE DISTRICT.
To accompany report by A.M.Macgregor to William Baird & Co. Ltd. 1952.
Geology by A.M.Macgregor and G.Bond, 1951.
Topography reduced in the Surveyor General's Office, Salisbury; from air photographs taken in 1951.
Matusadona Hills
NEBIRI
WB No.3
WB No.4
NENYUNGA
SIABUWA
Chizarira Tundazi Range
WB No.1
CHIREYA
MADZIWADZIDO
NEMBUZIA
WB No.7
WB No.2
GORODEMA
NEMANGWE
INYOKA ESTATE
CHARAMA
SAI
GOKWE
SANYATI
Sengwa R.
Busi R.
Scale 1:250,000
MOZAMBIQUE
ZIMBABWE
BOTSWANA
EXPLANATION
Kalahari Beds Pliocene to Recent
Basalt
Arkose
Red Beds
Conglomerate
Forest Sandstone
Escarpment Grits
Stormberg
Upper Triassic
Karroo System
Lower Beaufort
Permian
Ecca
Upper Sandstone
Black Shale & Coal
Wankie series
Lower Sandstone
Lomagundi
Transvaal
Granite and gneiss
Greenstone
Swaziland
Pre-Cambrian
Geological boundaries
Faults
Fossil localities
Shafts
Boreholes, William Baird & Co. Ltd.
Other
Native Department Camps
Native habitations
Trigonometrical Stations
Heights in feet above sea level
Roads
Tracks

– 10 –

My Valley

Much of the first three months of 1951 was spent collecting all the old records of geological work done in the entire Sebungwe region. Although it did not amount to much in total, it spanned a long period of time, beginning with Alford's ill-fated expedition in 1896 when he and his party only just survived a journey on foot from Salisbury to Nebiri, where coal had been reported by Trooper Hunt.[1] Molyneux had explored and even diamond drilled around the Busi river between 1898 and 1903. Parsons had drilled in the eastern Sebungwe in about 1913, and a mining company had explored and drilled in the central Sebungwe.[2] Mac himself had done a long traverse just before World War II from Gokwe on the plateau to Binga's village on the Zambezi, a distance of about 175 miles. Some fossils had been collected, and were known to me, but even adding it all up, the sum total of geological knowledge was meagre for an area of some 10,000 square miles.

It was arranged that I should go down into the Valley, through Gokwe, on 1 April and get things started. Mac would follow soon after, and later would come three diamond-drilling machines and their crews. We would jointly map about 4,000 square miles and, in addition to mapping, my job would be to log all the cores from perhaps ten boreholes.

For the first time, I had a Land-Rover, complete with four-wheel drive, in which to get around this difficult piece of country. It was not new. It had already done 20,000 miles according to the clock, and, to judge from its outward appearance, they had been hard miles. It was well battered, and had a length of light rail welded across the front bumper. It was a short wheel base,

[1] Charles John Alford was a British geologist and mining engineer and a Fellow of the Geological Society of London. He visited Rhodesia to investigate coal and gold deposits, which led to the publication a 'Report on an expedition in search of coal in the districts immediately south of the Zambezi' for the British South Africa Company (Salisbury, 1894).

[2] C.E. Parsons wrote 'Notes upon a geological section from Gwelo to the Zambezi River', *Proceedings and Transactions of the Rhodesia Scientific Association* (1903), 4, 48–52. See also 'The Kariba Gorge Attraction' in a PowerPoint presentation by Tim Broderick, *KARIBA: 60 Years Since Inception: A Geological and Geotechnical Review* (Kariba, November 2015), <http://www.geologicalsociety.org.zw/sites/default/files/news-attachments/5.%20T%20Broderick%20-%20Kariba%20review%20-GSZ%20-%20%20Kariba%2015.pdf>.

canvas-hooded model of early vintage, and was permanently in four-wheel drive, except for reverse. A small ring on the floor engaged all four wheels in reverse when needed, as it often was. A long scratch on the dull green paint on the bonnet looked like the word 'Susan', and Susan she became. I very quickly became addicted to Land-Rovers in general, and to Susan in particular. She took a severe thrashing in the weeks that followed and never gave the slightest sign of trouble. She towed five-tonners out of mud, hauled Mac's impossible old Morris 30 cwt truck out of sand, and saved my legs hundreds of miles of rough walking.

My field notebook records the journey into the Valley in three parts: the main road from Bulawayo to Que Que; the road from Que Que to Gokwe; and the Native Department's track from Gokwe to our base camp at Madziwadzido on the Sessami river, more or less the focal point of our Exclusive Prospecting Reservation, selected during our reconnaissance the previous December.

The main road was easy but slow; Susan objected to travelling at more than forty-five miles per hour. The distance from Que Que to Gokwe is only ninety miles, but the road was in terrible condition, having been cut to pieces by lorries. It took me just under six hours, and in places the fine pink dust was so thick that Susan's front axle ploughed through it, sending a cloud up and over the windscreen at any speed above five miles per hour. Sections on Kalahari Sand were so bad that it was easier to leave the road and wind about among the trees.

The next section of the journey, from Gokwe to Madziwadzido, about seventy miles, was a complete contrast and presented no problems at all. This track was usually used only once a year – by the Native Commissioner on the long patrol from Gokwe to Binga, along the river to the Sanyati confluence and back inland – and the only road fork, near Madziwadzido on the Sessami, a total distance of about 450 miles. These were the only roads in the whole of this huge district, and their maintenance was done section by section by conveniently placed villages, each year in time for the famous long patrol. The whole operation was paid for in raw salt, since salt was a rare commodity in

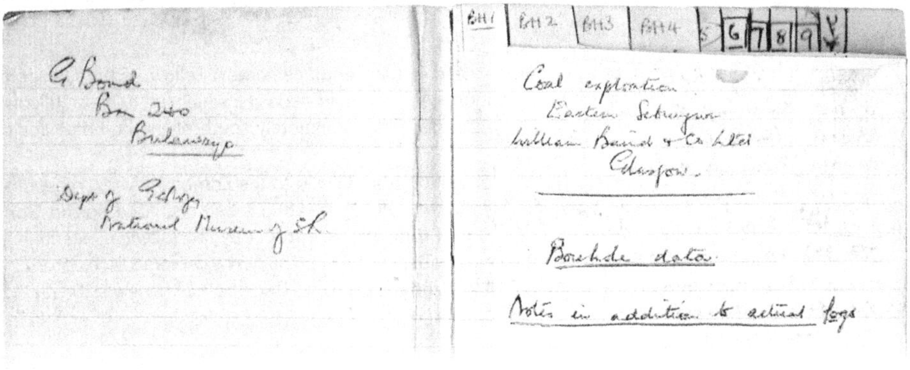

the Valley. The drifts across the rivers were a corduroy of poles and grass, but the poles were carefully pulled up and stacked before the rains to preserve them for the next year. No lorries used this road system, and between the rivers it stayed in remarkably good condition. After the horrors of the road between Que Que and Gokwe, travelling on it was a delight.

There would be few problems in getting drills to base camp, though I had some qualms about one piece of road, which went on bare sandstone outcrop for nearly a mile and ended in a steep drop of over a hundred yards, where the road dived back into the mopane woodland and was hard to see. If a lorry and trailer missed it, there would be trouble.

I had two or three weeks in which to explore for suitable drill sites before Mac and the drills were due to arrive. The main problem would be water, because diamond drills are thirsty machines and the previous wet season had been very poor in the Valley. Even after good rains, surface water is in short supply quite early in the dry season. Mac had indicated three areas for the first holes and only one was close to base camp. This was the easiest. It was at Bari salt pan near the Ume (Bumi or Omay) river, and an old shallow shaft put down for coal exploration some years earlier had plenty of salt water in it. The second, in the Ume–Kaonga valley, was more difficult, and the third, right in the north of the area, looked nearly impossible. It would have to wait for Mac.

Bari salt pan was one of the few areas where salt was traditionally won by the local population. It consisted of a low semicircular depression, about a hundred yards in diameter, in which there were several salty seepages. There were two ways of extracting the salt. The seepages spread out on gravelly areas and evaporated, leaving a crust of crystalline material. This consisted of common salt mixed with some Epsom salt and sodium bicarbonate. This efflorescence was scraped up, dissolved in water, and then evaporated until the common salt crystallized out before the other constituents.

The eyes of the seepages were surrounded by thick growths of dark-green rushes. These were cut, sun dried, and then burnt to ashes. The ash was then leached in water and again evaporated until the salt crystallized out. Both methods were laborious and time-consuming, and it was not surprising that the sparse local population would willingly be paid in salt for casual labouring. We took plenty of this valuable currency and paid for road-making with this rather than with real money.

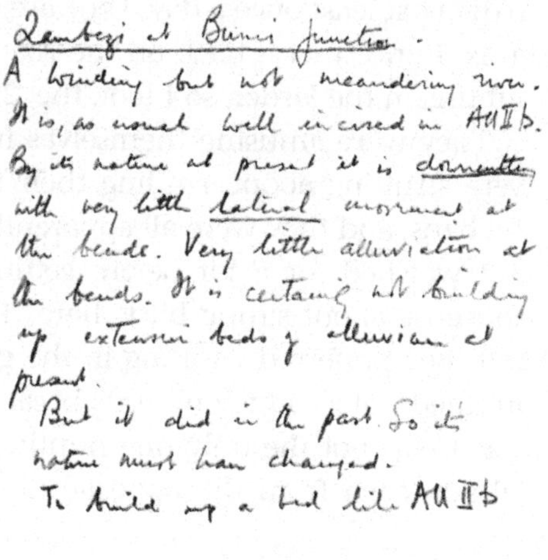
Zambezi at Bumi junction.
A winding but not meandering river.
It is, as usual, well incised in All IIb.
By its nature at present it is downcutting
with very little lateral movement at
the bends. Very little alluviation at
the bends. It is certainly not building
up extensive beds of alluvium at
present.
But it did in the past. So its
nature must have changed.
To build up a bed like All IIb

I enjoyed my time alone exploring the area we were to map and drill. The country was wild, very sparsely settled, geologically virtually unknown, and the solitude was there to be savoured to the full, but a diversion was soon to occur that took my mind off coal exploration for a time.

I am definitely not, and never have been, the mighty hunter type. Snakes and crocodiles I have shot without compunction, though I prefer to leave even these creatures alone. I once was teased quite unmercifully by my passengers in the Land-Rover when I carefully drove round a puff adder on a road through a mining camp instead of killing it. Looking back, it was rather stupid, because there were children and pet dogs living in nearby houses and puff adders are nasty poisonous things. But it was doing me no harm and, quite without thinking, I applied my usual practice.

Where the larger game animals are concerned, I much prefer to leave them undisturbed, and I have spent much time wandering about in their country armed with nothing more lethal than a hammer. In any case, the feeling persists that any rifle capable of knocking down an elephant or rhino would knock me over backwards with the recoil; and I have never fired anything bigger than a .303 in my life.

The antelope all looked so beautiful in their natural surroundings that it grieved me almost beyond bearing to shoot them. Sometimes it had to be done in the Valley for meat, but it was done with reluctance and as a very last resort. I preferred to buy a goat once a week, eat some while it was fresh, and live on tinned food or Guineafowl for the rest of the time. This attitude was regarded as sheer madness by the Batonka, but in the face of their scornful looks I remained stubbornly and unrepentantly a non-shooter.

Birds were a different matter, and so were fish. I cannot think why. Guineafowl were generally plentiful in the Valley anywhere near water, for they need to drink at least once a day. They are silly birds, but very quickly learn to avoid guns. I once met a flock on the road just after sundown. There was nothing but tins in the larder, so I took the .22, stopped the Land-Rover and got out.

They were amusing themselves in the way of Guineafowl. The cock birds were strutting about, ruffling their feathers and making silly little rushes at the hens, and they were all apparently blissfully unaware of me. I crept nearer and watched for a bit before getting near enough to shoot. There was no nonsense about sitting birds here. They were usually so difficult to get near, and they preferred running in the grass to flying away. This must have been an uneducated bunch of birds because they let me get quite close. Then I saw that I was not the only one hunting them. A jackal was making a very skilful approach from the opposite direction. As I had never seen this before I watched. Eventually, when the jackal was only a few feet from the nearest bird, he was spotted, and the whole bunch suddenly took off. I let the jackal go and went back to the Land-Rover. It would have to be a goat or bully beef after all.

On the whole, my attitude of non-aggression has been respected by the other side and I have had extremely little trouble from wild animals. But in May 1951, while I was at Madziwadzido, this peaceful state was suddenly shattered by lions. Looking back over the years, the very few times when I have had trouble with animals have generally been when Mac was around. It is quite a long story, though in essence it was simple enough: the lions were hungry, their leader was an old and almost toothless lioness, and man is the easiest animal to kill.

Our campsite at Madziwadzido was one used by the Native Commissioner on his journeys through his district. It was just a clearing with three huts, thatched rondavels of pole and mud. Because of the heat, they were open-sided, except for a low wall about three feet high, and there was a space, about two feet high, up to the eaves all round. They were built in a clearing round a huge and quite unclimbable thorn tree. This had a massive trunk about six feet in diameter and belonged to the kind called locally *muunga*.[3] It was a very pleasant spot and I was duly grateful to the NC for permission to use a ready-made camp.

When I arrived I had made my usual enquiries of the local people about game. Buck were said to be very scarce: just a few impala. Elephant had been around a good deal in the wet season, and there were some still within a few miles. Their spoor was everywhere, but none of it was very fresh. Lions were said to have been in the area some years before but, in view of the scarcity of buck, it was not surprising that they had gone. Rhino were not nearer than twenty miles, and buffalo seldom passed that way. The position seemed very satisfactory and there were plenty of uneducated Guineafowl. At first they could be shot easily with a .410 shotgun. After a week they could only be got at long range with a .22 Hornet. They learn fast.

About two days after arriving, I decided to make my first traverse up the Sessami river, which lay within a quarter of a mile of camp. There was a lot of old elephant spoor in the river bed, but nothing alarmingly fresh. I had only got about half a mile upstream from camp when I saw a spoor that I didn't like at all. It was dry sand and not very distinct, but it was too large for hyena and was distinctly cat-like. I had taken on an old local man called Mpofu (Eland) who was reputed once to have killed a lion single-handed with a spear. He had shown me the skin, with a slit-like wound, and I was inclined to believe him. I called him over and consulted the expert. He was distinctly doubtful, so I let

[3] A general name for an acacia with straight white thorns.

it pass. I found no more spoor that day and decided I was being worried for nothing. I carried on working by day and also collected the nucleus of a gang of men to help cut tracks to various parts of our allotted district.

Gradually, our team began to arrive, and with them came the paraphernalia of diamond-drilling apparatus, lorries and drums of petrol. The first to arrive was my geological colleague, Mac. Then came the three diamond-drillers and their teams of trained men. Getting their heavy equipment down into the Valley over the existing tracks was quite an undertaking in itself. In such a programme, we obviously needed an expert in the art of diamond-drilling who could supervise the whole problem of supplies, repair breakdowns and look after the whole of that side, leaving the scientific staff (all two of us) to get on with mapping and the examination of cores.

The appointed genius 'i/c drilling' did not arrive for a few days after the men and machines, but he was known by name. It was Grobbler, and to me this sounded ominous. I visualized a large, taciturn man, armed with every conceivable type of weapon for the destruction of the local game. I dreaded his arrival, and was prepared to dislike him heartily when he did put in an appearance.

I have never been so wrong in my life. Johnny Grobbler was certainly an Afrikaner, but of the very best type. He came armed with a light and heavy rifle and a double-barrelled 12-bore, but, as will be seen, they were much more for the protection of his young wife, Dot, whom he brought with him, than for any mass attack on my local game. Both Dot and Johnny were completely delightful. He was then about thirty-five, of medium height and build, and with a full naval beard. He delighted in the wildness of the country, and I think both Mac and I took to him at once. He insisted on camping about ten miles away from our base because he said he did not want to be in our way.

His story, which gradually came out during visits to his camp, was extraordinary. Before the war he had been a diamond-driller in South Africa, working for a company in Johannesburg. He had joined up on the outbreak of war and had been posted to North Africa after training. He had been taken prisoner by the Italians quite early on and was sent to a prison camp in Italy. There he found another diamond-driller from South Africa and the two had become firm friends.

When they were released after the war, they found themselves back in Johannesburg, each with a wife and a joint total capital of £400. The prospect of working for various drilling firms for the rest of their lives did not appeal to either of them, and diamond-drilling was booming because the new Orange Free State goldfields were just being explored. They decided to set up in business on their own account, Johnny as the technical brains and his friend as the business manager, but £400 must have seemed a very meagre sum with which to start up. However, they secured three small broken-down machines of identical type at scrap prices, and proceeded to build one really serviceable rig out of the bits and pieces. All this had to be done in their flats as they had no workshops.

They had no trouble in getting a contract and, being expert drillers, made it pay. From this very small beginning they had built up, in four years, a large and flourishing business, owning several machines of the type capable of going down 8,000–9,000 feet, as well as smaller rigs for lesser jobs. At the end of that time they could see that the Free State drilling was coming to an end, so they very shrewdly sold out - for about a quarter of a million pounds. This was not all theirs, because they had taken in partners to bring in capital as the enterprise expanded, but the two original members came out of it very nicely. Johnny decided to retire, but his partner stayed on as business manager for the new owners.

Johnny had one surpassing passion in life: wild animals in the veld. He had proved it by using his new capital to buy 20,000 acres of undeveloped land. It was a long, narrow strip adjoining the boundary of the Kruger National Park, and he made this his home. He called it a farm, but all he grew were a few pumpkins and some mealies. He built a house and lived there with Dot, fishing and watching the animals, of which he never seemed to tire. I need have had no worries on this score, for he was as ardent a game preservationist as anyone I have met. He had come out of retirement to take on this job only because the remoteness of the country appealed to him.

In all the four months he spent in the Valley, he shot one impala when meat supplies broke down, a few Guineafowl for his own camp, one cheetah, which he mistook in half-light for a leopard and for which mistake he never forgave himself, and half a lion.

I had mentioned the spoor that I had seen early in April to both Mac and Johnny. Mac was completely disinterested, having been among lions so often in the past; I don't think he believed in it any more than I did by then. Johnny was much more interested, but only, I think, because his wife would be afraid. They had lions on their farm, but there she had a good solid house, and here they were living in a tent. It was not many days after Johnny's arrival, however, before we had startling proof that there were indeed lions in the area, and that, furthermore, they were a nasty bunch.

My camp was on the route used by the Valley Tonga on their long walk up from the Zambezi to the plateau. The average distance of this walk, which they undertook when they wanted to go and look for work, was about 150 miles and was divided into regular stages with a recognized sleeping spot at the end of each. The distance varied because the night stops were determined more by the availability of water than by distance. One of them was within

200 yards of my camp at Madziwadzido. Both were determined by the availability of water in the Sessami river.

The work Mac and I were doing often involved staying away from base camp for a night or two, sometimes together and sometimes separately. I got back to base one evening after a break of this kind to find Mac still away but considerable excitement among our labourers. Fires blazed all round the camp; they had even lit small fires in the doorways of their own huts. As these were simply rough shelters of dry grass on a frame of light poles, this was an exceedingly risky business and indicated a quite unusual fear of wild animals in the occupants. I soon learned why.

The day before, a party of men had passed through on their way to the plateau and had camped for the night at the usual spot. As lions were not supposed to be around, they had not taken any special precautions and had let their campfires die down at bedtime. At about midnight, an old man of about sixty had been seized by the shoulder by a lion and pulled out of his blankets. His shouts woke up his companions, who very bravely seized logs from the smouldering fires and succeeded in driving off the lion. I asked about the victim and was told that he was very severely mauled and had been taken back to his village, about fifteen miles away, to die. I gathered that he was probably dead by that time, and, as it was already dark, there seemed very little I could do, so I went to bed.

Mac was still away, and I felt a bit uncomfortable and lonely when I put out the light. There was a good series of fires around the camp, but the low walls of the hut felt very insecure. I loaded a shotgun and a rifle and tried to sleep. The African night is full of noises, and generally they sound friendly enough. Without them I should feel ill at ease, and I can recognize most of them. There is the constant chirping of crickets, the croaking of frogs, nightjars, owls and a host of others. Normally I listen to them in the darkness after putting out the light, turn on one side, and go straight to sleep.

I slept badly that night and was glad of morning, but there were no noises that could be put down to prowling lions. I asked the men, but they had heard nothing. However, on looking in the dust of the roadway just outside the camp, it was disturbing to find fresh and unmistakable lion spoor. The local people confirmed my reading and added that there had been at least four individuals, three adults and one well-grown cub.

This was really alarming. We had been very closely inspected during the night by a family of lions who had made no noise at all and had the previous night shown that they were prepared to eat humans. Having been frustrated then, they must by now be even more hungry.

I wished Mac, with his experience, would come back. I even thought of moving camp – lock, stock, and barrel – to Johnny's place, but that would mean Mac returning, perhaps late at night, to an empty and unprotected camp. I had to stay.

Mac returned in the afternoon and I told him the position. He showed even less concern than I had expected. He appeared completely unworried and did not even mention lions during the evening. When we went to bed, all he said was that no one had ever been taken by lions from under a mosquito net. After which remark he got into bed, tucked in his net, said goodnight and was soon fast asleep.

I found it hard to settle down. It was not long before there was the most awe-inspiring series of grunts from quite close by, which soon had me out of bed with a shotgun and a powerful torch. I woke Mac up. He listened for a minute or two, heard the grunts, said 'lions' rather sleepily, and went off to sleep again. By this time, the men were chattering incessantly and stirring up their fires. I was sure that their huts would be burned down, but somehow they kept the flames just away from the grass.

I could not see anything from the hut with my torch, as the grass beyond the clearing was too tall, so I went out and stood on the bonnet of the Land-Rover to gain height. The grunting went on from various directions, but still I could not make out any eyes to shoot at, and in any case I could not hold both gun and torch. I fired a few rounds in the air to scare the brutes away. This produced silence from the lions but rather sleepy and peevish protests from Mac, who complained that I had woken him up. The shots seem to have had the right effect, for there was no more grunting, only deep breathing from the sleeping Mac. After a long time, I went back to bed, and there was no more disturbance that night.

Next morning, we learnt that the lions had gone off to a village about a mile away where there were only women and children and a few very old men. They had killed a couple of dogs right in the village and eaten them on the spot. The villagers were terrified, as well they might be, and I was asked to go and hunt for the lions. Never having done such a thing before, I was terrified myself, but I felt bound to do something. Quite apart from anything else, if this went on for long, all our labour force would disappear, and one could hardly blame them.

In the afternoon I took a .303 and some ammunition, screwed up my courage, and asked for volunteers with spears and dogs to go lion-hunting. To my surprise, I got quite a good response, and I set off with about ten men and a few dogs. I had no idea where to look, but we started operations from the scene of the last night's kill. We hunted around for some hours, but it became clear that the lions were too clever for my amateur efforts and we didn't even find fresh spoor.

That night was a repetition of the previous one. An early spell of grunting around the camp, much firing in the air by me from the bonnet of the Land-Rover, and not much sleep for anyone except Mac. In the morning, there was again fresh spoor almost among the fires and again a report from a nearby village of dead dogs and terrified locals. At this stage Mac had to leave to go to Salisbury for a conference on our progress and would be away for several days. I was left to deal with the situation as best I could.

I had one or two more unsuccessful daylight hunts, without seeing any sign of lions (thank goodness) and I consulted Johnny at his camp. He had neither seen nor heard any sign of lion in his area. He suggested that I should move camp to his place, but I explained why I must stay, and left him.

The nightly gruntings went on. I fired off more cartridges, got very little sleep, and in the mornings the same stories came in of dead dogs and terrified locals.

Then there was a diversion. A week after the first appearance of the lions, the local headman came in and announced that the first victim was still alive but extremely ill. Could I help him? This was a new situation, and the next day, instead of fruitless and aimless lion-searching, I took my stretcher bed, first-aid box, including M&B tablets, and, with my Museum skinner Donton and another man, set out to see the poor old man. We managed to get the Land-Rover within about eight miles of the old man's village and then had to walk.

Eight miles is quite a long walk. I had the local chief's 'policeman' as my guide and he was a very reluctant member of the party. He was obviously convinced that the old man was as good as dead and could see no point in doing the eight miles to the village and another eight back to the Land-Rover on a hot day. He needed considerable cajoling to get him along. After a while, however, he mentally shrugged his shoulders and walked ahead in stubborn silence.

This gave me time to wonder what lay ahead, and in what condition the patient would be when we got to him. Quite apart from his physical condition, there was the perpetual problem of local psychology to be faced. By the administration of aspirin and faith for vague pains and small ills, plus plasters and bandages on sores and cuts, I had built up quite a local reputation, but so far all the things I had treated were simple. I had probably hastened healing, but nature would have cured them all in time without my pills and plasters.

This was a totally different set of circumstances. If I treated him and he eventually died of his wounds in spite of my efforts, all the locals would regard me as responsible for his death. If he recovered, so much the better all round; if he was already dead when we arrived it would be easy. If he looked, to my untutored eye, as if he would die in any case, what should I do? The safest plan would be to refuse to touch him, but while there remained any chance at all, I felt duty-bound to try to save him.

We eventually reached his village. There were a few women and children around, but none of the signs of a recent death. I concluded that he must still be breathing, and began to ask where he was. No one seemed very interested, but eventually a woman pointed casually to a hut. I went in, and there was the patient. He was leaning back against the wall, with a filthy piece of rag round his shoulder. Sweat was pouring off him and he had a glazed look in his eyes. He was muttering almost ceaselessly, the only word I could catch being *mhondoro*, the local name for lion. He was old and grey and very skinny. I wondered why the lion had chosen such a poor-looking meal from the party sleeping by the roadside.

I took his pulse, more as a starting point than as an important piece of evidence. It was very fast and his breathing was shallow and rapid. I didn't like the look of his condition a bit; he looked and sounded extremely ill. I had expected extensive wounds on his body, but all I could find were two deep punctures on his left shoulder. These had been made by the lion's canines, and it looked as though they had been very blunt. It is generally old lions who take to man-eating, and this seemed no exception. Later events proved the truth of this surmise.

Although the wounds were not extensive – just two punctures, one in front and the other behind the shoulder – they were unbelievably septic, and his condition was due to serious general blood poisoning. Lion bite is generally a pretty septic thing, as their teeth carry so much bad meat between them. Wherever I pressed on his back or front, pus poured out of the punctures. I had now accepted responsibility for him and had to do the best I could. I pressed as much filth out as possible, and took out the M&B tablets. It was then that I realized I had forgotten to put the proper dosage on the bottle before leaving. I had been briefed by a doctor friend, but here I was with a dangerously sick old man, and no idea how much of the medicine to give. It seemed obvious that he would die of blood poisoning fairly soon without M&B; the only thing to do was to take a chance, give plenty, and hope for the best. I decided on four tablets right away and two every two hours. I discovered afterwards that this was just twice the proper dosage.

He was too sick to move at once, and that meant either staying at his village all night or leaving the proper number of pills, going back to camp, and returning the next day. I chose to do this, but another difficulty cropped up at this stage. How was the 'brother' who was staying with him to know when to make him take his pills without a watch? For the rest of the day it was easy. Time in the Valley was measured by the height of the sun, and I took the 'brother' outside and pointed to the elevation where I thought the sun would be at two-hourly intervals. That was easily comprehended, so it only remained to do the same for the moon, which luckily was at a reasonable stage.

There was not much more to do except go home and hope for the best, so I left instructions that he should be given water to drink and, if he could touch it, *maheu*, which is mealie porridge made so thin that it can be drunk like beer. I said I would come again next day and, if possible, take him back to camp. I wondered what I would find on returning to his village.

We started the long walk to the old man's village in good time the next day and arrived there about mid-morning. Again I wondered what I should find. Again there was no sign of mourning and no great interest evident among the few visible villagers, but there seemed to be a good deal of activity centred on the old man's hut. I hurried over and went inside. I was met by one of the miracles of modern medicine. The patient was sitting up, breathing easily and talking quickly and sensibly to his 'brothers'.

I felt his pulse; it was almost normal. The glazed look, the profuse sweating, and the babbling about lions had all gone in twenty-four hours under the influence of the rather excessive doses of M&B. The wound still looked very septic and the old man was very weak, but the transformation was little short of miraculous. Now was the chance to take him back to camp, give him a night or two in which to rest, and then send him to hospital at Gokwe, some seventy miles from our base. There would be a supply lorry due to leave about then. I had brought my camp bed, a very light steel-and-canvas affair, which is easily carried as a stretcher, and two of my own men. I could have brought more, but I more than half expected to find a corpse.

I went out of the hut and into the village, which had only about ten families in it, and called for volunteers to help to carry the patient. I reckoned that with eight men, six carrying and two resting in turns, we could easily get him back to the Land-Rover, eight miles away. Not one volunteer came forward. No one seemed in the least interested in giving the old man - one of their own villagers and, no doubt, related to most of them - the best possible chance to recover.

I was furious, and stormed round the village cursing them all to eternity and threatening all kinds of dire penalties. They knew as well as I did that it was all 'wind' and that my threats were idle. All one does by behaving in this way is to look foolish. I realized this soon enough, but I was too angry to resort to bribery. If they had not enough decency to help an old man from their own village, why should I care? But the old man had gone so far to recovery that I couldn't leave him now. He was willing to go to hospital, and that was quite unusual in itself.

After sitting down calmly and smoking a few cigarettes, the tab ends of which I destroyed so that these ungrateful people should not benefit from my cast-off tobacco, I tried again. With great reluctance, four more or less able-bodied but rather elderly men agreed to help. Perhaps they had intended to all along, but had merely been waiting until it was obvious that I would not pay them for helping their colleague. The stretcher was quickly assembled and the old man made as comfortable as possible on it with a blanket.

When we actually started, there was another surprise: nearly everyone in the village, mainly women and children, came with us. This was wonderful, as the women could easily take turns at carrying. We covered the ground fairly quickly, with frequent halts in the shade, during which I anxiously eyed my patient. If he suddenly collapsed and died on me at this stage, it would be bitterly disappointing. He stood the journey, including the Land-Rover ride, very well indeed, and I was so pleased that I relented and handed out cigarettes liberally to all who had helped to get him to the road.

The lions left us alone that night and I hoped they had gone. For the old man's sake, I was glad that they kept away. He seemed so much better the next day that I sent him straight off to hospital, sure by now that he would recover.

I have never seen him again. I was told that he made a complete recovery in hospital and was discharged about three weeks later. He joined a party walking back to the valley from the plateau, and must have passed through my camp on his way. But he never made any attempt to see me.

After sending the old man on his way to hospital, the lions came back with renewed persistence, and for a few nights I had very little sleep. Mac was still away. I tried cruising up and down the only road after dark, hoping to surprise one of them in the headlights, but with no luck.

Johnny tried it from his camp, too, after one of his drillers had had a most alarming encounter. This man had wandered off from the machine one morning to attend to the usual needs of nature. He had gone off about 200 yards along a path through a dense reed bed round an area in which there were salty springs. While strolling back to the machine, he had for some reason glanced back over his shoulder. There was a lioness walking behind him at a distance of only a few yards. He is said to have broken all records for the remaining hundred yards, and to have reached the machine in a state near hysterics and almost white with fear. I doubt if it was one of my lions, as the drill was ten or eleven miles away from my camp. It stirred Johnny into going out in his truck at night, and he did have a shot at a pair of eyes but with no apparent effect. Next morning, he could find nothing at the spot, and wisely he did not investigate too closely alone.

It must have been nearly a fortnight after the attack on the roadside party that news came in one morning that the lions had killed a donkey at the village about half a mile away, and that they had eaten only part of the carcase.

Previously there had been nothing worthwhile left of their kills of dogs and goats, but here there sounded to be half a chance that they might come back and finish their meal. If they did, there might be a possibility of getting in a shot or two. Donton, a quiet little man from Nyasaland, volunteered to go with me, and I put new cells in the torch and focused the beam. I took Mac's rifle (a .303 made in 1897!) and filed the tips off seven rounds of ammunition. There was no magic about the number seven – it was merely that I could find only seven. I had heard somewhere that for soft-skinned game it was better to blunt the bullets and prevent them going right through and doing relatively little damage.

So the innocents set out well before nightfall. Donton seemed quite unperturbed at the prospect of my amateur lion-hunting efforts. The kill was near a village, which lay about 200 yards off the road from my camp to Johnny's, and it was perhaps 200 yards into the bush on the side of the village away from the road. It was easy enough to drive the Land-Rover to the village, but it seemed wiser to leave it there and walk the rest of the way. Later that night, in the darkness, I rather regretted that daylight decision, but at the time I was still doubtful whether the lions would return to their dead donkey.

The carcase lay in a clearing with only a few scattered bushes near it. We did not go too close to inspect it, but I could see that the locals had been there before us, because there was a cunningly contrived trap gun aimed just above it. It was just as well that we did not inspect too closely or we might have got shot by this contraption. I was rather amused at this trap gun. The Africans in Southern Rhodesia were not supposed to own firearms, and I was fairly sure that this ancient fowling piece was completely illegal. It was an old percussion cap muzzle-loader, which must have been hidden away for years.[4]

I conferred with Donton about a suitable tree, and we selected a fairly stout mopane that looked easily climbed, but as it was still only dusk, we settled down comfortably at the foot to watch and wait. Darkness came, not in the way it is supposed to descend like a curtain in the tropics. It certainly does come much quicker than in higher latitudes, and the twilight is only short, but when one is sitting out of doors with no artificial lights, the eyes accommodate progressively to the falling night, and the change does not seem

[4] 'The earliest smoothbore firearms loaded with shot were the 'fowling pieces' that appeared in 16th-century Europe. In the early 17th century, the barrels were made as long as 6 feet (1.8 m) in an attempt to gain maximum accuracy.' 'Shotgun', *Britannica*, <https://www.britannica.com/technology/fowling-piece>.

Also known as a caplock, a percussion lock muzzle-loader was an innovation in 19th-century muzzle-loader development. 'Fulminate of mercury began to replace gunpowder as a priming agent. This explosive, housed in a metal container known as a percussion cap, ignites the black powder when the hammer strikes the cap.', 'Percussion Lock Muzzleloader', *Hunter-ed*, <https://www.hunter-ed.com/muzzleloader/studyGuide/Percussion-Lock-Muzzleloader/222099_88843>.

nearly so sudden. It is most abrupt at sundowner parties in town when the lights are bright. At one moment it is still light, at the next brandy it is dark.

We sat silently waiting, I at any rate inwardly frightened and not knowing what to expect, Donton as impassive as ever. Perhaps an hour after dark, we both heard a noise; we rose quietly to our feet. I pointed the rifle in the direction of the noise, Donton shone the torch over my shoulder down the barrel, and there was a pair of eyes. I fired. As I did so, the eyes vanished, and their owner, a jackal, made off in a great hurry, quite unhurt.

It was a good rehearsal, but it showed that, in my state of nerves, my aim was pretty bad; and it left only six bullets in the magazine. I reloaded and tried to relax. I thought I could see a faint smile on Donton's black face; at least his row of white teeth stood out clearly in the starlight. I consoled myself with the thought that the one shot had probably scared the lions away for good that night, and we settled down again.

How long we waited, I cannot say; perhaps another hour. Then we both heard a very distinct rustling as some animal moved through the grass near the carcase of the donkey. We jumped up. Donton shone the light down the rifle barrel and there, clearly visible in the beam not twenty-five yards away, were two lions, walking across a clearing towards the kill. I aimed at the lower part of the shoulder of the second lion and fired. It immediately reared up and spun round on its tail with a most fearful grunt and made off. So did we. We climbed that mopane tree in record time, taking rifle and torch up with us. Five rounds left, one lion wounded, how badly it was impossible to say, and two or three more prowling around.

Dreadful noises came from a patch of thick grass for a few minutes, but gradually they grew fainter and stopped. The silence was most disturbing; even the normal night noises seemed to have stopped. I looked down at the ground, about twelve feet below. Donton seemed fairly satisfied with our perch, but I remembered seeing films of lions climbing trees like cats and felt highly uncomfortable about the whole situation. I wished the Land-Rover was nestling in the grass at the foot of our tree, but it was more than 200 yards away in the village. It had not seemed far in daylight, but in the night it sounded a long way

It was then that I discovered that I had only three cigarettes and that Donton had none. We smoked one each. Occasionally we could hear grunts from one of the other lions, sometimes from one direction, sometimes from another. This was quite enough to deter us from making a dash for the safety of our Land-Rover. We just had to make the best of it, roosting in our tree, and wait till morning.

We had endured about two hours of this and it was nearly midnight when I thought I heard a motor engine. Soon it became plain enough. Johnny must have been to Gokwe, and was going to pass within 400 yards of our perch, but how could one hope to attract his attention? Shouting was useless, and I had only five rounds left. It was exasperating and tantalizing and I cursed our luck heartily as the noise of the motor got louder.

Quite suddenly it stopped and the motor was switched off. Before I had time to realize it and let off a shot to attract attention, a noise like a small war broke out, a real volley from a fairly heavy rifle. This could mean only one thing: Johnny had surprised a lion on the road in the headlights of his truck. As suddenly as it had started, the firing stopped. This time I wasted no time, and let off two shots in the air. Dead silence for a few seconds, and then faint shouts. I let off one more for good measure, and we both shouted as loudly as we knew how. There was an immediate answer, the engine started up, and soon a light appeared from the direction of the village. More shouting soon guided the vanette to our tree, and a very surprised Johnny and Dot peered into the branches from below. We wasted no time in getting down and into the cab, all four of us, windows wound up tight, and then back to the clearing of the village.

Johnny's story was simple enough. He was coming home late from Gokwe, and when he was level with the village, a lioness had appeared in his headlights. He immediately stopped and fired through the driver's side window. His aim was a bit wild and he emptied his magazine at the lioness, put her down, but then found he was out of ammunition. This was a most awkward situation, and to make it worse he had a severe bout of malaria and a high temperature. His wife was worried enough about him before meeting the lioness, and when it still lay alive in the road, she was nearly beside herself. It was while they were considering what to do next that they heard my shots. This startled them very considerably as they had, of course, no idea that I was out lion-hunting myself.

Now that we were rescued, the next job was to finish off the lioness. I got Johnny to drive as near as he dared, got out and crouched in the headlights. The old lady was also lying crouched, facing me on the road. She could raise her head and I had no idea how badly hit she was. I took very careful aim. She was a proud sight as she glared back at me. Those eyes were full of courage and defiance. Even a seriously wounded lion is a beautiful animal, but it was essential to kill this one cleanly. There might still be another wounded lion to deal with in daylight, and we could not afford to allow this one to make it two. I pulled the trigger, the lioness's head jerked down, and her front paws clawed convulsively once or twice. Then she rolled slowly on her side and lay quite still.

Just as Johnny and I were congratulating each other on this joint effort, a large lion with a black mane walked calmly out on to the road about fifty yards away. I had just time to take a quick aim and fire at him. Luckily it was a clean miss and it was the last round I had. He wasted no time in getting out of the lights and I scrambled into the cab equally fast.

We drove up to the lioness, who certainly looked dead, but we took no chances. Too many people have been killed or maimed by 'dead' lions. We left her there and I drove Johnny back to his camp, where Dot put him to bed, gave Donton and myself coffee and fresh ammunition, and we returned to the scene of the night's excitement. The lioness had not moved, so, with great caution, I got out. She was quite dead. My last shot had finished her quite cleanly. Johnny had hit her several times, none of them in very vital places, but a bullet far back in the spine had crippled her rear end.

We went into the village and eventually persuaded one hut to unbarricade its door. When they heard that the lioness was dead there was tremendous rejoicing and everyone came out. This was really very foolish of them as there was one unwounded lion and one that I knew I had hit not far away.

It took about six men to haul the dead lioness into the back of the vanette. After doing this I decided to have a drive round to see if I could find any sign of my lion. I drove to our tree and then down to the edge of a belt of tall grass into which I thought the lion had retreated. Not fifty yards from the tree, there he was, apparently crouching on the edge of the grass with his head resting on his paws. I stopped the vanette and watched for several minutes. He didn't move, so I drove cautiously up so that the lights were full on him. Still he lay without moving. It occurred to me that he might be dead, but, if so, it was certainly a lucky shot. Eventually I drove the car right up beside him and shone the torch on his face. To make absolutely sure I stretched out and poked the muzzle of my rifle into his eye. His head rolled over sideways and I was finally convinced that he was dead.

Another trip in triumph to the village, and this time no trouble in getting volunteers to pull the carcase into the back of the van.

There are several tailpieces to this story. We had the two lions skinned the next day. The one that I had shot was a mature young lion with a half-grown mane and in beautiful condition. My one lucky shot had found a spot that had proved fatal fairly quickly. Johnny's lion (though I claim half of it, hence the remark earlier about half a lion shot by Johnny) was an old lioness of considerable size. She was as thin as a rail, and her ribs stuck out through her short, rather scruffy coat. The skin of her chest and front legs contained a number of broken porcupine quills, showing that in her old age she had resorted to catching even these slow-moving animals; the resultant pinpricks cannot have improved her temper. But it was the state of her teeth that was more significant: all four canines were reduced to blunt stumps, either by breakage or wear, and nearly all the rest of her teeth were worn down, flush with the gums. Between us we had managed to dispose of a potentially very dangerous old lioness, far beyond her prime, who, in her declining years, was resorting to man-hunting and leading her family with her.

Although Donton had shown very little outward sign of fear during that night's escapade, he soon afterwards asked to be sent back to Bulawayo. There he gave notice and returned to Nyasaland. I have the feeling that he had had enough. For that matter, so had I - of lion-hunting. In retrospect, the whole performance smacks of crazy amateurishness, with beginner's luck firmly on one side, and I have no intention of repeating such folly.

Being secretly rather pleased with our joint success, I sent a note up to the Native Commissioner saying that between us we had killed two of the lions and that the other two had departed. I gave no details, but his bush telegraph must have been in good order for the reply. After sending his thanks and congratulations, he went on to say: 'I didn't know that part of a geologist's duties included roosting in a tree like a Guineafowl.' Very little happened in Ivor Cockcroft's enormous district without his full knowledge!

The lion that I had so fortunately missed with the last round in the magazine moved about forty miles and took up residence about halfway up the escarpment. Much to the annoyance of the migrant locals, he chose to live near one of their regular camping places because it was at the only permanent water for some miles. This made the place unsafe, and it must have upset the stages on the route very considerably. He lived thereabouts for two years, during which time he occasionally raided cattle and was once wounded by a Native Department Messenger, who put a .45 Martini-Henry bullet into him. They lost the blood spoor and the lion must have recovered, for he continued to raid the cattle belonging to the local people.

He eventually came to a very sad end. He killed a heifer, and the infuriated owner built a low brushwood fence round the carcase and set a snare made of steel cable in the only entrance. It must have surprised the owner when he returned in the morning to find a large black-maned, and exceedingly angry, lion firmly caught round the waist, ready to do battle with anyone or anything that came near. The man made great haste to Ivor Cockcroft at Gokwe, about thirteen miles away, and told him that the country was 'thundering', which must have been a fairly accurate description. Ivor took a suitable weapon and duly shot the lion. He said afterwards that he felt it was a great shame that such a fine animal should meet his end in such an ignominious way.

After this episode, our Number 1 borehole at Bari salt pan progressed rapidly but with no spectacular coal intersection down to 400 feet. I was logging core when the hole was at this depth and watched a most curious happening. The crew set about the routine of pulling up the rods and stacking them in 20-foot lengths against the drill derrick. When only sixty feet of rods and the 10-foot core barrel remained hanging in the clamp at the collar of the hole, the rods began to rise up out of the hole without being lifted mechanically. As the stem of the rods rose, the head man of the drill crew could hardly believe his eyes; he had never seen anything like it before.

He grasped the slowly rising rods and held on. He was slowly lifted off the ground and hastily let go. The remaining three 20-foot lengths were unscrewed as they came up the hole, and eventually the core barrel, full of rock, popped out of the hole like a cork, followed by a water spout that went higher than the top of the 30-foot derrick before falling in a deluge over the machine and the astonished crew. The hole had punctured an artesian aquifer, hitherto unknown.

We should have suspected it from the presence of the springs at Bari, close by, but could have had no way of predicting its depth. Subsequent analysis showed that the water was about five times saltier than sea water. It was quite useless for man, beast or irrigation, which was a pity, since the flow from a hole about two inches in diameter was at least 5,000 gallons per hour. Fifteen years later, on my last visit, it was still flowing. Several of our boreholes in the Ume–Kaonga valley were also strongly artesian, and the further south one goes, the fresher the water becomes until it is potable. Our Number 2 borehole was a case in point, and I lived on it for two or three months. It tasted lovely, with no sign of chlorides. A subsequent analysis indicated that it was thirty times higher in fluoride than the optimum for human teeth.

The site for this hole was beside a small stream, with a pool of standing water that we used for drilling. It was also the water supply for a small nearby village. The Headman was anxious about his water, and as we drilled, so his pool began to shrink. He was so worried that he walked the seventy-odd miles to Gokwe to ask the NC to stop the hole, but the NC was adamant that our hole should go on down, so he walked back. By the time he got home, we had struck artesian water, and his pool had not only filled up with beautifully fresh water, but had started to overflow.

We soon finished at that site, pulled up the casing and moved five miles up the valley to our next site, which had no water. This was to be carted in bowsers to the drill, using the artesian flow we had found, but the hole caved in and blocked off the flow. This alarmed our local Headman, who promptly set off for Gokwe again, this time to demand that the hole should be reopened.

By the time he got back, we had done just this because we needed the water; this time we promised to leave the valuable casing down the hole. He was delighted by the thought of a private artesian well, and we suggested that we deserved some reward – an ox, perhaps. He considered our suggestion and replied that he was too poor to afford an ox (there were none in the area because of the tsetse fly) and would we accept a chicken? But somehow he never came along with that, or even with an egg.

The whole coal-exploration programme was rather inconclusive, but it provided me with the most valuable geological information about my favourite rocks, and was the foundation for the vast amount of later research work in the Zambezi Valley. For one thing, we had discovered unmistakable evidence that, 270 million years ago, there was extensive continental glaciation there, which filled a puzzling gap in the knowledge of this famous event in the southern hemisphere. It had been stated quite categorically that the great Dwyka Ice Age was not represented in Southern Rhodesia; we now had evidence that it was present in the eastern Sebungwe. It has subsequently been found in many parts of the middle Zambezi Valley, and also in the Limpopo Valley in the south-east of the country.

The artesian basin that we had accidentally discovered is still unexploited, although there is evidence that it stretches from about as far east as the Sanyati river west-south-west to the Gwai river, a distance of about 130 miles under country which has a desperate need for water. Salinity is a problem, but the southerly part of the aquifer could surely be put to some use.

By combining results from our artesian boreholes with structural observations, it was also shown that all the hot springs of the region were in no way connected with volcanic activity, as had been previously suggested. The region was last volcanically active about 170 million years ago, and it had seemed to me rather a long time for any late-stage manifestation to survive. The springs are artesian, like the boreholes, and the water is rainwater falling in the southern part of the region and running down the dip of permeable rocks to depths of over 7,000 feet. This is enough to raise it to boiling point and, if it comes quickly up an open channel such as a fault, heat loss is minimal. Even the Binga spring, which squirts up into the air as boiling water, is of this origin. The gases and the salts are picked up during the long passage of the water underground and released at surface.

In our one field season together, Mac and I mapped about 4,000 square miles by our crude survey methods. When air photographs became available the next year, the fit of our map was remarkably good. We had learned a great deal about the rock succession, the fossils contained in various quantities in the rocks, and laid the foundations for a great deal of subsequent work. I could have had no better teacher than Mac – either for the business of living in difficult conditions or for doing geological work in unknown country. Incidentally, when we started making the map, our sheet of paper was blank except for the position of two primary trig beacons, and they were 72 miles apart.

Some years later I was asked by the Atomic Energy Authority to sample gases from our artesian holes and the natural springs. This was done with large glass bottles, first filled with water and then by inverting them over the springs, driving out the water and collecting the gases to be analysed later for their helium content. To the local people, this looked like taking samples of absolutely nothing and finally convinced them that I was quite mad, a suspicion that I think they had entertained all the time I worked among them. Their reaction was uncontrollable and unconcealed mirth at my expense.

Scientifically, the 1951 season was very satisfying, but there were other interests. We made our own topographic map as we went along. The names of the major rivers were well known, but the smaller tributaries had to have names on our map for easy reference, and we consulted the local people as the only source of information. In the north of the area, I asked the name of one small river and was told it had the odd name of Matanke. While traversing up it, I came across an old 500-gallon galvanized-iron tank, probably left there in 1913 by Parsons, and this explained the unlikely name. Another small river was the Mawadzi, which drained north-east of a minor watershed. I later found that a river rising on the other side of the divide and running south-west was also called the Mawadzi; this was a bit confusing, but a translation of the word might help.[5]

I began to be fascinated by the Batonka and their next-door neighbours, the Mashankwi,[6] and this grew with the subsequent time I spent among them. They live in a harsh and dangerous environment but are completely adapted to it. Left to my own devices and shorn of all my familiar technology, I would die in the Valley very quickly. They live in it permanently without any of the life-support systems I need. The converse is not quite true, but they are, to me, highly educated people for the life that they have chosen, whereas I am a novice. Their honesty was total, nothing ever disappeared from my camps, however long I left them unattended.

[5] *Mawadzi* is not a Tonga name; there is no *-dzi* in Tonga. It does mean 'water' in Nyanja/Chewa.

[6] These are probably the Shangwe/Vashangwe people. Nkomo Thabiso Macmillan says: 'Shangwe is a pejorative term used by immigrants to name the indigenous people of Gokwe. Pius Nyambara notes that the term is associated with primitiveness and backwardness. The indigenous people object to be called as such and insist that the term Shangwe describes the place in which they stayed in Chief Nemangwe's area rather than who they are. In essence Shangwe refers to the Korekore-speaking people in Gokwe.' 'The Displacement of the Shangwe People of Chief Njelele in Gokwe: An Examination of Ethnic-based Socio-economic Exclusion of the Minority, 1951–2015' (Gweru: Midlands State University, dissertation, BA Hons History, 2015).

This was well illustrated while Mac and I had our base camp at Madziwadzido. By Whitsuntide that year, we had not heard or seen any sign of any motor vehicle other than our own. Yet, in the very early hours of Whit Sunday morning, I could have sworn I heard a light motor car go past the camp. I was so sure that I got up just after dawn and inspected the track, and there was the spoor of a narrow-tyred vehicle going towards the Zambezi. It must have come from Gokwe, but this seemed strange. Perhaps it was some enterprising fisherman.

I put the matter on one side, but two days later I had to go down the Zambezi road for some forty-five miles to visit our furthest-west borehole. Some twenty miles from camp, I came to a particularly steep dip into a stream crossing and, as I approached, a blond hatless head came out of the dip. I stopped to talk to its owner, who was a young European, with no luggage, walking in the direction of Gokwe, all alone. We exchanged polite greetings and I asked him where he was going. The answer was indeed Gokwe, about 100 miles away. Then his story came out.

He and a friend had come out from England in March to join the technical staff of the Post Office. They had bought a light, second-hand vanette and had decided to spend their Whit weekend going to the Zambezi 'to see crocodiles'. They had managed the track from Que Que to Gokwe in some way, but at Gokwe they had met Ivor Cockcroft, the Native Commissioner. When he heard their story, he flatly forbade them to go any further into his district and turned them back. But they went back just a few miles and slipped through Gokwe in the dark.

As Ivor Cockcroft had warned them, their vehicle had inevitably broken down and almost fallen to pieces. This was the car I had heard in the night, and it had eventually died thirty miles down the road. This young man had volunteered to walk to Gokwe for help, while his friend stayed to look after the dead vanette. When we met, he had covered ten miles and was already very tired and thirsty. He climbed into Susan and we set off to find his friend. We towed the vanette back to camp, dumped it on the side of the road, and fed the two adventurers. Our supply truck was due to go to Gokwe that day, so they were lucky. Ivor gave them a terrible tongue-lashing, put them on a lorry going to Que Que, and they got back to work only a day or two late.

Their truck and all their belongings stayed completely untouched for three months, until they could arrange for it to be towed away. Such security of property is rare in the modern world. They could probably have seen a crocodile in the Hunyani only a few miles from Salisbury – and have had something lifted from their car while they were doing it.

That episode illustrated one facet of the Batonka, but there were others which impressed me. People who can neither read nor write are usually good storytellers. This applied to the Zambezi Batonka, and, as they sat round my fire in the evening, they told long yarns. Certainly it sounded like it, and there was a wealth of dramatic gesture, but I have never learnt their language. I am sure that I missed a great deal of interesting gossip as well as their traditional stories. I had to rely on translations by my cook, Million, and naturally he abbreviated a good deal to give me the gist of things.

Some of the stories I heard in this way made it quite plain that the Batonka had their own traditions of a Creator and stories of the beginning of things in the world around them. One of their stories has a nice little twist of human psychology; another concerns that grotesque tree, the baobab. The first is about cats and women, and may be a sly dig at their matrilineal society. The story explains why cats always live in the kraals with the women. Cats were not commonly kept by the Batonka, but I have seen them and, as one would expect, they did live in the villages.

The story goes right back in time to the beginning, an interesting sidelight on Batonka mythology, as it involves both a Creation and a Creator. I will call the Creator 'God', and in the beginning he made all the animals and put them on earth. At first they all lived quite separate independent lives, including the cat. After a while, the cat was lonely and cast around for some animal of about his own size with which he might make friends. His choice finally fell on Kalulu, the rabbit. The partnership worked very well for some time, but it ended suddenly and tragically when Kalulu was killed by the jackal, and eaten. This upset the cat very much, for now he was alone again. Furthermore, his confidence in Kalulu was so shaken that he would never dream of making friends with another of his kind. He reasoned that the jackal was a much stronger animal and so, catlike, he changed his affections and made friends with the jackal.

The story can really only be told round the fire, by one whose invention is such that he can describe the first, and subsequent, tragedies in detail, for the jackal was killed by a hyena, and the cat promptly transferred its affection to the stronger animal. And so it goes on. The hyena was killed by a cheetah, the cheetah was killed by a leopard, and the leopard was killed by a lion. By this time the cat was adept at changing his friends and has no trouble in going along with the lion.

This seems to have lasted for some time, certainly it did in the original telling. The lion was a strong animal, but one day a man came along with a spear and, after a great fight, the lion was killed. True to form, the cat transferred to the man, as the lion's conqueror, and they began the walk home to the village in triumph. When he approached his own hut, however, his wife came out in an exceedingly bad temper, and, with insults and abuse, chased him out of the village. The cat still lives in the woman's hut because it has never found anything stronger since.

I camped for quite a long time one year near the village of a chief called Nenyunka. This has nothing whatever to do with cats, although I first heard the cat story near there, but it does concern wives, and so I include it at this stage. When I first knew him, Nenyunka was perhaps in his early fifties, but it is always hard to tell the age of a Batonka. The girls mature quickly and marry young. After this, repeated child-bearing quickly spoils their young figures and, by the time they are thirty, they look like old hags. For other reasons, the men seem to age equally fast.

Polygamy was the rule among the Batonka. The women do a great deal of the cultivating, so the more wives, the more food can be grown. The children also do useful work, including herding goats. On the purely economic side, therefore, a bachelor would be at a great disadvantage, and a large family should ensure a prosperous and comfortable old age.

Nenyunka's main village contained perhaps 1,000 people, but there were many other villages scattered over some miles of country that were under his rule. According to the Native Commissioner, he was a model chief. He settled disputes justly, and his people were well behaved, contented and no trouble to the authorities. He had five wives, and no doubt a large number of children. He could be counted as a successful and prosperous man, and he was extremely polite. He never failed to come to the camp when I arrived in his territory, and if I stayed many days he was sure to come to see me two or three times more. Since he spoke Chilapalapa, we could converse fairly easily, and sometimes I would encourage him to stay in the evening and talk round the fire. He told me a great deal about the tribe to which he belonged. One night I could not resist a little gentle leg-pulling about having so many wives.

'Nenyunka,' I said, 'I have quite a lot of trouble trying to live peacefully with one wife. How do you manage to cope with five?' He replied, 'You Europeans are stupid to have only one wife. When she is in a bad temper, she has no one to quarrel with but her husband. When any of my wives are in a bad temper, they fight among themselves so much that they leave me alone.' It was an angle of polygamy that I had never considered.

Another story I summarized in my notebook concerns the baobab or cream of tartar tree, which is surely one of the most grotesque of trees. It has a squat trunk of enormous diameter. In one field notebook, I found an entry that a baobab under which I once camped was 63 feet in circumference at the height of my shoulder. It was by no means a monster of its kind, and I can only assume that I measured it one evening because I had nothing else to do. I am sure I have seen larger ones.

The trunk generally goes straight up without branching for fifteen or twenty feet and then the whole tree splits up into tapering branches with very little central trunk. It looks like some gigantic root vegetable turned upside down, with its head in the ground and roots in the air. They are slow-growing, and the heart wood of a 15-foot diameter specimen felled during the bush-clearing on the floor of Kariba lake was aged by the Carbon-14 method in the Gulbenkian Laboratory of the Chemistry Department in the University of Rhodesia. It came out at about 1,000 years, and a 15-foot diameter gives a circumference of about 45 feet.[7] My 63-foot example must have been considerably older than this.

Although they conform to a general plan, no other tree seems to be so infinitely variable in form. Many are so odd that I once thought it would make an amusing hobby to collect photographs of them. For half of the year they are leafless. The bark is an unhealthy purple colour, rather like an 'expensive' complexion of the typical stage colonel. They are useless for timber as the wood is just a fibrous mass, though it is said that they make good pulp. The root system is shallow and old ones sometimes fall over. When they die, they just collapse in a horrid fibrous heap, as though they had melted.

When the rains come, they are transformed into lovely green mushroom shapes, with large fig-like leaves and beautiful white exotic blossoms several inches across. These last only about a day and then drop. Their fruit is bomb-shaped and about six or more inches long. It has a hard, green shell, which can only be broken by hammering. These fruits hang on long, slender stalks all over the outer branches. The trunks are often hollow, and one that I used to pass on the road into the Valley was regularly used as a hut in which travellers slept the night. The entrance was easily closed with poles and it made a lion-proof lodging. When an archaeologist friend of mine was doing an extensive 'dig' in the Limpopo Valley, he converted a hollow baobab into a darkroom.

It was an important tree to the Batonka. They made very good string and rope out of the fibrous stuff just below the thin bark, and hardly a tree was without the scars they made on it. They collected the fruits in large numbers and boiled the soft, dry, pithy interior into a porridge-like mass. The pithy stuff contains cream of tartar, and it is surprisingly pleasant to chew, even

[7] These were reported on in A. Robins and E.R. Swart, 'Southern Rhodesian Radiocarbon Measurements' *Radiocarbon* (1964), 6: 32. 'The Gulbenkian Radiocarbon Dating Laboratory in Salisbury came into operation in October 1962. [It] suffered numerous setbacks in its brief history, its counter proving particularly irksome. The laboratory produced a whole suite of dates in the early 1970s that were either directly refuted by other laboratories or considered too young by the archaeologists who submitted samples. By 1978, the laboratory was closed.' Peter Robertshaw, 'Radiocarbon Dating and the Prehistory of Sub-Saharan Africa', in Ervin Taylor, Austin Long, Renee S. Kra, *Radiocarbon after Four Decades: An Interdisciplinary Perspective* (New York: Springer Science & Business Media, 2013), 340.

when one's mouth is dry with thirst. The slightly acid feel (there is no real taste) makes the saliva run and gives the impression of thirst-quenching.

In Zimbabwe, the baobab is practically unknown above 3,000 feet, so it is never seen on the developed plateau. When I pass the first one on the journey, I really feel I have left town behind me.

The Batonka are very observant of things about them in the bush, and it is not surprising that they had a story to account for the 'upside-downness' of the baobab. This story, like that about the cat, goes back to the Creation, but, instead of animals, the characters are the trees. They were all asked to choose the kinds of places in which they would like to have their permanent homes. The baobab chose the lowveld so as to avoid the cold nights in the winter on the plateau.

During the first summer season, when the lowveld was burning hot and the cooling showers were few and far between, the baobab felt hot and uncomfortable. It longed for the comparative cool of the highveld plateau. When the Creator walked among the trees, the baobab complained bitterly that the heat was unbearable on the lowveld. Wouldn't it be possible to move on to the plateau to escape from the scorching sunshine by day and the hot stuffy nights? Anything, said the baobab, would be better than being scorched, even the cold and frost on the plateau. So the move was arranged, and the baobab basked in the more gentle sun of the plateau summer, where the days are quite hot, but the nights are pleasantly cool.

Summer passed and, as so often happens on the highveld, winter arrived almost overnight. This is a strange feature of the highveld climate, which I have noticed myself. One morning, generally in mid-April, there is suddenly a different feel in the air. The sun still shines from the same cloudless sky as yesterday, but then it was a summer sun and today it is different; winter has begun. Sudden drops in temperature are felt because of the relative change more than because it is actually cold. After many days or weeks of 85°–90°F in the shade, 65°–70°F feels distinctly chilly.

The baobab shivered, and this was only the start of winter; the nights would get colder still. Quite severe frosts are not uncommon in May and June on the plateau. As the winter wore on, the baobab began to long for the warmth of the lowveld and to regret the change. It was perhaps better to swelter in the lowveld summer in the shade of one's own leaves than to shiver naked in the cold of the highveld winter.

When the Creator came to visit the highveld dwellers, the baobab again complained bitterly, and demanded to be taken back to the lowveld again. This must have been irritating, even to the long-suffering Creator. The baobab's request was granted, and they were all taken back to the lowveld, but they were planted upside down as a punishment and to prevent them from ever complaining again. There they have remained ever since, silent and with their root-like branches spreading in the air.[8]

In the days before I had even been in the Zambezi Valley, I used to look at maps and wonder what this vast area of uninhabited territory was like. I knew much of it was completely without human beings - even Africans, partly because of the tsetse fly and the risk of sleeping sickness. Most of it was included in a large square on air-navigation maps. This square was labelled 'Prohibited Flying Area'; it seemed to imply that it was desolate country, in which anyone who was unlucky enough to force-land had very little hope of survival. The maps themselves were a bit scanty. There was very little topography shown, and the rivers ran in suspiciously straight lines, sometimes even dotted. I suspected that the maps were the vaguest of generalizations, and I later found out, sometimes at the expense of much sweat and toil, just how misleading they could be.

There was one name, which appeared even on maps in the *Times Atlas*, as well as on larger-scale ones published in Southern Rhodesia, which fascinated me. It was Piccaninyemba, and it was written in the middle of nowhere in type of the same size as the smaller towns on the plateau, where there is a sizeable white community. Yet there was no road shown going anywhere near Piccaninyemba, even the dry-season track from Gokwe to the Zambezi did not pass nearer than about eight miles of it. For three or four years I could do nothing about it, but I was quite sure that there was no European living within miles of it. The 1951 field season exploring for coal provided the chance to find out more about this mysterious place, which was fairly near our base camp at Madziwadzido.

Naturally we had to get around our area fairly extensively, and I began to ask local people more about Piccaninyemba. When Africans have something they do not wish to talk about, they can be very evasive, and on this subject they seemed more evasive than usual. However, it seemed certain that, whatever else it was, Piccaninyemba was not even an African village. Apart from this, I couldn't find out if it was a place, a hill, a valley, a group of rocks or even exactly where it was.

[8] More stories by the Tonga people can be found in Tim Matthews, *Tales of the Secret Valley* (Harare: Baobab Books, 1988) and Pamela Reynolds and Colleen Crawford Cousins, *Lwaano Lwanyika* (Harare: Colleen Crawford Cousins in association with Save the Children Fund (UK), 1991). A 'sound collection of proverbs and sayings, rich in Tonga history' was compiled by Isaac Mumpande, *Tonga Proverbs: Tusimpi* (Harare: Silveira House, 2001).

One of our jobs was to make access tracks to parts of our district so that we could get further in our Land-Rovers before we had to start walking. For this purpose, we had collected a 'road gang' of about twenty local men, and I concentrated on them in efforts to find out more. In time and with patience, I found one man who would talk; his name was Cabbage. Even he was reluctant to talk much, and would never point in the direction with his finger, but would nod his head vaguely. This was an important clue, as it suggested that the name referred to a hill and was one of those haunted hills at which the local people would never point. To do so is to invite disaster.

Eventually it was possible to piece some of the story together. Piccaninyemba was indeed a haunted hill, about ten miles from our base camp, and its position, surprisingly enough, was correctly shown on the maps. The name is difficult to translate with any certainty, and seems to have nothing to do with the haunting. *Piccanin* is Chilapalapa for anything small. The rest is not so easy. *Nyemba* could mean bean, or it might be a corruption of *nimba*, the Tonga name for the little red-brown antelope called Sharpe's grysbok, which lives in the district.

Cabbage told me some of the details of the manifestations. No African would spend the night there because it was impossible to sleep for the noise. Throughout the night 'impis' marched past, clashing their spears, beating drums and shouting war slogans. After them came the women and children, wailing and singing. The procession was said to go on all night, yet the next morning there would not be a sign or footprint to be seen on the flat ground at the base of the hill. No wonder no one would go near the place, if that was indeed what happened at night.

Cabbage's description has been confirmed by other local people since, and I was even told of one European who would vouch for the truth of it. He was a man called Cameron, who lived in the Valley for some years. He used to have a small store and did occasional road-cutting jobs, but although the remains of his building still stood in 1951, Cameron seemed to have disappeared. I should have liked the opportunity to talk to him, because he is said to have spent one night on Piccaninyemba with a man from another district. My local men swore that he left the accursed place at dawn after spending a completely sleepless night clutching his rifle and listening to the marching multitudes passing below. In the morning there was not a spoor to be seen.

All this was very convincing second-hand evidence, but I wanted to try it for myself. Only Cabbage would volunteer to go with me, and I arranged to go on the first available Saturday afternoon. When the day came, I called Cabbage after the morning's work, during which he had seemed perfectly normal, but on being told that the time had come and we were going to Piccaninyemba for the night, there was nothing doing. His reply was, '*Mina maningi sick.*'

(I am very ill.) Nothing I could do would change his mind; not even bribery would cure his sudden illness. I was definitely not going by myself, so I never did go to find out for certain what happened.

I discovered later how the name got on the map. In the early days, some Native Commissioner had decided that there should be a police post there and actually got as far as building the foundations of a few small buildings. It is not recorded why the project was abandoned, but I think I can guess.

The pages that follow are taken from Geoffrey Bond's field notebook dated 19 and 20 August 1955. They describe some days at Walker's Drift, which was flooded when the Kariba Dam was built.

It was named for Hugh Felix Walker, who, Jonathan Waters says, 'was in charge of the transport for Captain John Carden's party of BSAP, who came to establish a post at Monze. Carden brought about 16 wagons which were ferried across by lashing canoes to their sides at Sijoba, a point that became known as Walker's Drift.' J. Waters, *Kariba: Legacy of a Vision* (Harare: New Zanj Publishing House, 2020), 32.

It was the point where the road to Northern Rhodesia crossed the Zambezi.

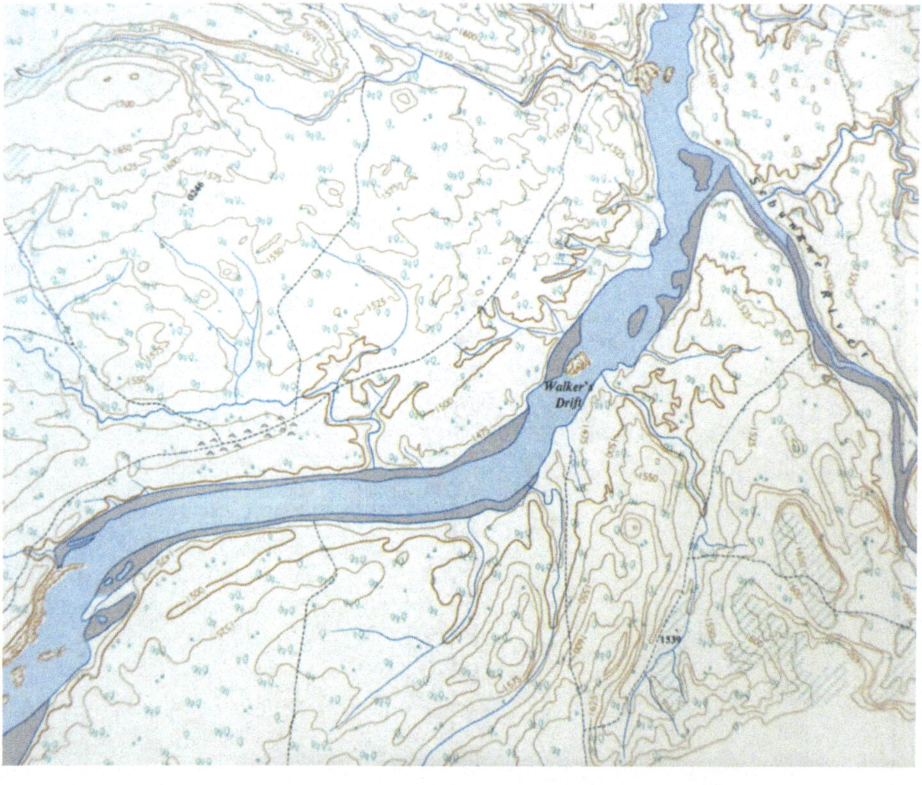

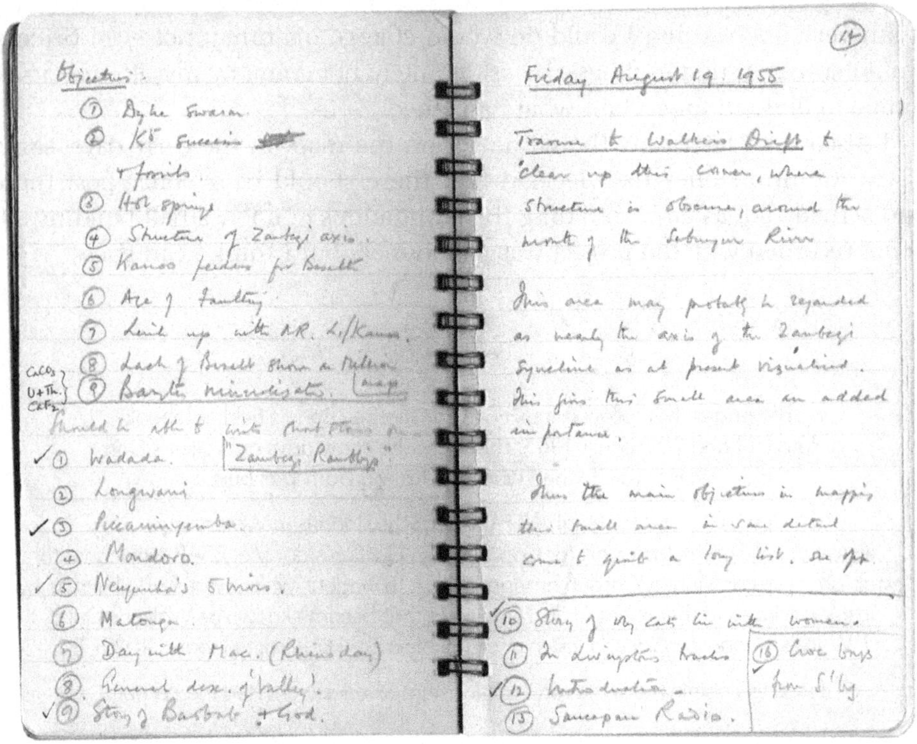

Objectives:

① Dyke swarm
② K⁵ succession + fossils.
③ Hot springs
④ Structure of Zambezi axis.
⑤ Kanoo feeders for Basalt
⑥ Age of faulting
⑦ Link up with N.R. L/Kanoo.
⑧ Lack of Basalt shown on Bulletin
⑨ Baryte mineralisation. [map]

Should be able to write short stories on

✓① Wadada. "Zambezi Ramblings"
② Longwani
✓③ Picaninnyemba.
④ Mondoro.
✓⑤ Nenyamba's 5 wives.
⑥ Matonga
⑦ Day with Mac. (Rhino day)
⑧ General desc. of valley.
✓⑨ Story of Baobab + God.

(14)

Friday August 19 1955

Traverse to Walker's Drift to clear up this corner, where structure is obscure, around the mouth of the Sebungwe River

This area may probably be regarded as nearly the axis of the Zambezi Syncline as at present visualised. This gives this small area an added importance.

Thus the main objectives in mapping this small area in some detail come to quite a long list. See opp.

✓⑩ Story of why cats live with women.
⑪ In Livingstone's tracks
✓⑫ Introduction.
⑬ Saucepan Radio.
⑯ Croc boys from S'by

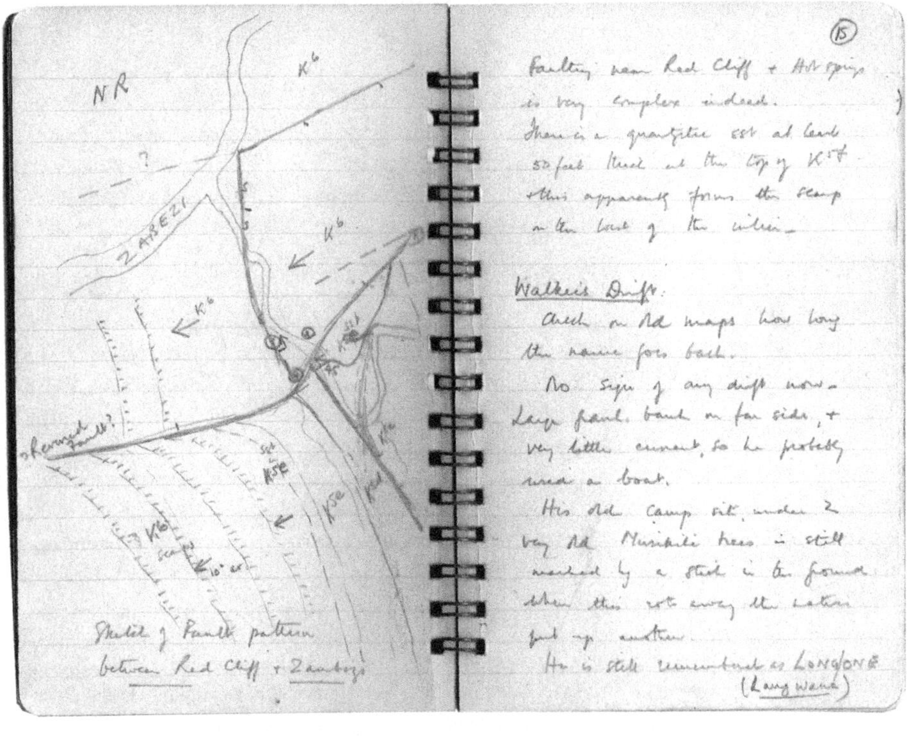

(15)

Faulting near Red Cliff + Hot springs is very complex indeed. There is a quartzitic sst at least 50 feet thick at the top of K⁵ + this apparently forms the scarp on the bank of the river.

Walker's Drift.

Check on old maps how long the name goes back.

No sign of any drift now. Large sand bank on far side, + very little current, so he probably used a boat.

His old camp site, under 2 very old Musikili trees is still marked by a stick in the ground. When this rots away the natives put up another.

He is still remembered as LONGWONE (Longwane)

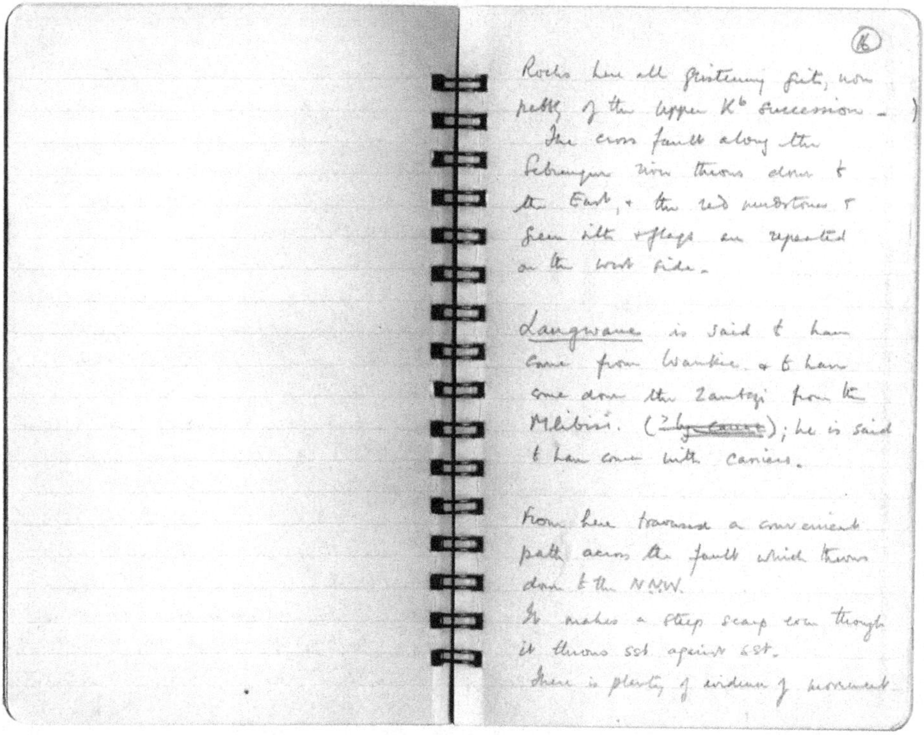

(16)

Rocks here all glistening grit, non pebbly of the Upper K^b succession. –
The cross fault along the Sebungwe River throws down to the East, & the red mudstones & green silts & flags are repeated on the west side.

Langwane is said to have come from Wankie & to have come down the Zambezi from the Mlibizi. (? by canoe); he is said to have come with carriers.

From here traversed a convenient path across the fault which throws down to the NNW.
It makes a steep scarp even though it throws sst against sst.
There is plenty of evidence of movement

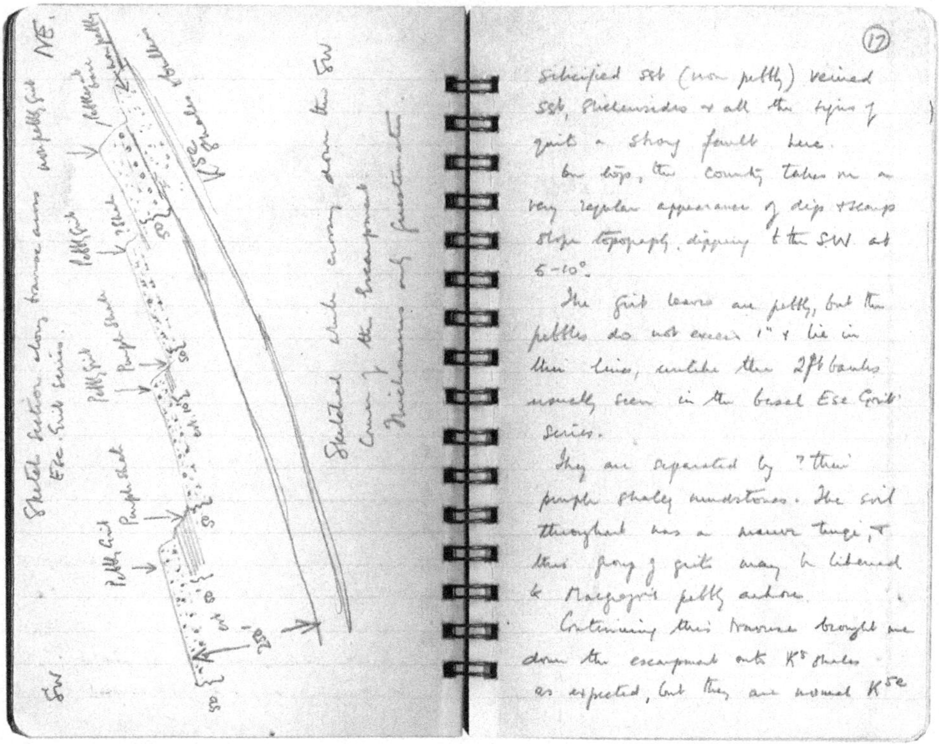

(17)

silicified sst (non pebbly) veined sst, slickensides & all the types of grit a strong fault here.
On top, the country takes on a very regular appearance of dip & scarp slope topography dipping to the SW at 5-10°.
The grit lenses are pebbly, but the pebbles do not exceed 1" & lie in thin lines, unlike the 2ft bands usually seen in the basal Ese Grit Series.
They are separated by ? thin purple shaley mudstones. The grit throughout has a mauve tinge, & this group of grits may be likened to Mazgozi's pebbly arkose.
Continuing this traverse brought me down the escarpment onto K^5 shales as expected, but they are normal K^{5c}

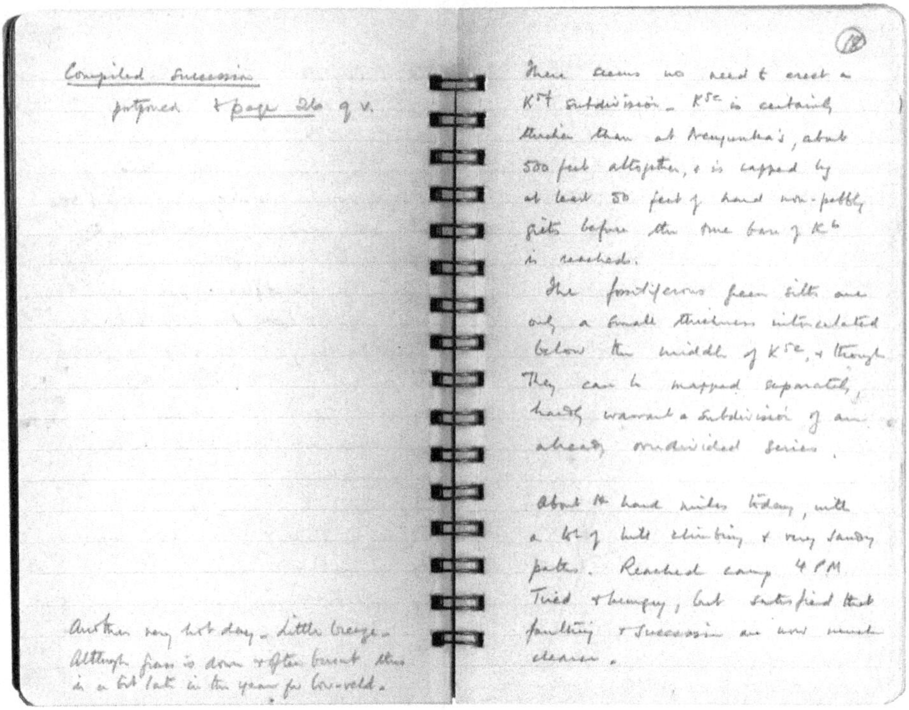

Compiled Succession
postponed to page 26 q.v.

Another very hot day – little breeze. Although grass is down & often burnt this is a bit late in the year for low-veld.

(18)

There seems no need to erect a K5d subdivision – K5c is certainly thicker than at Nenyunka's, about 500 feet altogether, & is capped by at least 50 feet of hard non-pebbly grits before the true base of K6 is reached.

The fossiliferous green silts are only a small thickness intercalated below the middle of K5c, & though they can be mapped separately, hardly warrant a subdivision of an already overdivided series.

About 14 hard miles today, with a lot of hill climbing & very sandy paths. Reached camp 4 PM. Tired & hungry, but satisfied that faulting & succession are now much clearer.

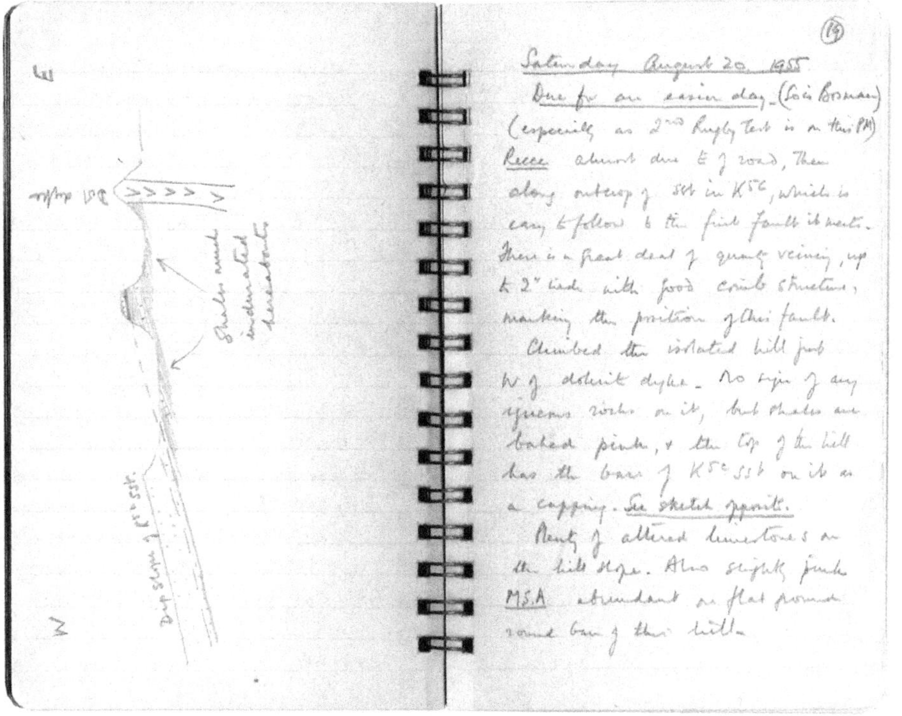

(19)

Saturday August 20 1955

Due for an easier day (Sid's Bosman) (especially as 2nd Rugby Test is on this PM)
Recce almost due E of road, then along outcrop of sst in K5c, which is easy to follow to the first fault it meets. There is a great deal of quartz veining, up to 2" wide with good comb structure, marking the position of this fault.

Climbed the isolated hill just W of dolerite dyke – No sign of any igneous rocks on it, but shales are baked pink, & the top of the hill has the base of K5c sst on it as a capping. See sketch opposite.

Plenty of altered limestones on the hill slope. Also slightly pink MSA abundant on flat ground round base of this hill.

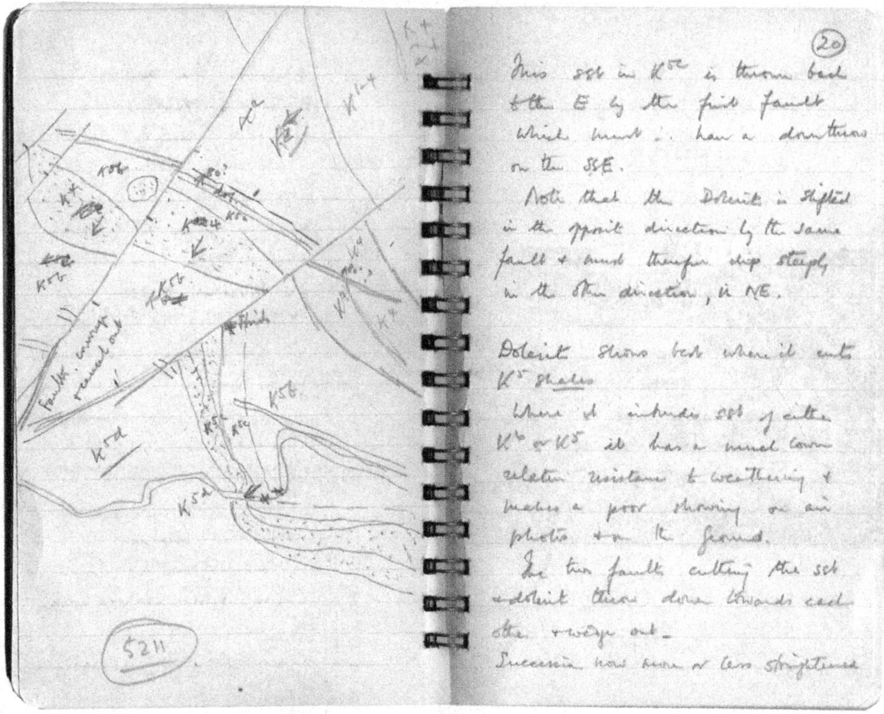
5211
20
This sst in K5c is thrown back to the E by the first fault which must ∴ have a downthrow on the SSE.
Note that the Dolerite is shifted in the opposite direction by the same fault & must therefore dip steeply in the other direction, i.e. NE.
Dolerite shows best where it cuts K5 shales.
Where it intrudes sst of either K4 or K5 it has a much lower relative resistance to weathering & makes a poor showing on air photos & on the ground.
The two faults cutting the sst & dolerite throw down towards each other & wedge out.
Succession now more or less straightened

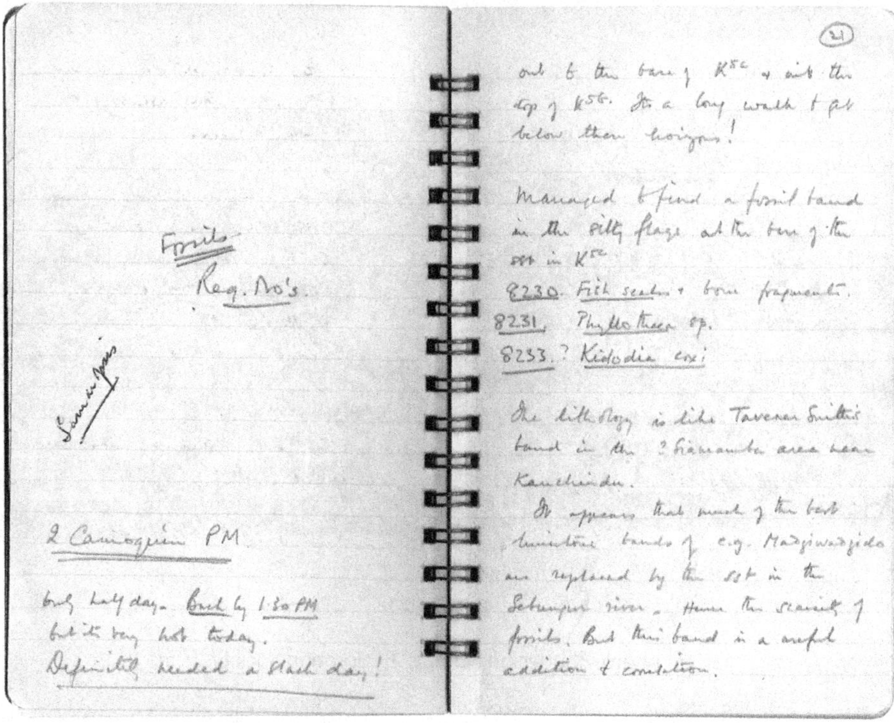
Fossils
Reg. No's
Samples Nos
2 Camiorzgiwe PM
Only half day. Back by 1.30 PM but it's very hot today.
Definitely needed a slack day!
21
out to the base of K5c & into the top of K5b. It's a long walk to get below these horizons!
Managed to find a fossil band in the silty flags at the base of the sst in K5c.
8230. Fish scales & bone fragments.
8231. Phyllotheca sp.
8233. ? Kidodia coxi
The lithology is like Tavener Smith's band in the ? Siamambo area near Kanchindu.
It appears that much of the best limestone bands of e.g. Madziwadzido are replaced by the sst in the Sebungwe river. Hence the scarcity of fossils. But this band is a useful addition to correlation.

RHODESIAN SCHOOLS EXPLORATION SOCIETY

REPORT

Upper Kariba Expedition

MAY, 1956

RHODESIAN SCHOOLS EXPLORATION SOCIETY

PATRON-IN-CHIEF

The Honourable Sir Robert Tredgold, K.C.M.G.

EXECUTIVE COUNCIL

Elected Members:
F. Gordon Deedes, Esq., Chairman
G.E. Fuller, Esq., Treasurer
The Reverend Father J. Fitzsimons

Representatives of the Rhodesia Teachers' Association:
T.R. Craig, Esq.
H.A.B. Simons, Esq.

Representatives of the Heads of High Schools' Association:
The Reverend Father Desmond Ford
V.W. Milward, Esq.

Representative of the Federal Ministry of Education:
G.H. Tanser, Esq.

MATABELELAND BRANCH COMMITTEE

H.A.B. Simons, Esq., Chairman
A. Ainsworth, Esq., (Technical High School), Hon. Treasurer
D.C. Barbanell, Esq., (Milton School), Hon. Secretary
Dr. G. Bond
C. Conolly, Esq.
D.L. Carley, Esq.
B. Haddon, Esq. (Northlea School)
D.V. Houlding, Esq. (Falcon College)
J. McCrudden, Esq.
Brother O'Connor (Christian Brothers' College)
C.W.D. Pagden, Esq.
R.H.N. Smithers, Esq.
R.F.H. Summers, Esq.
Dr. G. Wright

Group photo, Rhodesian Schools' Exploration Society, 1956

– 11 –

My Valley Shared

No account of the mid-1950s would be complete without mentioning the activities of the Matabeleland Branch of the Rhodesian Schools' Exploration Society,[1] and of Harry Simons, its Chairman.[2]

When it first began to operate, I confess to having been sceptical of its worth and in consequence probably adopted an attitude of lofty indifference. But after one or two experimental ventures not too far from Bulawayo, they approached me for suggestions on sites for future exploration. I began to be impressed not only by their enthusiasm but also by the quality of their scientific leadership. The reports of their early efforts were good. I suggested the western end of the middle Zambezi Valley as their next target, and this was accepted for the 1955 expedition.

Since I had been there, it naturally fell to me to select a campsite, which I did with completely selfish motives to suit my own research. There was, however, plenty of scope for the other disciplines, but the site chosen, on the banks of the Sebungwe river, sixty miles beyond Kamativi and twelve miles short of the Zambezi at Siachalaba's village, was no easy one for a convoy of lorries to reach. Consequently, the selection of boys to take part in it was done with an eye to how many boy-horsepower they could be expected to apply on tow ropes when lorries got stuck in either sand or mud. We selected thirty out of about eighty applicants. My reconnaissance in late April confirmed that the track from Kamativi, used only in the dry season by a few venturesome fishermen, was in very poor shape. It was even difficult to find in places. The convoy would have a hard time in May.

The planning for such a venture is a major undertaking. The whole party would be beyond the reach of supplies and had to be completely self-sufficient

[1] A history of the Society appears in A.H. Siemers, 'Rhodesian Schools' Exploration Society', in I.P. MacLaren (ed.), *Some Renowned Rhodesian Senior Schools, 1892–1979* (Bulawayo: Books of Zimbabwe, 1981), 317–22.

[2] Harold Arthur Bertrand Simons spent twenty-five years in the Ministry of Education in Rhodesia, of which nineteen years were as a headmaster. 'His contribution in building up the museum and sense of tradition at Baines School is worthy of special commendation. His greatest contributions have all been associated with what can be broadly referred to as the educational field.' Citation: 'The Society's Gold Medal: 1975 Presentation', *Rhodesiana* (1976), 34: 53.

for food, medical and scientific equipment. To look after their health, patch up minor injuries and ensure against major catastrophes required a full-time doctor on the expedition, preferably one used to primitive conditions. Dr Louise Westwater was the one who volunteered.[3] A quartermaster who was also a schoolmaster was in charge of feeding us, and the police lent us a radio transceiver for daily communications. Harry Simons, then headmaster of Baines School, was the leader, and various other Biology masters were to look after the groups dealing with scientific disciplines other than Geology, which naturally fell to me.

The transport consisted of two three-ton lorries, lent, rather rashly I thought, by Bulawayo industrialists, to which was added the doctor's Land-Rover, my Museum short-wheelbase Land-Rover, and one other. Stores and boys were to travel on lorries and staff in Land-Rovers. The journey of 200 miles from Bulawayo to Kamativi mine was eventful only on the last twenty miles, the section of the main road that goes through the Gwai gorge. If the majority of the expedition thought this was difficult, I wondered what they would think of the sixty miles from Kamativi to the Sebungwe river drift.

Naturally, the whole party spent the night in Kamativi and were well entertained. We set off before 7.00 next morning. Sixty miles and twelve hours later, we reached camp. The lorries were in constant trouble - either in the mud in the vleis, in rocks in the river crossings, or in the heavy sand sections in between. Even the Land-Rovers got stuck more than once, and it was a tired party that made camp that first night in the Zambezi Valley. The elephant had been particularly conscientious since my earlier recce, and the road was strewn with fallen trees.

The scientific work was done by dividing the boys into parties of five or six under expert leadership, and it proceeded smoothly. Hopefully, this kind of

[3] A New Zealander who served in the BSAP between 1965 and 1969 met Louise Westwater: 'In the short time ... perhaps an hour ... I spent in Louise Westwater's company I knew I was in the presence of 'A real Africa woman.' From her dark, sun-burnished, and wrinkled-with-age face shone two eyes that carried the conviction that White and Black could find common ground when commonsense and good-will prevailed over fear and its resulting racism and bigotry. ... Using a staff of two Black assistants and a Land-Rover modified to suit her work she worked in sometimes remote African populated areas trying to reduce the effects of diseases ... malaria, bilharzia and glaucoma ... that were endemic amongst African people living in the rural areas.' Gordon 'Johnny' Johnston, *So I've Done My Three in the BSAP* (2012?), 167, <http://johnnybsap.weebly.com>.

In December 1976, she married David Reid-Henry, a bird artist, who died less than a year later at the age of 58. She was with him when he died. 'Obituary: David Morrison Reid-Henry (1919–1977)', *British Birds* (December 1977), 70: 539–41.

Between 1984 and 1997, she owned a two-bedroomed cottage, Stancro, in Rapness, Westray, Orkney Islands, Scotland, and visited there occasionally. 'A doctor of an Edinburgh family, she became Medical Officer for Health for Matabeleland ... She died suddenly in a hotel in London on 4 August 1997, leaving all her many belongings at Stancro.' 'Stancro', *FindGlocal*, <http://www.findglocal.com/GB/Kirkwall/1126295224089053/Stancro>.

exposure would encourage some to take up scientific careers; even if one or two did so, there would be a long-term benefit to the country. The short-term benefit was ensured, since the collections and observations would be recorded in the Expedition Report by the section leaders.

My group of five had a hard time learning the rudiments of geological mapping in rather harsh terrain, but in two weeks we made a tolerably good geological map, and I stored up much information on a new area of my favourite rocks, from which we collected quite a lot of fossils. These were later identified and added to the Museum's collections.

The long-term effects on at least two of my party were entirely negative geologically. Angus Strover distinguished himself by discovering a hitherto unknown hot spring. He spotted a small patch of intensely green vegetation among the otherwise brown scrub, walked on it, and fell through it up to his middle in hot water. Luckily it was not hotter than a normal hot bath and he was none the worse. But eventually he became a Doctor of Medicine. For a time he was a lecturer in the Faculty of Medicine in our own university. Rob Summers, a Bulawayo boy, did not make any such spectacular discovery, but went on to university in South Africa, did Pharmacy, and later became Professor of Pharmacy in Salisbury. The fate of the other three is unknown to me, except that none became geologists. Perhaps other party leaders were more successful at recruiting than I was.

There was only one medical excitement on the expedition. No one was bitten by a snake, stung by a scorpion or chased by an elephant, but at about 7.00 p.m. on a Saturday night, a large boy reported to the doctor that he was feeling a bit unwell, whereupon he folded up in an agonized heap at her feet. Her diagnosis was acute appendicitis, in need of immediate surgery. This posed a real problem. The nearest hospital was at Kamativi mine, where there was a resident doctor. His wife was also a doctor, but Kamativi was sixty miles away, on a track that had taken the convoy twelve hours in daylight. Being a Saturday, the medical staff at Kamativi might be away for the weekend. Wankie Hospital was a further sixty miles away, and this appendix was acute.

One Land-Rover was sent immediately to alert Kamativi, the patient was dosed with pain-killers and loaded into a second Land-Rover on a stretcher bed, carefully lashed down, and set off with the doctor. I brought up the rear with ropes, shovels and axes in case the 'ambulance' got stuck. By 11.00 p.m. we had got the patient about halfway to Kamativi and I had a puncture. The time taken to change the wheel left me some way behind, and I came to the worst section of road without catching up with the patient.

This section, the Tinde vlei, began with a bad stream-crossing and then continued in a dense reed bed for several hundred yards, through which there was no well-defined track. I was astonished to see Land-Rover headlights weaving about aimlessly in the reeds. The vehicle with the doctor and patient had got hopelessly lost in this reed bed, and were just about to emerge from it, heading back towards camp instead of towards Kamativi. I led them through it on to the road again and we reached Kamativi at about 2.00a.m. to find the operating theatre ready for the patient but wondering where he had got to. The angry appendix was removed in time and all was well. The expedition returned safely and vowed that the next one in 1956 would be still further into the Zambezi Valley.

So, in April 1956, I joined forces with a medical party in two Land-Rovers who were to go through Kamativi to the Zambezi. At the start of the previous dry season, when I had reopened this road on my own, it had taken me all day to cover the first fifty miles. This year, with three Land-Rovers, it would be easier, since, if the leading vehicle got stuck, there were two more to pull it out backwards, and more hands to remake the really bad bits. I hoped that we should reach the Zambezi, at Siachalaba's village, by the evening of the first day.

The doctors were doing a mass vaccination programme against smallpox, and hoped to do about 4,000–5,000 Batonka within a week. My job was again to reopen the road and choose a campsite for the next expedition of the Rhodesian Schools' Exploration Society, as far downstream on the Zambezi as possible. My own view was that it would be impossible at this time of year to cross the Logola river, if indeed it proved feasible to reach even that point by lorry. As the rainy season had been a heavy one I expected that the 'road' would be in rather poor condition and would need the attention of several local gangs before it would be possible to take the expedition's lorries even as far as the Zambezi.

There were three main obstacles on this road. The first was the escarpment down into the Valley, which was reached some twenty-five miles after leaving Kamativi. It was generally easy enough to get down but quite another matter coming up again. It was not particularly steep, but the surface was deep, loose sand. In four-wheel drive it presented no particular problem, but to ordinary vehicles it was a menace. The long pull in sand made them boil, even if they did not bog down and need digging out.

The second bad patch was the Mlibisi river drift, not far beyond the foot of the scarp. This was a particularly nasty crossing. The approach was a steep bank of loose sand, renewed every year by the floods of the wet season. Again it was easy to get down, but I had vivid memories of trying to climb it in a Land-Rover the year before, when it took ten attempts, stops and retreats backwards into the river before it finally clawed its way on to the top. Later in the year I had it dug out, but that would all have been destroyed by the next floods.

The bed of the river was only about fifty yards wide at the drift, but it was full of massive round sandstone boulders, through which it was possible to clear a narrow track by rolling out the largest. The Land-Rover then had to bounce around on the rest in about a foot of water. The climb out on the far bank was not difficult.

The third bad patch was the Sebungwe river drift, eleven miles short of the Zambezi itself. It was like the Mlibisi, but nearer a hundred yards wide, and the water was usually deeper at the same time of year. By August, both rivers are practically dry, but their condition in April depends on the lateness of the rains. That year the rains had persisted into April and it could be expected that both rivers would be flowing fairly strongly.

As I knew the road well from previous years, I led the party. Dr Louise came next, having been down the previous season, and G. came last, not having been that way before.[4] We left Kamativi in good time on a brilliantly fine sunny morning, but there was still the feeling of summer in the air, and as the day progressed clouds built up over the Sijarira hills. There was a distinct chance of rain later in the day and, as both the Mlibisi and Sebungwe rivers rise in these hills, this brought with it the chance of flash floods, which could hold us up.

The first few miles were surprisingly good. Places where it was reasonable to expect difficulty were easy. Vleis were dry. Small streams were low and the drifts undamaged and we made really good progress, averaging ten miles an hour. There were occasional trees across the track that elephants had pushed over, but it was possible either to drive round them or to move them aside. Elephants seem to delight in making obstructions of this kind across any kind of man-made track. They have thousands of acres of virgin bush in which to push over trees, and indeed they do pull them down in large numbers, but roads seem to fascinate them. They often walk miles along a man track and, as though to preserve their private haunts from intrusion, they leave a trail of fallen branches and trees to obstruct the passage of trespassers. These were our only obstacles, and I had visions of reaching the Zambezi without trouble, but there were still the two major river crossings ahead of us.

We reached the bank of the Mlibisi without any mishaps at all, and here we met our first snag. The river was running strongly and was a dirty brown colour, indicating recent rain in the Sijarira hills upstream. It looked exceedingly difficult, if not impossible, to cross. I called a halt and instructed my cook to make a fire and some tea while we explored. It was then about

[4] 'G' must be Dr Gavin Wright, who was Government Medical Officer at Nyamandhlovu at one time, and was a member of the Matabeleland Branch of the Rhodesian Schools' Exploration Society.

2.00p.m. and we had covered about half the distance from Kamativi to the Zambezi.

While the tea was brewing, I waded across the river, feeling the boulders and testing the depth of the water. The cleared track from last year was visible, but the water was more than knee-deep in much of it and I could feel the boulders as I slithered about. It looked and felt a nasty problem to get three Land-Rovers through without sticking. If I had been by myself, I doubt I should have attempted it, but with two spare vehicles to pull it might be possible. My own was brand new, and the running-in period had only been completed on the main-road run from Bulawayo to Kamativi. As usual it was grossly overloaded, and it seemed sheer cruelty to decent machinery to take it over this drift.

We had our tea and a snack and discussed it. Then we all set about moving such rocks as we could find in the brown and turbulent water. At last we had moved all we could, but there was one huge rock that half blocked the way at the deepest and softest part of the crossing. It made a nasty kink in the most awkward place, but there was nothing that could be done about it.

I climbed into Susan (all my Land-Rovers have been called Susan), engaged the lowest gear ratio in four-wheel drive and slithered cautiously down the sandbank into the water. It came in through the door and all but covered the foot pedals. The exhaust pipe blew bubbles behind me, but I kept my foot down and the engine revved happily enough. There were still plenty of boulders under the water and Susan bounced madly from side to side. The fan splashed water up through the bonnet and out over the flat wings as I crawled and bucked towards the other bank. I had been too lazy to take off the fan belt, and just chanced it.

In spite of the ill-treatment I was handing out to a new Susan, I began to enjoy myself. I reached the kink at the big rock and, by sheer good fortune, bounced past without touching it. Steering as such was hardly possible, because every bounce of the front wheels jerked the steering wheel one way or the other. If it had jerked the wrong way at the kink, I should have had to stop, reverse and try again. To my great surprise and relief, Susan crawled out of the water and up the far bank on to dry land, without even misfiring once. I drove up on to level ground, put on the handbrake and, leaving the engine running at a fast tick-over to dry out plug leads, climbed out and waved to the others.

The psychological effect of getting one vehicle over a bad and unknown quantity is considerable, and Louise began her crossing with a broad smile while I stood on a rock and waved. She went well as far as the kink, and then stuck. I waved, intending to imply that we were all coming to push her forwards. She mistook my waves for instructions to try reversing, which she

Crossing the Mlibisi river, 1956

proceeded to do. Before we could reach her, she had backed about a yard, put the Land-Rover in low gear forwards again, and cleared the obstruction; she also finished the crossing unaided. After this, G. made no mistakes and we were able to go on our way.

The river appeared to be still rising and it was more than likely that we were now cut off from retreat, but, as this crossing is usually the worst, we felt fairly confident that we could now reach the Zambezi. The twenty miles to the Sebungwe river was easy going for our type of vehicle but would have been a hard slog for anything else. It is level but sandy most of the way. The sand gave us no trouble and the few rocky patches were easily crawled over. We reached the Sebungwe riverbank at about 5.00 p.m. The trucks were all parked on the bank while I waded in to try the depth. To my intense disgust, the water was waist deep and running strongly, so much so that I had trouble keeping my feet. It was obvious that even Land-Rovers would not cross it.

There was nothing for it but to camp and wait. Perhaps the river would go down enough in twenty-four hours to allow at least one truck to cross. We were now eleven miles from the Zambezi and twenty-six from our final destination, but time was precious to the doctors if they were to get through their vaccination programme. I stuck a small stick in the sand at the edge of the water. Time drags at an enforced stop of this kind. We had camped at my old site of the year before.

Before breakfast next morning I went to examine the flood-marker. The river had risen nearly a foot. I moved the marker stick to a new position at the water's edge and returned gloomily to tell the others that there would be no chance to cross at least until evening. We were not cheerful. The Mlibisi would now certainly be impassable behind us, and the Sebungwe was equally impassable in front of us.

To pass the morning, we took fishing rods and walked down to the Zambezi-Sebungwe junction, about seven miles; the fish were not helpful. On the way back it rained hard, and we had visions of the river rising even more. Under these conditions, waiting would be useless. It might be several days before we could drive over. The only other way was to raise carriers and walk. As I had travelled before with carriers and was known to the local people this, in theory, presented no problem.

Having decided that we must do the journey with carriers, it was necessary to decide how many were needed and to cut our baggage to the bare minimum. After conferring with the others, we decided we could manage with ten in addition to our two medical orderlies and Million, my cook. I went to Bosman's village and soon found him. I explained that I wanted ten men to carry us to Siachalaba's village eleven miles away, and he set off cheerfully enough to collect them. He was soon back with the required number, some of whom I recognized; they sat around and looked cheerful.

Then came the vital question, that of payment for the journey. I opened the bidding at 2/- each and was prepared to go as high as 2/6. They opened with a firm demand of 10/-. This was a scandalously high figure and I expected them to come down very quickly indeed to near my figure and then haggle for a bit in splitting the difference. However, surprise no! They refused point blank to do that journey for less than 10/- each.

It was soon obvious that they just did not want to go, and no threat or promise would move them. This was probably because there had recently been a funeral in the area and much beer had been brewed and drunk. Perhaps there was still some left, and perhaps the after-effects of heavy drinking and all-night dancing still lingered, but the plain truth was that they did not want to work. Rather than offend us by saying so outright, they preferred the indirect method of sitting firmly on their backsides and demanding a price that they knew I would never dream of paying.

This was a situation for which I was quite unprepared, and they knew that without carriers we could not possibly do the journey. The Batonka on his own ground can often hold a strong hand and he knows well how to play it. I had, however, one trump card left. Their chief, Tschabi, lived near the Mlibisi crossing and could be reached by the road on which we had come. The chief was still a power among these people, but he was paid by the government and the two doctors were on government business. It should therefore be possible to get the old man into the awkward position that he would have to order his people to work for the government at government prices. I took Million and Bosman and drove as fast as I could to Tschabi's village.

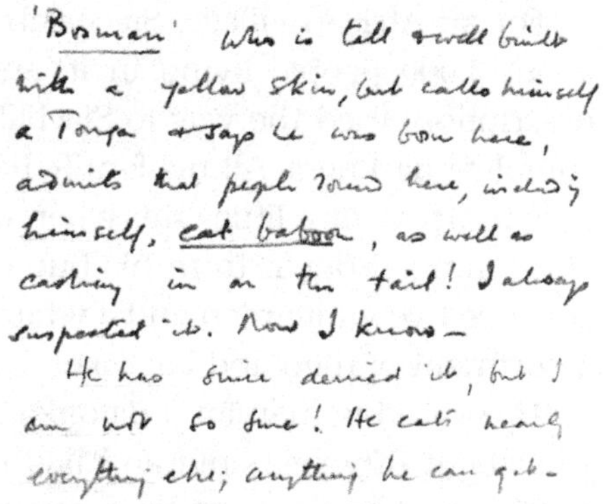
'Bosman' who is tall & well built with a yellow skin, but calls himself a Tonga & says he was born here, admits that people round here, including himself, eat baboon, as well as cashing in on the tail! I always suspected it. Now I know –
He has since denied it, but I am not so sure! He eats nearly everything else; anything he can get –

If the old man was away on a journey, we would be sunk, but if we were lucky enough to find him at home, I felt all would be well. As luck would have it, he was in. At first he flatly refused to believe that any of *his* people would behave so badly. They must come from beyond the Sebungwe river and belong to Siachalaba's people. This excuse held no water, and even Bosman could not uphold it. Then he expressed his amazement that they should ask such an exorbitant price for the job. He volunteered the suggestion that 2/– each was a fair wage, ordered his own 'policeman' to get into the Land-Rover and see that his orders were carried out, and we departed.

The same group of men were lolling about our campsite when we returned. They listened to the words of Tschabi, as relayed by his policeman, and behold, we had the same ten carriers who had refused to work for less than 10/– quite ready and willing to do it for 2/–. The loads were soon ready. After some preliminary wrangling for light or easy-looking loads, they were hoisted on to heads and off we set, just after dusk, to walk to Siachalaba's village.

It was a lovely, fine, cool night with a moon, and nothing could have been more pleasant than walking without a load. The carriers seemed perfectly contented. They went on ahead, with strict instructions where to put their loads down and wait for us. The track was in good condition and easy enough to follow in the moonlight. It is used as a footpath by the Batonka during the wet season. They walk on the wheel tracks, sometimes using one, sometimes the other, which makes driving tricky to begin with when the road is again used for motor vehicles. It was exasperating to find how easy it would have been for the Land-Rovers if only we could have crossed the Sebungwe river.

We struck a fast pace, which was very stupid, because none of us had done much walking since the previous dry season. Louise, the youngest member of the party, positively galloped; G. and I, both on the wrong side of forty, did our best, but it was hard going. We covered the eleven miles in three hours, finally overhauling our carriers, who were resting, before they could reach Siachalaba's village.

For an African village, Siachalaba's was quite large: there must have been about 3,000 people living in it, and 'groups of villages' would be a better description. I led the way to Siachalaba's family group of huts, which was in complete darkness. All his family had gone to bed early that night, for it was only 9.00 p.m. or a little after when we got there. After some shouting, a sleepy chief climbed down from his hut, complete with old bush hat, to find to his surprise two white men and a white woman in his presence, together with an assortment of men and baggage.

He was a big man for a Batonka, probably nearly six foot and well built. He was about fifty years old and had one good eye and an empty socket. What he saw that night must have confirmed his feelings that all white men are mad, but he took it calmly and politely. The usual greetings were solemnly exchanged and the position explained. The doctors had come to vaccinate his people, and I wanted at least ten strong and able-bodied men at the camp by 8 o'clock the next morning to carry us to Siansali's village, about fifteen miles down the river. The rate was agreed at 2/6 per person, and I had no doubt that the required men would appear at the appointed time. I don't know what the authorities thought of Siachalaba as a chief, but in my dealings with him he had always been helpful, polite and effective.

His village was typical of the large Batonka kraals on this part of the Zambezi. It lay on the inside of a great bend where the river swung round in a half circle. The land it enclosed was some two miles across the ends of the bend, and was bounded by the river and backed by a long escarpment about 500 feet high. The bend marked an old floodplain of the Zambezi when it ran at a level about 100 feet above the present and was covered with a fertile alluvial soil. This combination of arable soil with abundant water for people and goats allowed these river people to live in much larger communities than was usual. There were more than 1,500 huts in the village, arranged in little family groups of perhaps a dozen buildings. The groups lay in long discontinuous lines, each one on the top of a low terrace feature marking a stage in the recent down-cutting of the river from its old flood plain to its present level.

The huts themselves were circular with thatched roofs and were nearly all built on stilts. I have heard many reasons put forward for this practice. It has been claimed that it was a protection against floods, but as the villages were all built on terraces well above the level of any possible flood, this could hardly be true. The people themselves could not give any good reason. They merely said that they had always built this way.

The ground below the hut was always in shade at the heat of the day and the women did much of the sit-down domestic work there, including the cooking. Whether this was the idea behind the design it is impossible to say,

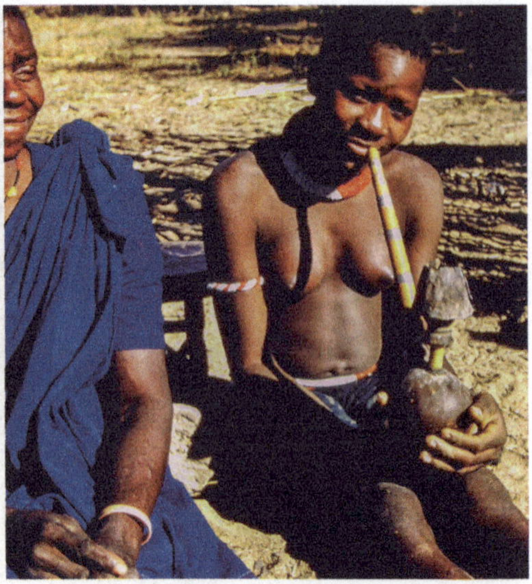

but it seems as good a theory as any. Siachalaba's was typical of many of the Tonga villages in the valley: wherever the conditions were similar, big kraals were to be found.

This, then, was our campsite for the night, though we were on the fringe of the town, under the shelter of a group of evergreen musikili trees, right on the high bank of the Zambezi.

True to his word, Siachalaba was there before 8.00 the next morning, with more than enough carriers from whom to choose our party. After a mad rush from the Sebungwe river drift the night before, I had one or two tender spots on my feet. G. was so badly blistered that it was useless for him to try to go further. It was decided that he should stay behind and vaccinate all the people over whom Siachalaba ruled, and that Louise and I should go on to Siansali, fifteen miles downstream, and deal with the people there. Louise seemed to have no foot troubles at all and to be looking forward to the remainder of the journey.

We did a further subdivision of baggage, food and cooking pots, taking for ourselves only the barest necessities and the medical supplies. This allowed us to reduce our carriers to eight, none of whom would have to carry a very heavy load. We estimated that we ought to be back at Siachalaba's in four days' time. After remarkably little preliminary nonsense from the carriers, we were off to a very good start, this time in broad daylight.

Our route lay on the narrow strip of flat country between the meandering Zambezi and the long straight escarpment that forms the last step down into the inner Valley. For most of the journey we should be in sight of the river except when it would be hidden by thick bush. As there was only one track and the destination was the next village, of a size comparable to Siachalaba's, there was no chance of the carriers going to the wrong place. I let them make their own pace, and kept near us only our own men who were carrying the tea things and a little food for lunch. The tender spots on my feet got nicely numbed after the first mile or so, and it was as pleasant walking as it could be.

As the miles went by, the sun got hotter and hotter and we had a tea break after some five or six miles. After twelve miles I was tired, and my feet were getting sore again; the sun was boiling hot and I was more than ever annoyed with the Sebungwe river for being too full to cross, for, again, the track was quite feasible for the most part. Louise showed very little sign of weariness, but she did not refuse a good tot of neat whisky after lunch. We reached our campsite in a clearing near Siansali's village at about 3.00 p.m., and soon had Louise's small tent up under a tree. We had not brought beds, only bedding and mosquito nets. A good layer of cut dry grass makes comfortable lying, but it does leave a nasty feeling that anything that crawls, from ants to snakes, can too easily get into the bed.

I still had to see whether the Logola drift would be passable for the forthcoming expedition, and I think it was more bravado than anything else that made me suggest to Louise that we should walk on to look at it in the two or three hours' daylight that remained. My feet were decidedly sore, but I wouldn't confess to a woman, especially one who seemed quite impervious to heat or distance.

I could see the deep gorge in the escarpment where the Logola breaks through to join the Zambezi; it could not be more than three miles at the most. Siansali, like Siachalaba, lies on the inside of a big bend, and must be about the same size. Leaving all the carriers in camp, we strolled off and took our time over the walk. One look at the drift was enough. There was a broad, placid stretch of water fully 300 yards wide where the road should have been, with hippo spoor coming up the bank. It would be weeks before the river fell enough to allow anything on wheels to get across.

We sat on the bank, resting until the sun reached the horizon, and then turned back for camp. Darkness falls with a fair amount of speed, but in country that I knew well it seemed to make very little difference. It was not long, however, before I realized that I didn't know this bit around Siansali as well as I thought I did. In fact, the further we went, the less I seemed to know it. In the end I had to confess, and it was a galling confession to have to make, that I really had no idea of the exact spot where our camp lay. I knew roughly, and I knew it was in a group of musikili trees, but after investigating several, all of which ought to have contained our camp but didn't, it was all too plain that we were temporarily lost. There was no danger at all, only the feeling of foolishness that made me furiously angry with myself.

Any local of five or more years would have led us in the darkness straight to our objective. There were plenty of potential guides but, out of sheer stubbornness, I tried for over an hour to do without asking. Louise was very patient, but she must have been dead tired after twenty or more miles on foot. However, there were no complaints as we stumbled on, sometimes in tall grass, at other times over old mealie fields, falling in concealed holes, taking paths that led nowhere, and generally behaving in an idiotic manner.

At last I had to swallow the remnants of pride and head for the nearest hut with a fire. From that point, a small boy set out with no hesitation at all, along a perfectly good path and we were in camp within ten minutes. As soon as I could see the light of a storm lantern, I dismissed the guide with a liberal tip and we strolled into camp as though nothing had happened. I doubt if I fooled Million, who seemed to be smiling more broadly than usual, but I said nothing and he very tactfully remained silent on our late arrival.

The next two days were spent vaccinating and treating a great multitude of ailments. I arranged with the chief for the carriers to reappear on the third morning. At 8 o'clock sharp they were waiting, and I was surprised to find that they were all complete strangers. I had intended Siachalaba's men to wait and carry us back. They had had other ideas, and it was explained that they could see no point in waiting, even though they had not been paid, and had gone home after arranging for eight of their friends from Siansali's to carry us back. I didn't like this at all, and I didn't like the look of the men who had taken their places either. Furthermore, I had a strong feeling that, even at this early hour, some of them were drunk. One in particular I mistrusted. He fancied his English, which poured out non-stop and was mostly quite incomprehensible.

In spite of this change in personnel, we got off to a good start soon after 8.00a.m. The loads were lifted on to the men's heads and they set off at a tremendous pace. Louise and I followed with Million and one of the medical orderlies, each with a small load. We passed our last outward-bound stop, where we had had a reviver of neat whisky, and after about three miles reached the small village of Siamunga. There we were met by our loads, neatly piled by the side of the track through the village, but no sign of any of the carriers. There were plenty of local women and children gathered around but a significant absence of males.

Enquiry among the locals soon produced the disturbing news that our carriers had gone to look for reliefs to take over the job of transporting our luggage to the next village. They had no intention of going any further themselves. When they had found enough men, they would go back to Siansali. This was rather the kind of trouble I had expected. As all the men had vanished, there was nothing to do but wait.

Louise wandered around taking pictures while I sat and smoked, and the day wore on. The sun climbed higher and the temperature with it. It must have been two hours before any of the missing carriers returned. So far they had failed to find more than half the number of men, and the chances of getting a full team seemed remote. I cursed them soundly, and swore that they would not get paid a penny unless they completed their job, but money meant little to these people and no combination of threat and bribery would alter their resolve to abandon us and go back to Siansali. This they proceeded to do, leaving me fuming.

After rampaging round the village, I persuaded two half-grown lads into joining us. I picked up one load myself, and Million and the medical orderly picked up two more, and, with much cursing, I got the convoy moving slowly and reluctantly towards Siachalaba. They began to get quite cheerful after the first half mile, and I began to think that perhaps we might after all manage

Friday April 27th
Siachelaba – Sebungwe Drift + 'main camp' –
Ab. 70 miles in 6 days.
MLW is marvellous – Never complained
Saturday April 28th
Radio with Fighter 0645 hrs
messages for ZEF3. (to Siavonga)
1/ Recce to Lusitu completed.
2/ ~~to~~ Lusitu Drift hopeless.
200 yards of deep water.
3). Road to Sianzali OK except for one 'drift' which needs 2 or 3 days work for jeeps.
~~4). Suggest camp~~
4) Sebungwe river going down well
5) Suggest camp site between Siachelaba + Sianzali or/
6) at Siachelaba.
7) Final selection to be made next

(36)

Monday May 21. 1956.

Left Kamativi midday – Arrived back at Camp 4PM.

Remainder of evening spent clearing up –

There had been rain during the early afternoon!

Million had done well in my absence – Clothes all washed + camp in good order – Apparently nothing missing – But I forgot to bring sugar! Too tired –

Million at the camp

to cajole them into going all the way. At the next village they threatened to dump their loads to look for more 'brothers', but by some means we got them through it. I heaved a sigh of relief, as the village looked even more deserted than Siamunga's had done.

There were still about eight miles to go and the day was now at its hottest. We were approaching a spot where Louise had treated a very old woman with terrible leg ulcers on the way out. This old woman lived in a very small village, of no more than six huts, a few hundred yards off the track. I knew that Louise would want to stop and see the old girl if possible, but I feared that our carriers might try to run away if I told them to put their loads down and rest. If they did desert us, it would be highly inconvenient. Million would be left on guard, and Louise and I would walk on to Siachalaba's. Once there, I knew the chief would soon send out men to fetch our baggage. It would mean a hold-up of twenty-four hours, with no proper bedding, and we should have to share the remains of G's rations.

When we came to the nearest point on the road to the patient's village, there she was, sitting by the roadside. How she knew we would pass that way on that day I cannot say; as far as I knew, she had been given no hint of our proposed return date. Perhaps she had sat there each afternoon. She was much too old and decrepit to do work of any kind, so this was as good a place as any in which to sit, whiling away the days. There she was and, medical instinct being strong, there was no alternative but to stay and re-dress her sores; penicillin internally and externally had already done wonders. These sores, which are almost tropical ulcers, were a curse among the older people particularly. They became chronic and deeper and deeper until the poor old victims could hardly move. The flies crawled over them perpetually. The only local treatment seemed to consist of applications of dung and a rough covering of green leaves tied on with strips of bark. These were so much better after four days that there seemed a good chance that one further treatment would mend them sufficiently for nature to do the rest.

The men made a fire for tea and lunch while Louise gave a fresh injection and re-dressed the sores, packing them with gauze soaked in penicillin and putting plaster puttees round her legs in the hope that the dressings would stay on for several days. The porters seemed less sullen, and none of them had so far run off into the bush. The sun grew even hotter, and we were too far from the river to see the coolness of the water. In spite of the heat, it was peaceful in the shade and I began to hope that the rest of the trip would be uneventful. After an hour or so we decided to go on. The loads were lifted on to the heads of the carriers and we set off. My load seemed very heavy, but it was not as large as some of the others. I sympathized secretly with them and kept them all in sight.

After an uneventful mile or so, I began to hear sounds of distant revelry, singing and shouting; we were obviously approaching a beer-drink and my spirits sank. There could be nothing more calculated to upset our carriers than to find a large party of their own kind, men and women, sitting in the shade drinking traditional beer. We soon reached the source of the noise. There were some thirty Batonka, closely grouped under the shade of a large evergreen tree, right beside the Zambezi. It was a lovely place for a picnic, well chosen from every point of view except ours. The temptation was too much for the carriers. They flung down their loads and rushed into the crowds.

This was the last straw. I could recognize only about half of them, and once they had mixed with the others I had no chance of sorting them out. My instructions could hardly be heard above the noise of their singing and chatting, carried on as always at the top of their powerful voices. There seemed to be no hope at all of getting them started again. All my shouting did was to produce roars of laughter. To them it must have been a huge joke to see two angry Europeans, one a woman, stranded in their home country.

It was no joke to us, and I got more and more angry with them. The more I fumed at them, the more they laughed. I felt ridiculous and undignified, but Louise seemed to be quite unmoved, except to subdued mirth. After more unprofitable shouting, I withdrew to cool off and talk it over with her. We were only some six or seven miles from Siachalaba and she could easily have gone ahead with the medical orderly and a message for Siachalaba, but that meant giving in to these ill-disposed carriers, and I was reluctant to be beaten by them.

There was another angle to it as well. All these people were due to be moved out of the Valley, which had been their home for generations, when Lake Kariba began to fill.[5] I knew that the Native Department had spent patient months among them, gently persuading them to move peacefully. By their skilful diplomacy they had secured almost 100 per cent goodwill and co-operation from the various chiefs and their people. If by a tactless move on my part I caused resentment, it might undo some of the good that their efforts had produced; I should then be extremely unpopular with the authorities as well as with the locals.

It was, therefore, a situation of some delicacy in the wider sphere as well as one of immediate inconvenience to Louise and myself. Their behaviour was, however, quite intolerable. Here was Louise, a government doctor, who,

[5] 'In 1958, the Federal, Northern and Southern Rhodesian governments set up a coordinating committee [the Kariba Lake Coordinating Committee] to plan and execute projects for the development of the lake. ... On May 8, 1962, the KLCC ... announced an official name change to Lake Kariba from Kariba Lake [having] finally accepted that, as had been pointed out, 'Nyasa, Bangwela, Tanganyika ... are all prefixed by the word Lake and Kariba should follow suit.' Jonathan Waters, *Kariba: Legacy of a Vision* (Harare: New Zanj Publishing House, 2020) 94, 153.

GEOLOGICAL MAP WEST SEBUNGWE EXPEDITION MAY 1955

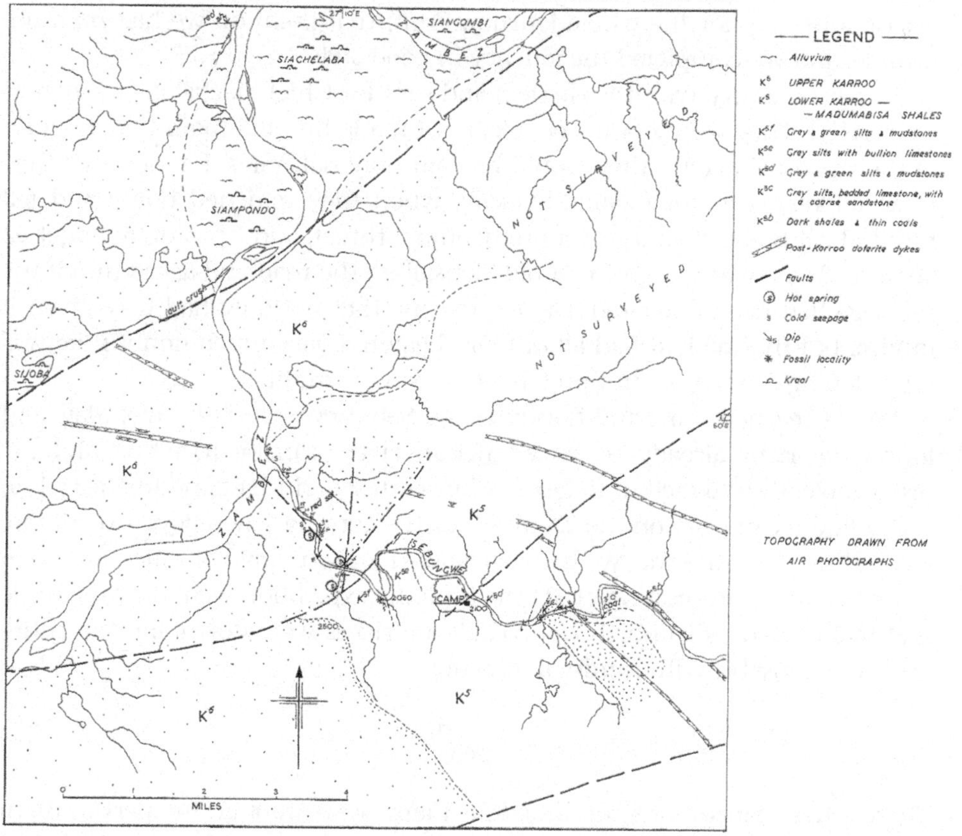

at considerable inconvenience and no little discomfort, had treated literally hundreds of their people for all manner of painful illnesses and afflictions, as well as vaccinating them against the scourge of smallpox, and they had not even the decency to help carry her bags. That alone annoyed me. Added to this was all my previous experience in dealing with them, in which I had always found them honest and helpful. That they should choose such a time to shatter my faith was galling to me.

I eventually collected our original porters, but it proved quite impossible to get any more out of the beer-drinkers. So it was still necessary for me to carry a load. They were sullen, and although they were carrying their loads, I still felt that we might have further trouble. The party proceeded in silence for the next mile or so until we came to a small one-family village. When we were level with it, a tall, well-built Tonga man came out to greet us. He seemed quite horrified to see a white man actually carrying a load and took it from me immediately. I do not know what he said to our carriers, but it sounded extremely sarcastic. He assured me in Chilapalapa that he would certainly go all the way to Siachalaba's with us, and I mentally vowed to give him at least double pay for the whole journey. From that point on we had no more trouble, and we completed the journey in good style.

I was still angry with the carriers and sent for Chief Siachalaba as soon as we arrived. They were not his people, but I made him tell them that, because of their bad behaviour, they would be paid only half rates. I added that they must all report to their chief, Siansali, when they returned home and tell him to beat them. It sounded a pretty empty threat, and I promptly forgot all about it. Some weeks later I met old Siansali on the road and asked him if the message had been delivered. He assured me that not only had they given it to him, but that he had had all of them beaten. I very much doubt if he was really telling the truth, but I pretended to be very gratified.

We all reached our Land-Rovers at the Sebungwe river the next day, and found the river already passable. African rivers can be irritating and unpredictable. G. returned to Bulawayo, but Louise and I had one further job to do, to find a campsite on the Zambezi for the coming schoolboys' expedition, so we drove back the way we had just walked for so many hot miles, and soon settled on an appropriate spot. By the time the expedition went over the route that had caused us so much trouble, all the rivers were down and the entire track was passable with minimal difficulty.

These school expeditions continued for many years, though my part in them faded out. My main target area was the middle Zambezi Valley, which had been visited twice by them. Their report series is valuable scientific and historical literature. Museum collections have been enriched, and surely some of the participants have become scientists. True exploration in Zimbabwe is now over. The country is well mapped, and air photographs and satellite pictures have done their worst. The Bush War has come and gone. Surely it is time to revive such ventures. Even if they cannot truly be called 'exploration', there is still plenty of scientific work to be done, and we need all the embryo scientists we can enlist so that development, without destruction, can go ahead in this country.[6]

[6] The Rhodesian Schools' Exploration Society ran over sixty expeditions between 1954 and 1971 (Alex Siemers, 'Schools' Expeditions', *Heritage of Zimbabwe* (2009), 28: 137–47). It would seem that the security situation in the country would have at least suspended the expeditions. Siemers's piece in MacLaren's *Some Renowned Rhodesian Senior Schools* (see fn. 1), published in 1981, ends with: 'the Society is poised to resume its activities as soon as the situation in the country returns to normal'. There was apparently an attempt to restart them after Independence but it did not succeed, mainly because, it was said, the new generation of teachers were surprised to find that the teachers had given up their own holiday time and paid their own way; they were less willing to do the same.

– 12 –

Limpopo and Makarikari

Not all my research and consulting time in the 1950s was spent in the Zambezi Valley, though it absorbed a goodly proportion of it. By 1952, the decision had been taken by Rhodesia Railways to build a new line linking this country to another port on the east coast, Lourenço Marques ('LM', now Maputo). This new line, the South-East Connexion (SEX for short), was to start from Bannockburn near Shabani, and run to the border, east of Pafuri, and on across the Limpopo to LM. Our engineers were to build the line from Bannockburn to the border while Portuguese engineers were building the rest.

By June 1952, the centre line was more or less decided but had not yet been cleared and pegged on the ground. Davies, the railway surveyor assigned to the task,[1] had traversed the full length on our side of the border, and his comments on the general aridity caused doubt on the feasibility of operating this line by thirsty steam locomotives needing to drink 5,000 gallons every twenty-five miles. Don Macdonald, the railway geophysicist, decided that a water recce was vital and asked me to join in. Davies would be our guide.

The three of us converged by Land-Rover on Bannockburn and set off across country. It was dry, and we were soon dry, too. Our first stop, for lunch, disclosed a peculiar lapse in our planning for the trip, which we had estimated would take a week. There was a mass of food, and things to eat it with, but somehow each of us had relied on the others to bring the cooking things. Between us we could not even raise a kettle or a teapot, let alone saucepans and frying pans. We decided to improvise. Tins can easily be heated in the fire, water can be boiled in empty tins and tea brewed in them, but all our efforts to fry eggs and bacon on shovels or in my prospecting pan were dismal failures.

We had no difficulty in reaching the border, though it was a slightly odd feeling to have seen only one road in a week: the main road to Beitbridge. We came out of the bush on to it rather unexpectedly, having miscalculated our position slightly, crossed it at right angles and dived back into the bush.

[1] P.H. Davies, Rhodesia Railways Assistant Engineer, Surveys.

Engineer Davies's gloom about water was fully justified. Surface water was very scarce. We saw some pans in the sandy part, where the centre line crosses what is now the Gonarezhou game reserve. My notes reflect my astonishment that these beautiful pans, in this remote area with such abundant water, appeared to attract no game. There was hardly a spoor to be seen. Such pans could not be relied on for engine watering, and attempts to find a site for a major dam on the Nuanetsi river were unsuccessful. From our geological observations, combined with the known low rainfall in the whole area, the chances of underground water seemed also to be poor. Our report to the planners, therefore, was pessimistic, and we suggested that the line should be operated from the start by diesel-electric locos. Needless to say this was ignored. Water would turn up somewhere!

I charged no fee for this week's work, which involved more than 1,000 miles of travel, on the understanding that I would get a free first-class ride to LM and back when the line was built. I have not yet been to Lourenço Marques and back by train.[2]

Although my main interest was firmly centred on the middle Zambezi Valley, being based in the National Museum meant opportunities to join in other people's expeditions, even as a driver, and to see other disciplines at work. One such trip in 1952 was to the Haroni–Lusitu river junction at the south end of the Chimanimani mountains.

I had already spent about two months in these mountains in 1950, where the Musapa river cuts a deep valley as it runs eastwards into Mozambique, near the north end of the range. More than 5,000 feet of relief on the flanks of this valley allow the geologist to see right inside the structure of the range. My work there had suggested that here was an example of an Alpine type of fold mountain range, a rarity in Africa, consisting of great slices of folded and even inverted strata, thrust over each other from the east. To go to the south end of the range would perhaps help to confirm these earlier observations. The main objective was to collect birds in the thick forest at this low-lying point, not much more than 500 feet above sea level. So I joined Reay Smithers's party.

To reach this remote forest area entailed going through pine forests at an altitude of about 4,000 feet, which we did by Land-Rover. However, we could get no further on wheels than the steep edge of the Lusitu valley. The local people are the Ndau, and we had arranged for carriers to meet us on the edge of the

[2] A detailed story about this line appears in R.D. Taylor, 'The South-East Connection', *Heritage of Zimbabwe* (2005), 24: 81–93.

valley to transport our gear eight miles down the scarp to where we intended to camp. To our surprise, all the carriers turned out to be women, some with babies on their backs. There was only the merest suggestion of a rough path down the scarp, and all these women had walked up it that morning.

They turned out to be incredibly strong, and carried our loads all the way to the campsite without any sign of fatigue, stopping only occasionally when a baby needed to be fed. The Ndau are a sturdy people and are mighty hunters, using bows and arrows. Their arrow points were as big as the spear blades of the Batonka and the bows so powerful that they hunted buffalo with them in the forest. There were buffalo skulls in the village to prove it, and I was assured that it was possible to drive an arrow into the flank of a buffalo bull with such force that the point of the arrow came out the other side. I expressed no interest in a practical demonstration, feeling that hunting buffalo this way was not for me.

The young Ndau who came with me each day had an uncanny knowledge of the forest, which was so thick that the light was too poor for photography. It was soon evident that there was little geology to be seen, and the dreaded buffalo bean (*Mucuna coriacea*) was rampant. This stings like ten thousand nettles, and the pain lasts for days. My Ndau guide had a keen eye for it and treated it with care, but claimed that the tip of his nose was immune to its stinging hairs. I refused to let him demonstrate in case his illusions were shattered.

The Ndau had developed a unique method of fishing in the big pools of the Haroni and Lusitu rivers. They made very light Y-shaped nets with the mesh stretched like a triangular butterfly net between the arms of the Y. Holding the handle of the net, the fisherman dived in and chased and netted the fish by swimming under water. They were strong swimmers, and both the women and men stayed under water for astonishing periods of time. Having caught a fish, they surfaced, trod water, killed the fish, put its head in their mouths, and disappeared under water again. I tried it myself, but in the peaty brown water I could see no fish. Since the rocks were so poorly exposed in the forest, I hunted butterflies for George Arnold. The bird people had a tremendous expedition and made very significant additions to the list of species known to occur within the borders of this country.

When the time came to go back to our Land-Rovers, the same group of women carried our belongings up the path, with the same apparent ease, while we all struggled along panting and resting. Three thousand feet of climb is a long haul even without a load on one's head and a baby on one's back. The Ndau were, indeed, a powerful breed.

The ornithologists, led by Reay Smithers, had evidently had enough of mountains, for the next collecting trip they organized was to the Makarikari Lake in Botswana.[3] This time they would also collect small mammals, rodents and cats, as well as birds. I went along as geologist-driver again.

There could be no greater contrast than between the rugged mountains of the Haroni-Lusitu area, with its gloomy forest cover, and the incredible flatness of the Makarikari country, with its glaring sunlight and sparse vegetation. Reay chose October, one of the hottest months and the driest time of the year, to do this collecting trip; perhaps it was something to do with migrants, but I would have preferred June. Our campsite was right on the bank of the Nata river, about seven miles from its confluence with the Makarikari Lake. It overlooked a long pool of moderately fresh water, which we shared with the game that drank there. These included hundreds of wildebeest that drank mostly at night. I counted more than five hundred going down to this pool one moonlit night before I fell asleep.

There was a small stone hide on the pebbly bed of the river on the other side of this pool, which was periodically occupied by a Masarwa Bushman, armed with an ancient muzzle-loader.[4] We examined this ancient gun, which had at some time in its long history been modernized from flintlock to percussion cap. It fired more or less spherical lead balls, propelled by home-made gunpowder. The proud owner showed us one bullet that he claimed had killed three wildebeest 'this year'. Gunpowder was necessarily expendable, but to lose a bullet was a calamity. We never saw him let off this fearsome weapon, but it was presumably done at such close range that to miss was virtually impossible.

Nothing was wasted from these wildebeest. The meat was dried into biltong and traded for grain, along with the bones and the hide and locally mined salt. Additional hides and bones were collected from the many wildebeest that died from drought and poverty in that desolate region. They lay in grotesque attitudes, in varying degrees of decay, where they had perished. An artistic member of our party kept me busy for two hours, moving my Land-Rover around such a casualty while he got the right angle and the right lighting for his photograph. He never showed me the result.

Desolate it may be, but, if you have an adequate support system to stay alive and can stand the October heat and the tremendous drop in temperature during the night, the Makarikari country is fascinating. Except for a shallow arm about five miles wide, the lake was practically dry while we were there. There were flocks of flamingos, estimated by the ornithologists

[3] Now known as Lake Makgadikgadi.

[4] See Chapter 10, fn. 4.

to number hundreds of thousands. Pelicans were plentiful, but what they fed on was a mystery; the water was too salt for fish. Like many waterfowl, they were careless with their eggs, and fresh ones were easy to find resting on the bottom of the shallow water. Reay and I splashed happily in water half way up our shins and collected half a dozen, which we took back to camp. Our colleagues were very suspicious of them, claiming that they must be bad (although they did not float) and that if they were fresh they would taste of fish. The yolk was like a ripe tomato, but we made very good, though rather odd-looking, scrambled egg from them.

Game included large herds of springbok and many ostrich. On one occasion, we surprised a cock ostrich on the mud flat and chased him in the Land-Rover. He managed 35 m.p.h. quite easily but went straight for the water, which effectively stopped the Land-Rover. He went right across the five miles of water to the far shore and we watched him all the way through binoculars. The incredible flatness of the region was brought home to us by the fact that the water never came up to his hocks. Chasing springbok on the same mud-flats showed that they could manage 40 m.p.h., which was as fast as I could persuade the Land-Rover to go on the slightly yielding surface, and still 'pronk' in beautiful great leaps beside us, showing no signs of distress.

The dried-out lake flats were dazzlingly bright but generally firm enough to drive on for miles. There were, however, soft areas, where it was possible to bog down even in four-wheel drive. With experience, it was possible to spot these patches, which were slightly less reflective, and avoid them. One day I took three of my colleagues so far out on the lake floor that the low shoreline was lost in the haze. When we were about ten miles offshore, I failed to spot a soft patch, but I got out of it without trouble. I have never known three people decide so quickly and positively that it was time to go back to camp for lunch. Back-tracking on the light spoor of the Land-Rover was easy.

Bird-collecting is a mornings-only occupation so that skinners can preserve the day's collection before it goes off. On such bird-collecting trips, the custom

was to keep a large three-legged iron pot permanently on a low fire. Into it went nearly all the carcases of the birds as they were skinned. About the only exceptions were birds of prey. This pot, known as ornithological stew, was a main item of diet.

Night-shooting in the Kalahari was largely under my control. The hunters were armed with double-barrelled 12-bore shotguns, and they used lamps on their heads to pick up eyes. I took the party out after supper along the only track, which ran north to south, parallel to, but a mile or two east of, the Nata river. The rough track we had made from this road to camp made a T-junction with it, but was quite hard to see, even in daylight. My job was threefold: first, to take the hunters to what might be a likely spot for small cats; then, by light signals from the Land-Rover, to prevent them getting too far away and lost in the night; and thirdly, to find the way back to camp when they had done enough hunting.

By keeping an eye on the stars, particularly the Southern Cross, navigating back to the 'main road' was not too difficult. Finding the track to camp was much harder. Thinking back, I cannot imagine why we never put up a marker of some kind at the junction. Since the hunters were never quite sure what size cat their 'bulala' lamps would pick up,[5] one barrel had a .410 adapter in it with a small shot, and the other was loaded with the heaviest buckshot available, just in case one of the really big cats showed up. Their spoor was common.

My sleeping on this trip was hampered by an incomplete camp bed. It had only three legs, one having been mysteriously missing from the start. I abandoned it and slept on the bare sand. This had one advantage: the nights were very cold because of the clear sky, and I could use nearly all my bedding on top to keep warm. Being on the ground made me vulnerable to exploration by small crawling things, and I woke one night when what I took to be a large beetle was crawling up inside my pyjama leg. I put my hand over it and it was soft and warm and furry. In the light of my torch it proved to be a mouse, but a remarkably small and very beautiful mouse.

It was pale chestnut above and pale cream below, with great black eyes and a mass of whiskers. It seemed to be an adult, but its teeth were so small and its mouth so weak that when it tried to bite my finger I felt nothing; the skin was too tough for it to break. I put it carefully in a spare live trap for inspection by the zoologists next morning. They were delighted. It was a specimen of a Kalahari mouse, *Mus minutoides*, perhaps the smallest mammal in the world, and almost impossible to trap by conventional methods.[6] Obviously

[5] A battery-charged head lamp used in hunting. From the Nguni verb *-bulala*, to kill. *Dictionary of South African English*, <https://dsae.co.za/entry/bulala/e01346>.

[6] Also called the African pygmy mouse.

setting a geologist to sleep on the ground was the answer. I got several more in my bedding in the remaining time in that camp, and the zoologists caught none in their traps.

To begin with, our camp roused the interest of large numbers of Pied Crows. They irritated us, and mostly congregated on the dry bed of the river. Their harsh voices disturbed the afternoon sleep of the zoologists, who claimed they needed a siesta before night-hunting. I shot at them occasionally, but missed consistently with a rifle that I knew well, and I was puzzled by my apparently bad shooting at a range I estimated at fifty yards. I hit one eventually at their usual meeting place, where they were discussing us so noisily. Immediately there was silence. They walked up more or less in turn, inspected their fallen brother, took off and vanished. They never came back, but I paced out the distance. It was 150 yards, but in that clear air and total flatness my estimate had been about a third of the real range.

Probably to reassure me that my shooting was really very good, the camp gossip had been that these crows were a special species that was already so full of holes that a few more did not matter. They gave them the Latin name of *Corvus perforatus*, the perforated crow.

This was a very successful excursion for me, and I found the geology of this large inland basin, with no outlet but evaporation, very revealing. All inflow is fresh water, mainly from the Nata river, which rises near Plumtree on the Zimbabwe watershed, and flows in a great anti-clockwise sweep to discharge due south into the Makarikari Lake. How long it has taken to concentrate the minute amount of salts in fresh water into this strong brine is unknown. There are even thin beds of pure salt (sodium chloride) in the banks of the streams forming the Nata river delta. These were exploited by the local Masarwa Bushmen; they may even have been an attraction for Middle Stone Age man, perhaps 30,000 years ago. We found a factory site of this age in the Nata delta, and I recalled the concentration of stone tools of the same general age round the Bari salt pan in the middle Zambezi Valley. In such a hot climate, salt has always been a commodity vital to humans and other mammals.

My second - and, as it turned out, my last - visit to the Makarikari was under very different circumstances - to help in the search for a lost twin-engine civil aircraft that had taken off one early December day to fly to Maun and gone missing. Because the Museum staff was known to have operated in the Makarikari area, the Botswana authorities asked us to take part in the ground search. December can be as hot as October and, if the plane had force-landed in the desert, the occupants would soon die of thirst.

We were assigned to the area about midway down the east shore of the Makarikari Lake and were in position within twenty-four hours of getting the message. Reay Smithers and Robert Halstead led a party of three Land-

Rovers, hastily stocked up with water, food and petrol. Our only radio was my old 'saucepan' receiver. After a good deal of persuasion and talk to the control tower at Francistown airport, they reluctantly agreed to send us a message on the progress of the search on an agreed frequency every two hours in daylight. We heard every message plainly, and listened to the rescue when it took place several days later, after the aircraft had been found by a South African Air Force plane.

The pilot of the missing plane had missed Maun in the haze, run low on fuel and landed safely on an island in the Okavango swamps. They had plenty of water, a rifle with which they shot meat, and were still surviving when they were spotted. A light aircraft landed beside them, refuelled their plane, and both took off safely.

They were lucky to survive. If they had been forced down in the waterless area like ours, it was doubtful if they could have lived forty-eight hours. The heat was intense. Deserts and semi-deserts have always fascinated me, and the Kalahari had intrigued me even on my first train journey through its margin. Such country may be fascinating, but it must be treated with the utmost respect, or disaster will surely follow.

This same area had claimed the lives of two pupil pilots in World War II, who got lost when they missed Plumtree on one leg of a cross-country exercise from Khumalo air base at Bulawayo. The ensuing search by many planes, on one of which I had flown as a passenger, found the plane undamaged on the floor of the pan, but the pupils, against all instructions, had left it to walk out of the desert. They were never seen again.

On my last trip, on our return, we came across the makings of another tragedy. Just north of the Makarikari and well off the track, we happened to notice a lorry, more or less hull-down in the sand. We soon reached it in four-wheel drive. It was deeply embedded and its driver was a young man of 21, Juan Phillips Montero, whose Cockney accent flatly contradicted his Iberian name. He had been trying to dig his lorry out for three days. He was a Colonial Office astro-fixer and his business was to fix the ground control for an aerial photographic survey. Apart from making his ground fixes, his only other instruction seemed to be to find a cable office once a month. The Colonial Office only paid his month's salary after receiving his cable as 'evidence of life'. We got him on his way again and restocked his dwindling water reserves. We wondered how many more cables his colonial employers would get. No one should have been operating in the Kalahari with only one vehicle, and that with only two-wheel drive.

Altogether, working for the National Museum was proving to be stimulating and varied, and my experience was being broadened by every new facet of the work that came my way.

– 13 –

Radioactive Prospecting

One question that the geologist is asked quite often by the layman goes something like this: 'So you are a geologist. You must often come across wonderful new mineral deposits. Do you ever peg them for yourself?'

The answer, of course, is that the geologist never just 'comes across' mineral deposits like that. Generally, he doesn't even look for them; that is the province of the prospector, amateur and professional. Even if the geologist did look for such things, he probably wouldn't find anything because he looks where he thinks things should be found. Nature is not always like that. The prospectors look everywhere, and eventually one of them finds something. Also, there are many more prospectors than geologists, so, purely by weight of numbers, they have a better chance than we have. When something interesting is found, the finder generally gets a geologist to tell him why it's there – as though that mattered one iota so long as it is there. Modern scientific mineral searches are big-team affairs, certainly not the province of the lone, wandering geologist any more.

However, there was one occasion when I did find something, and the manner of its finding was, to put it mildly, odd. The fact that it cost a large mining company a certain amount of money to prove that it was, after all, quite worthless, makes no difference to the story. Such a thing could only happen to a geologist. If it had been a prospector's find, it would certainly have turned out to be an enormous money-spinner. Nature is like that. I am not bitter because it failed to make a fortune for me. I had my pipe dreams in the weeks that followed the initial discovery, and I woke up so gradually that the blow was well cushioned. Subconsciously, I probably expected no reward; luck like that is not my line.

In preparation for the movement of people out of the Kariba basin, a new road was cut into the valley in 1956, and a temporary Native Department station was built at a place called Manjolo Springs. This new road made accessible a piece of country that previously it had been possible to get into only on foot with carriers. I had only been on the fringe of this area before, so, in June 1956, I decided to go into the Valley by the new road. The Native Commissioner at Manjolo was my old friend, Ivor Cockcroft, and so it was natural to stay one night with him in passing through to the Zambezi.

I arrived soon after dark and found him living in a very temporary corrugated-iron prefab of three rooms with a concrete floor. His son was also living there as a Land Development Officer, and, as he was an ex-miner, I was not surprised to find that he had been doing some prospecting. He was away in the Valley but had left some samples for me to examine when I passed through. Mining and prospecting is an incurable disease, and although this particular victim had been badly hit in his only mining venture, which was a failure, he still had the urge to find something. He was particularly anxious to know whether these samples were radioactive, uranium being one of the fashionable minerals that year.

On a journey of that kind, into country of the geological nature of this part of the Zambezi Valley, I would not normally have gone armed with a Geiger counter. It happened, however, that, a few days before, I had been provided with a very nice new transistorized model by the Atomic Energy Authority and had taken it with me. Here was a heaven-sent opportunity to show if off. I went out to the Land-Rover and came back proudly waving my new toy.

In using these instruments there is always a certain background level of radiation that shows on the dial. It varies from place to place within narrow limits, but it is a good thing to get a steady reading for this background before doing any measurements, particularly in a new area. After I had shown it around and explained what a wonderful little machine this particular model was, I put it down on the concrete to reach a steady background reading while I examined the samples by eye. They looked extremely dull and unpromising, but I was determined to test for radioactivity to use my new machine.

As I was sitting on a very low couch and the Geiger counter was on the floor at my feet, the easiest way of testing the samples was to put them beside it on the floor. None of them made the needle even tremble. They were obviously as dull as they looked, but I complained a little that the background reading seemed remarkably high. In fact, it was three times normal. The significance of this didn't strike me at first; it seemed only to make it more difficult to detect a low level of activity in the samples I was supposed to test.

Testing over, I picked up the machine and prepared to switch it off. It seemed, however, that the reading went down to about normal when it was picked up off the concrete. This was distinctly odd, so, instead of switching off, I put it down again. Up went the needle – not far, but distinctly. That chance observation started a succession of events that could almost be called a chain reaction. Why should a slightly higher level of radioactivity be recorded when the instrument was on the concrete? The floor looked like any other cement floor with a red polish. It obviously consisted of commercial cement, which could hardly be expected to have any radioactive content, and this must have been made up with local broken stone and sand. A few more tests in other parts of the room soon proved that the first reading was no illusion. The whole floor was radioactive, and quite strongly so in some places. Testing the bare sand outside showed the contrast very plainly indeed, for the reading was exceptionally low.

My performance had been watched with interest, and I suspected not a little quiet amusement, by the NC and his wife. When I finally asked where the sand and stone came from, I was acutely aware of it as he told me casually that there was quite a large heap of this broken stone lying in the bush not sixty yards away. I was provided with a torch and given the directions. The heap of broken stone was easy to find, and I put the machine on the middle of it. The result was startling - a reading of twenty times background - and wherever the Geiger counter was put on this innocent-looking heap of stone, it gave equally high readings. In theory all that remained to be done was to trace the source of this broken rock in daylight the next day.

In practice, there was much more to it than that. Manjolo lies on a sand plateau of considerable size, and I could guess that the stone had been collected by a gang of labourers with a lorry and driver. Under such conditions it might have come from widely scattered places. I slept badly that night, wondering what would come out of the next day's prospecting.

My ex-miner friend turned up the next day in response to an urgent radio message. He was sure he knew where the stone came from: they had cut a rough track along the foot of the first stony ridge bounding the sand plateau, and simply told the gang to pick up the loose rock lying on the slope. The track turned out to be about a mile long, and I searched the outcrop on the ridge for most of its length before getting even a mild 'tick' or seeing any rock like the obvious vein stuff occurring in the heap of concrete aggregate.

At last I could see that the end of the track was not more than a hundred yards away, and there were still no signs of radioactivity; the needle stayed stubbornly on the background reading. Then, suddenly, there it was. A reading about 300 times background. I spent the day tracing it along strike for about a hundred yards, and in width it seemed to be up to fifteen feet. Radioactively, it was red hot. More work in the next few days disclosed several more such bodies, scattered along this low escarpment. I did the whole thing alone so that no one except the NC and his family should know where I had been.

I had one particularly nasty moment. The outcrop of these reefs was on the steep slope of the little scarp above the sandy plain, and was generally covered in large boulders up to a ton in size. In stepping from one of these to the next I slipped, and a large rock rolled on top of my left leg, pinning me down. It took all my strength and a good deal of time to get free, luckily with nothing worse than a few bruises and abrasions. On the way back to my Land-Rover, I nearly trod on a buffalo asleep in the long grass. Altogether it had been quite a day.

The question now arose what to do with 'our' discovery. Being on the staff of the statutory body, and therefore not a true civil servant, I was the only one who could get a prospecting licence and we decided to form a small syndicate to peg the discovery. However, before I could start on the journey back to Bulawayo to get a licence, the Minister of Native Affairs, Ben Fletcher, turned up, and his first question to me was whether I had found anything radioactive in the area.[1] This really worried the syndicate, as it looked as though our find had already been leaked. In fact, it turned out later that he was only being conversational and knew nothing at all about it. We had an urgent meeting of the syndicate and decided that we had better 'come clean' and invite him to join us and, furthermore, to fly back to Salisbury with samples for analysis by the Atomic Energy Authority. He agreed to join us, and was duly sworn to secrecy. He went off with the samples the next day, and I set off for Bulawayo and a prospecting licence. More trouble developed on the way to Kamativi from Manjolo.

At a particularly bad spot in the road I met a three-ton lorry heading towards Manjolo. It was stuck, and naturally I stopped to help. It contained the Chief Government Mining Engineer, Frank Elliot, and again the first question was had I found anything radioactive. This was terrible. Frank professed to be on leave and heading for the Zambezi for a fishing trip, but he was going straight for the area in which we wanted to peg our find. It was hard to believe that his question was pure coincidence and that he really was going fishing. It was tempting to leave him stuck in the mud, but I pulled him out and fobbed him off with stories about Stone Age implement sites and unmapped Kalahari Sand areas, and hoped for the best. Again, it came out later that he was totally unaware on our discovery and was indeed only going fishing. At the time it was hard to credit that he knew nothing and was only being conversational.

I did the double journey in fast time, and we duly put up the discovery notice, the only one I have ever pegged. The problem then was: What next? Claims pegged and registered were safe for a year, but we were all anxious to

[1] Sir Patrick 'Ben' Fletcher, 1887–1966. 'A parliamentarian for 21 years, he held several Cabinet posts at various times. He was Minister of Agriculture, of Public Health, of Native Affairs (1951–1958), and of Lands. He was also President of the Bulawayo Agricultural Society, and of the Central African Trade Fair Association. Fletcher High School is named after him.' *A Concise Encyclopedia of Zimbabwe* (Gweru: Mambo Press, 1988).

get on with the job; the only thing to do, as none of us had any money, was to offer our claims to a mining company, which we did, on a free option.

The company concerned duly acquired an Exclusive Prospecting Order on the area, and sent in a team to investigate. It did not take them long to confirm the radioactive reefs and to have samples analysed. Then came the let-down: the radioactivity was genuine, but it was entirely due to thorium, and there was no trace of uranium. No one wants thorium, so the claims were abandoned and the syndicate dissolved. None of us is richer and no mine was established to spoil the peace and quiet of the Manjolo escarpment. But if anyone ever does want thorium, and the assays showed up to 16% thorium oxide, we can tell them where to find it.

The Manjolo episode was not my only brush with radioactive prospecting. Once reasonably portable gamma-ray detection apparatus became available, the game was on for our prospecting fraternity, and one of the keenest was one-armed Laurie Byerley. As a child in Hartley, an accidental discharge of a firearm had cost him his left arm just above the elbow, but he had overcome the handicap quite remarkably.

He had been a small-scale miner and prospector all his life, and had wandered over vast areas of very wild and unknown country. He quickly caught the uranium bug, and began bringing samples in to my office at frequent intervals for testing in the Museum. To begin with, my own equipment did not include any gamma-ray detector, but uranium can be detected in a very small quantities by a simple combination of chemistry and ultraviolet light. The test consists of powdering a small quantity of sample and fusing it into a sodium carbonate bead on a loop of platinum wire. The bead is then exposed in the dark to shortwave ultraviolet light, the kind used in scheelite prospecting.

As could be expected, most of Laurie's samples were entirely negative, but on one occasion they gave positive results. Like most prospectors, he was cagey about where he had been, but eventually he was prepared to talk – after he had put up his pegs and secured his claims. The locality was just outside Beitbridge, in a borrow pit dug by the Roads Department for gravel. By the time Laurie allowed me to visit his claim, I had a Geiger counter, and together we gridded the claims and recorded the results on a plan.

It was not encouraging: nowhere was the radioactivity very strong, but it was certainly there, and it seemed to be associated with a black titanium mineral, sphene, and another black mineral, allanite, which often contains thorium in small amounts. The country rocks were gneisses of the Limpopo mobile belt, and the radioactivity seemed to be restricted to small pegmatite veins in the gneisses. It did not look promising as a commercial proposition, but we held out hope that it would be possible to 'date' the rocks by one of the radioactive isotope ageing techniques. Both forecasts were correct. Laurie's claims did not make him a fortune, but an age determination was eventually done on these rocks.[2] It has been superseded, of course, by more modern methods, as the importance of the Limpopo mobile belt has been realized, but it was a start.

In the early 1950s, the term 'Energy Crisis' had not been invented and oil was still cheap. The rate of discovery was keeping pace with the rate of extraction, but far-sighted scientists could see, even then, that this could not go on forever. There would come a time when these finite oil reserves would approach exhaustion, and a new major source for the world's expanding technology and population would be needed. The peaceful use of atomic energy was an obvious solution, but to supply the projected host of atomic reactors, vast new reserves of uranium would have to be discovered. The prospecting could be done either from the ground or the air, but relatively cheap and portable detectors had to be evolved. By the mid-1950s, such instruments were, thanks to solid-state devices, being made, so the search was on.

The Geological Surveys for British Overseas Territories and the Atomic Energy Research Establishment at Harwell were in the forefront of the programme to encourage uranium prospecting and set up an office in Salisbury to foster the search in Rhodesia, which was considered to be worth a major effort but was unknown in terms of uranium potential. The office was under the direction of Ken Branscombe, a geologist trained in uranium prospecting, and the instrumentation was dealt with by Bill Munnoch, from Harwell.[3] Bill had already designed portable transistorized Geiger-Müller

[2] 'Born in 1897 in Southern Rhodesia [Laurie Byerley] lived and helped his father on the latter's gold mine near Bulawayo. At the age of eight he had his left arm shot off at the elbow by a crazed mine worker. ... He died in 1984 at the age of 87, while mining for gold in the Filabusi district.' Chris Pargeter, 'Laurie Byerley', *Magnum* (September 2015), 40(9): 58, 69.

[3] The Directorate of Colonial Surveys was established in 1946 and was later renamed Directorate of Overseas Surveys. See Martin Hotine, 'Overseas surveys', *Journal of the Royal Society of Arts* (1958), 106(5026): 785–98; F. Dixey, *Colonial Geological Surveys 1947–1956: A Review of Progress during the Past Ten Years* (London: HMSO, Colonial Geology and Mineral Resources, Bulletin Supplement No. 2, 1957); 'The Atomic Energy Research Establishment, Harwell', *Nature* (1948), 162: 317–20. Ken Branscombe was seconded by the UK Atomic Energy Authority in December 1955 to set up their Salisbury office. W.C.T. Munnoch was the resident electronic engineer in Southern Rhodesia throughout the Authority's field programme.

counters and scintillometers, suitable for hand-held traverses on foot, Land-Rover traverses with continuous recording ('cargo'), and airborne work. The office launched a major prospecting drive which was to last for several years. A few oddities of radioactivity were already known, such as Laurie Byerley's Beitbridge occurrence, and occasional refractory uranium minerals from pegmatite belts such as Bikita.

It was natural that I should get in touch with this new activity, and I lost no time in so doing, and getting basic instruction in this relatively new art. Branscombe and his staff were generous with their time, and, even more important, fitted me out with ample equipment for both laboratory and field operations. In a sense, the Geology department in the Museum in Bulawayo became a sub-office of the Atomic Energy Authority in Salisbury. This was natural, as it had always been an extension, albeit unofficial, of the Geological Survey, who had no office in the city at that time. Wherever I went by Land-Rover, I traversed the tracks, particularly in the relatively unknown regions of the middle Zambezi Valley, which I regarded as my own preserves.

After their initial period of orientation in Rhodesia, Branscombe decided to mount an ambitious airborne operation to survey a large tract, which could be roughly described as the Lomagundi region. This was to be flown on straight-line flight paths at a height of 500 feet, using a twin-engine aircraft, equipped with a vertical camera for control of the accuracy of the traverses, a scintillometer with continuous paper recorder, and a crew of four. The cost was estimated at around £7.00 per line mile, and the results would be plotted up in the Atomic Energy offices under the greatest secrecy.

Maps would then be prepared on which any anomalies found would be clearly shown, and, at a given time on a specified date, these maps would be published so that all interested parties, mining companies and private individuals alike, would start equal. In a way, this was the modern equivalent of the old 'diamond rushes' in South Africa, at which prospectors lined up and raced on foot to peg claims on ground opened to prospecting. I missed the start of this 'uranium rush', but I gather that all sorts of elaborate preparations had been made – including prospectors on motorcycles armed with licences, who were dispatched by their employers to rush off and peg.

The maps certainly disclosed some anomalies, and some of these were pegged and investigated, but no uranium mines resulted. To misquote an enthusiastic early Land Development Officer, the virgin fields of Rhodesia did not appear to be pregnant with uranium possibilities. The rocks chosen for Branscombe's survey were rather similar to those of the Copperbelt of Zambia, which contains some traces of refractory uranium minerals associated with the copper ores.

This was a setback for uranium prospecting in this country, but the Atomic

Energy Authority's offer to buy any tonnage of uranium ore going over a certain minimum grade still stood and was still an incentive for prospectors. A small vein of pitchblende, the primary uranium ore mineral, was discovered near Umtali and did, in fact, declare a small production before it apparently petered out. This material was brought to Salisbury and dumped.[4] It is not known, to me at any rate, how many tons were carted to this ore dump. It never moved beyond this point, but it was still there in 1961, when I was building up my new Department of Geology's teaching collection. I managed to locate it and samples are preserved in the Department's collection. It would be interesting to know what has happened to the rest, and to the site itself. Imagine the excitement, and dismay, if it is rediscovered some time in the future, in the midst of Salisbury's industrial area. It could be thought to be a new discovery in an area where pegging is prohibited, or to be a potentially dangerous source of radiation!

The samples in the Geology department, University of Zimbabwe, in 2021

Ken Branscombe was not unduly deterred by the relative lack of success of his major airborne survey. The anomalies had certainly shown up, and given him faith in the method. Equipment in those days merely detected total radiation, which could have come from any radioactive element. These, in nature, are uranium, thorium and potassium. Modern instruments can distinguish between the radiation emitted by these elements, which is a great step forward. Potassium is a weak emitter but is an essential element in all our granitic rocks, which have, in consequence, a higher background radiation than other rocks, particularly than sediments. Anomalies are to some extent masked by it, especially if they are not very strong. Modern equipment is much more sensitive and discriminating.

Ken decided to mount one more airborne survey but to modify the technique. We discussed the plans at length. His idea was to use a small, four-seater, single-engine aircraft with only a pilot and a navigator, who would be

[4] The source of the bulk sample brought to Salisbury was the Cripmore claims located in the Zimunya area south of Umtali (Mutare): R. L. A. Watson, *The Geology of the Country South of Umtali* (Salisbury: Southern Rhodesia Geological Survey, Short Report 38, 1962).

a geologist and, in addition to navigating, would handle the instruments in the air and plot up all the results. He chose me for this job, and we hoped to get the cost of the survey down to ten shillings per line mile without any appreciable loss of data and accuracy.

After a lot of discussion, it was decided to fly an area of Karoo sedimentary rocks between the Deka and Masumo rivers, which included Kamativi tin mine located in a narrow belt of much older rocks within the area. The northern boundary was to be the Zambezi river and the southern one the Victoria Falls main road. The height was to be 250 feet and line spacing the standard one mile. The area would need 3,000 line miles of primary coverage, with some leeway for extra miles if anything exciting showed up.

This form of excitement would be additional to that generated by the flying itself. Flying at 250 feet in a piece of country with nearly 2,000 feet of relief in the month of October, when thermal turbulence would be severe, would be exciting enough, but in addition it was 1958, the year when bush-clearing and -burning in the Kariba basin would ensure minimum visibility. The prospect was intriguing. I knew the country fairly well on the ground, having passed through it often with my 'cargo'-equipped Land-Rover, but I rated the chances of finding much in the way of anomalies as pretty poor, in spite of the thorium veins at Manjolo, which were younger than the sediments we were to investigate. I was seconded from the Museum for a month to do this special job. As it was largely experimental, no publicity was given to it, and no maps were to be issued. There would be no uranium rush this time.

The chartered aircraft turned out to be a Piper Tri-Pacer, single-engined, with a fixed tricycle undercarriage, very slow and manoeuvrable.[5] Our base was to be Victoria Falls, using the strip which is the base for tourist sight-seeing trips. Ken and I flew up with Andy Rybicki, the pilot. Andy was an ex-Fleet Air Arm turned civil pilot, and had done over 600 hours on single-engined jet fighters based on carriers.[6] He was obviously keen on this chance to do something unusual in the air, instead of just flying straight and level from A to B. Bill Munnoch went by road with all the equipment.

My terms of secondment were quite carefully drawn up. The Museum would keep my October salary. I would be given the rank and pay of an Assistant Director of Overseas Geological Surveys; for living at the Victoria

[5] 'Piper Tri-Pacer', *Aircraft Owners and Pilots Association*, <https://www.aopa.org/go-fly/aircraft-and-ownership/aircraft-fact-sheets/piper-tri-pacer>.

[6] The Fleet Air Arm is one of the five fighting arms of the Royal Navy. 'Fleet Air Arm', *Royal Navy*, <https://www.royalnavy.mod.uk/our-organisation/the-fighting-arms/fleet-air-arm>. Andy Rybicki is listed as one of the pilots who flew for Wenela Air Services: Ken Fuller, *Africair Ndege ya Wenela* (Germiston: South African Airways Museum Society, Digital Books, 1998), <http://www.saamuseum.co.za/digital-books/100.html>.

A 1955 Piper PA-22-150 Tri-Pacer

Photo licensed under Creative Commons

Falls, there was a climatic allowance, and for flying about six hours daily there was danger pay. Altogether, this added up to more than double my Museum salary, and, since I was an Inspector of Monuments, Bill and I would live rent-free in a Monuments Commission chalet in their rest camp at the Falls.

October is hot at Victoria Falls, and we intended to take off at 6.00 a.m. daily (except Saturday and Sunday of course), fly until about 11.00 a.m., and land before it got too rough. It would be too hot to plot the day's work before the sun went down, so the afternoon could be spent, profitably, fishing for supper. Consequently, our grocery bill looked odd in some respects, because the one shop supplied not only all our groceries but fishing worms as well, and the item '100 worms' appeared regularly among bread, butter, bacon, tea, etc. Our fishing was singularly unsuccessful, but the time by the river was considered well spent.

Bill installed the instruments; tests showed that everything was working, and Ken returned to Salisbury to await results. Since there were no reliable maps of the area, navigation was to be done from vertical air photographs. A mosaic had been laid down and stuck together into a huge aerial map, many feet square, covering the whole area. On this, flight lines had been drawn in bold, black ink, one inch apart. The whole was then cut into strips about one foot six inches wide, which were rolled in such a way that they could be progressively opened during flight, no easy matter in the rather cramped space available for the geologist-navigator-instrument-basher in the right-hand front seat. The instruments were behind us with the recorder and other switches reasonably conveniently placed. We did some trial runs, and things seemed to work tolerably well as planned. So we began our survey.

The Victoria Falls at 6.00 a.m. are a lovely sight, and we were constantly asked to photograph them for other people and did so – many times both at

6.00 a.m. and on returning at about 11.00 a.m. My camera was in Bulawayo, but there must be a number of tourists with rather special shots that we took on those morning flights.

Each flight began with an instrument check over a chosen piece of the main railway line, at a strict 250 feet above Jafuta siding, and the test was repeated on our return. The siding staff must have wondered what went on during these twice-daily regular flyovers. To be certain that all Bill Munnoch's black boxes were working properly, we had to do the test flights more than once on some days, but his instruments never let us down.

As expected, Andy turned out to be the ideal pilot for such work: he was meticulous and safe. Under such conditions it was only to be expected that we had our moments. Trying to keep at 250 feet in very corrugated country meant clipping the edges of escarpments, diving though valleys, and climbing steeply up the next escarpment. Diving over one scarp, the starboard wing dipped suddenly, nearly into a tree, but when I reproached Andy, he merely shrugged it off as a freak air current over the lip of the scarp.

I could see his instrument panel clearly from my seat. After turning at the main road near St Mary's Mission and heading north on the next line, there was a particularly high and steep scarp to get over. At our turning point we were well below the crest of this scarp and I was not surprised to see full throttle and the aircraft in a steep climbing attitude, but as the scarp got nearer the vertical speed indicator read 500 feet per minute *down*. We certainly were not going to make it. It was not surprising that Andy did a full power stall turn almost off the face of the cliff, and merely said, 'Let's try again a bit higher this time.' So we did, and got over at the second attempt.[7]

Navigating with my rolls of air photos was not easy. Even at about 100 m.p.h. at a height of 250 feet, features come and go remarkably fast. The carefully chosen kinks in streams or other salient points could easily be missed or misidentified, but we managed, and nightly plots showed very close agreement between projected flight lines and those we actually flew. We were allowed a maximum error of 300 yards either side of the line, and we only re-flew three lines. Turning points were closely identified, but the haze was so thick that it was frequently impossible to see the last turning point when we flew over the next on the fresh line.

The results were plotted each night in the comparative cool after sunset and with the ever-present roar of the Falls even louder when the sounds of the day had died down. Little anomalies were common, but for some days nothing startling turned up. However, one evening the chart showed not

[7] The stall turn is an evasive manoeuvre dating back to World War I, when it was used as a method to get behind the enemy or to escape pursuit. 'Stall Turn', *How To RC Plane*, <http://www.howtorcplane.com/stall-turn>.

just a little peak but a section where the needle had shot up to the top of its travel, stayed there for some distance, and then gone down to the well-known average background level. This was as unexpected as it was exciting, and my main worry was whether my system of linking the recorder to the flight path on the air photo strips was exact enough to re-locate this anomaly next day.

Six a.m. saw Andy and myself airborne and heading for the spot where we thought the anomaly should be, a hill called Katete on the east side of the gorge by which the Gwai river joins the Zambezi and a few miles upstream from the confluence. Its slopes were well wooded, but the summit was an open grassy area, yellow in the October drought and heat. We turned north so that we could do an instrument run along the line on which the massive anomaly had appeared on the traverse the day before. If the anomaly did not appear exactly where I had plotted it, surely some closely spaced runs would be bound to locate it - unless, unthinkable thought, it was just an instrumental gremlin.

We flew steadily north towards the yellow summit of Katete; the level on the recorder, which I had turned in my seat to watch, hovered around the average background count. We crossed the deep gorge of the Bidida river, a right-bank tributary of the Gwai, and Andy began to climb to keep at approximately 250 feet over the Katete itself. Without warning, the needle shot steeply up to its limit, stayed there over the crest, and then, as suddenly, dropped to the normal background. We were elated, and did several closely spaced runs north-south and south-north over different parts of Katete until we had fairly closely defined the extent of the radioactive area, which was more or less circular and confined to Katete hill itself. It did not even cross the Gwai or the Bidida rivers. The rest of the day's flight was routine survey on the predetermined lines. Nothing showed up, and for the rest of the survey period, nothing; but we had found what might prove to be an exciting prospect.

The last day's flying came along. We had no more prospect of excitement, not even of being severely buffeted by turbulence, which had twice got so bad that Andy had feared for the airframe, abandoned the flight, and climbed as fast as possible out of the worst of the rough air before heading for Victoria Falls.

On our last final approach, Bill Munnoch wandered out over the strip to wave to us. Andy put on full power and put Bill flat on his face on the tarmac, after which he did a full-blooded RAF 'beat up' of the landing ground for about fifteen minutes - to the amazement of the next group of tourists waiting to take off for their flip round the Falls.[8]

[8] To 'beat up' is to make a high-speed, low-level pass in an aircraft.

Katete was the only worthwhile anomaly discovered, but the whole exercise had proved Ken Branscombe's contention that such a survey could be done relatively cheaply and still provide useful data. The ground follow-up to Katete was still to come, but the rains set in before it could be mounted, and it was not possible to start before the following March. I set off with a young field assistant and two Land-Rovers to make our main base at Kamativi tin mine. Katete is only twenty miles roughly to the north-west of the mine, but there was no known track and we would have to make our own. It took us five days of chopping and digging to reach the south bank of the Bidida gorge, where we had to stop and make camp. It was a hard slog each day into the area of the anomaly, which was the summit area of Katete hill itself.

On the ground, the anomaly was very disappointing, with only a few moderately warm spots, but at 250 feet the scintillometer had picked up a cone of radiation that was the summation of all the small anomalies. What we had found was a previously unknown and unsuspected volcanic plug of a rare kind of rock called carbonatite. Subsequent work by a mining company confirmed our findings, and it later became the subject of research by one of my own graduate students and part of his doctoral thesis.[9] By the time he worked there, it was easily accessible from a new road, with a high-level bridge over the Gwai and tracks leading round the hill, but the hill itself was still very rough ground.

This old volcanic plug is, geologically speaking, quite young, being only seventy million years of age. It is the central manifestation of a rather weak episode of mineralization, probably contemporary with my thorium reefs at Manjolo and some weak uranium mineralization near Wankie. But, as an economic proposition, Katete has no interest. There is some uranium, some thorium, a little manganese, and some reasonable concentrations of lanthanum (a 'rare earth').[10] There is no phosphate of any consequence, as there is at two other carbonatites in the Sabi valley - the famous Dorowa and Shawa ring complexes, found by F.P. Mennell and now the source of our

[9] 'Professor G. Bond established the presence of the Katete carbonatite in 1958 when investigating an airborne radiometric survey anomaly for the United Kingdom Atomic Energy Commission. Following photogeological studies of the Zambezi Valley, the Anglo American Corporation pegged the complex for base metals in the late 1960s. ... [C.A. Lee, 'Petrochemistry of the Katete, Shawa, and Dorowa Carbonatites, Rhodesia' (University of London, unpublished Ph.D. thesis, 1973)] later investigated the petrochemistry of the carbonatite during the tenure of an Anglo American research grant.' B. Barber, 'Phosphate resources of carbonatites in Zimbabwe', *Fertilizer Research* (1991), 30: 247–78.

[10] 'The rare earth elements (REE) are a set of seventeen metallic elements. These include the fifteen lanthanides on the periodic table plus scandium and yttrium. Rare earth elements are an essential part of many high-tech devices.' 'What are rare earth elements, and why are they important?' *American Geosciences Institute*, <https://www.americangeosciences.org/critical-issues/faq/what-are-rare-earth-elements-and-why-are-they-important>.

phosphate fertilizers. This survey and the discovery of Katete was just another interesting geological exercise, without any economic pay off.

By 1959, the Atomic Energy Authority had more or less written off Rhodesia as a potential uranium province and turned its attention to assessing our resources of beryl, an ore of the very light metal, beryllium.[11] For this it was again possible to use a detecting device, but beryllium itself is not radioactive. It can be made to emit radiation if it is bombarded by a strong source of radiation, and a device called a berylometer had been evolved for this purpose. It was portable, but the bombarding source had to be applied within inches of the material being scanned. The source itself was so strong that it was highly dangerous and had to be handled from a distance of about six feet. So it was a cumbersome piece of apparatus and needed expert handling. I never used it.

My own 'cargo'-traversing continued whenever I went into remote areas, though little of interest came to light, but the apparatus did get put, briefly, to another use. At that time, a great deal of money, much of it in notes, came by post from South Africa for lottery tickets, and it was suspected that some of it was being stolen en route.[12] The most obvious possibility was in the Bulawayo sorting office, and I was consulted about possible scientific ways of detecting such thefts.

The simplest method seemed to be a radioactive trap, using a prospecting scintillometer and an alarm circuit. At dead of night, we experimented with this gear hidden in a passage leading to the sorting office. Using quite weak sources, it was impossible to pass the trap without triggering the alarm bell and a red light in the superintendent's office. The next step would have been to get the Harwell boys to impregnate paper with a radioactive solution, probably cobalt nitrate, and make up a dummy mailbag, using genuine envelopes with stamps and South African franking. However, since this dummy mail would be radioactive, it was thought wise to consult 'higher authority' about the possible health hazard. The answer was absolutely negative, so the trap was never used. That something strange was afoot must have somehow leaked out, and thefts virtually ceased forthwith!

[11] K.C. Branscombe, 'Beryl in Southern Rhodesia', *Chamber of Mines Journal* (1962), 4(7): 32–7; K.J. Ackermann, K.C. Branscombe, J.R. Hawkes and A.J.L. Tidy, 'The geology of some beryl pegmatites in Southern Rhodesia', *Transactions of the Geological Society of South Africa* (1968), 69: 1–38.

[12] 'The origin of the state lottery sweep in Southern Rhodesia', *Central African Journal of Medicine* (1959), 5(10), 563–4. There are photographs of the lottery tickets here: 'Southern Rhodesia State Lotteries: Lottery Tickets, 1950–59, *Rhodesian Study Circle*, <http://www.rhodesianstudycircle.org.uk/southern-rhodesia-state-lotteries-lottery-tickets-1950-1959>.

Interest in uranium prospecting dwindled away, but in the last year or two of the 1970s there were persistent reports in the papers of uranium discoveries on the Zambian side of Lake Kariba. In the first half of 1980, just after Independence and the lifting of sanctions, this spread to our side of the Lake, and a German geologist in charge of the Zambian operation came to see me about the geology on our side.[13] He was understandably vague about his operations in Zambia.

Modern airborne detection apparatus is much more sensitive than the machines available in 1958, and can discriminate between gamma radiation from different elements as well as logging the total count. Perhaps, after all, something may be found, but hopefully not in the beautiful game areas on our side of the river. Mana Pools is already under threat from a proposed new dam in the Mupata Gorge.[14] Surely the world is not so uranium-hungry that a huge, low-grade, opencast operation needs to be added to the threat. My guess is that they will find nothing, and I hope and pray that I am right. Only time will tell.[15]

[13] Saarberg-Interplan Uran GmbH were granted exclusive prospecting rights in 1981 (EPO 578) over a massive area stretching from Bumbusi west of Hwange to Mukumbura north of Mount Darwin: P.M. Oesterlen, *Exclusive Prospecting Orders, No. 501–650* (Harare: Zimbabwe Geological Survey, Bulletin 102, 1998), 46-9. A number of radiometric anomalies were realized, and 90% of the original area was relinquished. However, the company focused only on the Kanyemba Claims close to the point where the Angwa river leaves Zimbabwe. This uranium mineralization with vanadium accumulation, small by world standards, passes south-east from Zimbabwe, across the right-angled boundary with Mozambique and back into Zimbabwe. (Tim Broderick)

[14] 'The scheme is currently not being developed but has been reserved as one of potential sites for future development.' 'Mupata Gorge Hydroelectric Scheme', *Zambezi River Authority*, <http://www.zambezira.org/hydro-electric-schemes/mupata-gorge-hydroelectric-scheme>.

[15] 'Dorowa Minerals undertook extensive sampling and testing of its resources and found low-grade niobium, thorium and uranium together with some rare earths, but all at too low a grade to be economic.' J. Hollaway, 'Minor Minerals and Metals', in Martin Prendergast and John Hollaway (eds.), *Mining in Zimbabwe: From the 6th to the 21st Centuries* (Harare: Chamber of Mines of Zimbabwe, 2019), 552.

Making a road to the camp from Matope, Malawi

The same road a few weeks after use

Photos courtesy of Jeremy Burden

– 14 –

Zambia, Malawi and Crocs

My work in the middle Zambezi Valley had settled down over the years to two major objectives. One was to produce a reasonable geological map of the country bounded on the west by the Deka river, on the north by the Zambezi, on the east by the Sanyati, and on the south by the line roughly east-west through Gokwe. This area of about 15,000 square miles included EPO 14, on which Mac and I had mapped together in 1951, and my subsequent work was an extension of that joint effort.

A second major objective was to extend what Desmond Clark and I had started in the Valley in 1949 on the Stone Age archaeology and its associated Quaternary river terraces. There was increasing urgency about this because it became clear that Kariba dam was going to be built. The lake would flood the entire area in which the Stone Age deposits occurred and they would be lost forever.

Desmond Clark continued his pioneering archaeological work in Northern Rhodesia, and on one of his journeys into the extreme north of the territory he paid a visit to Kalambo Falls. He claims that his visit was pure sightseeing but that while he was there he fell into a concealed donga. The floor was covered in coarse gravel eroded from the walls. He searched this gravel and found a few beautifully made hand axes and cleavers belonging to the Old Stone Age. The gravel from which they had eroded proved to be old swamp or lake deposit, laid down in the valley of the Kalambo river, which today meanders quietly through the little basin surrounded by hills above the lip of the Falls. These Falls are spectacular not for the volume of water going over it, for this is quite small, but for the height of the single drop below. This is over 700 feet into a deep ravine between high quartzite cliffs on which Marabou Storks nest in complete safety. This magnificent gorge is cut into the Rift Valley fault scarp, which is the eastern wall of Lake Tanganyika.

The flat floor of this little basin was the scene of extensive excavation in 1955 and 1956, and in the second year I was invited to join Desmond's party of archaeologists to provide geological back-up. The Kalambo site had already proved to be stratified and to extend in time from the Early Stone Age to the Iron Age Bantu occupation. There is even evidence of the site of a

German camp from World War I. It was indeed of major significance. So were Desmond's excavations.

I calculated that, from his three major pits dug in 1956, more than 1,500 tons of material had been removed. This would be no great tonnage from a prospecting operation, but for Stone Age archaeology it was immense. It had to be taken off in layers no more than a few inches thick, sieved, and finds recorded on a grid. When a gravel layer was reached, it was exposed, leaving every stone in place. A complete plan was then prepared showing the exact position of each. These gravels were in old lake beds or swamp beds, it matters not which, on which old Stone Age parties had camped at seasons of low water. Their tools were scattered liberally on these floors, and even the foundation stones of rough semi-circular shelters were still in place. Even more surprisingly, and because of the acid nature of the deposits, logs of wood, twigs and leaves occurred on these floors; and, which made the site almost unique, there were crude wooden tools, such as roughly pointed sticks, probably used as spears. The next flood sealed the deposit undisturbed.

There are many other stratified living floors of the same age in other parts of Africa, such as Oldoway Gorge in Tanzania and Olorgesailie in Kenya, excavated by the Leakeys,[1] but they are devoid of wood, being in alkaline conditions. Wood and bone need different pH conditions. Wood is preserved in acid situations that destroy bone, and vice versa. Kalambo, therefore, had a very special significance, being rich in wood and pollen grains but completely devoid of fossil bones and teeth. Analysis of the pollens from successive layers allows inferences to be made of changes in regional vegetation, controlled by changes of climate. Altogether, Kalambo was a most important discovery, and its geology was complex and hard to understand. At least it has been mapped and recorded. Whether my tentative interpretations will stand the test of time remains to be seen.

My work at Kalambo was in October, and it was hot. Desmond suggested a Sunday outing down to Lake Tanganyika at Mpulungu. On the way we would call on the Miss Gamwells, who farmed near Abercorn, hoping to get to their homestead by about morning teatime. There were two Gamwell sisters, who had remained spinsters all their lives. They had emigrated from Britain to the Abercorn farming area in about 1925, and by 1956 must have been around

[1] The Oldoway Gorge is now known as the Olduvai Gorge. Robert Curley, 'Louis Leakey: Kenyan archaeologist and anthropologist', *Britannica*, <https://www.britannica.com/biography/Louis-Leakey>. 'Artifacts have been excavated at Olorgesailie since 1942. Olorgesailie is unusual because of the large number of hand axes found there, along with other bifacial tools (flaked on two sides to create an edge).' 'Olorgesailie, Kenya', *Smithsonian Institution*, <https://humanorigins.si.edu/research/east-african-research-projects/olorgesailie-kenya>.

The Gamwell sisters talking to
Senior Chief Tafuna of the Lungu
Photo courtesy of Colin Carlin, www.abercornucopia.com

sixty.[2] Desmond had known them for years and I knew a little of what to expect.

When they had selected their farming area from rough maps, it had been plain that the road system of the Northern Province was rudimentary. They had all learned to fly light aircraft before leaving Britain but had never owned an aeroplane. However, their enterprising attitude to life had paid off in other directions. For instance, on their large farm they had discovered a new species of *Aeollanthus* (*A. gamwelliae*, named in their honour) which contained a valuable essential oil.[3] They distilled and marketed it, adding a new export product to the territory's meagre list of exportable commodities. We were duly invited to tea.

The two sisters were rather alike. Both were pear shaped and invariably dressed in khaki longs and bush shirts. Their leather belts, suspended mysteriously round the maximum circumference of the pear, carried bush knives. Lanyards ended in police whistles that were used, by a complicated code of signals, for giving instructions to the staff. The whole tea party was conducted in the manner befitting an English country manor house. Both of them had volunteered for service on the outbreak of World War II, and one had risen to command the First Aid Nursing Yeomanry (FANY), in which she served with great distinction.[4] They impressed me greatly.

[2] The story of the Gamwell sisters is told online by Ian Singer, *Marian and Hope Gamwell: Two Extraordinary Lives* (version 02/10/17), according to which they 'found themselves in Abercorn, not far from the southern end of Lake Tanganyika on the 19th December [1928]. They meant to stay for just a few days to replenish their stores but were persuaded to stay at the Tanganyika Victoria Memorial Institute for Christmas and then to have a look around the area with a view to farming there as it was envisaged that the area was going to become a famous coffee growing centre. They dug eight foot test pits all over the district and chose the area which they were to make their home for the next thirty-five years. The homestead was a collection of high-thatched buildings snuggled together to make a home with accommodation for visitors.' <http://www.nrzam.org.uk/Site%20Resources/Abercorn_Documents/Gamwell%20Sisters%20v02-10-17.pdf>.

[3] Although originally described as *Aeollanthus gamwelliae*, this plant is now known as *Aeollanthus myrianthus* Baker subsp. *gamwelliae* (G. Taylor) Ryding. See 'Plants of the World Online', *Royal Botanic Gardens, Kew*, <http://powo.science.kew.org/taxon/20008017-1>. *Aeollanthus* is a genus in the mint family, *Lamiaceae*, which includes many culinary and medicinal herbs.

[4] FANY is still active as the Princess Royal's Volunteer Corps. 'The First Aid Nursing Yeomanry (Princess Royal's Volunteer Corps), founded in 1907, is an all-female voluntary organisation which deploys multi-faceted rapid response teams to support civil and military authorities in times of crisis.' 'Our History', *First Aid Nursing Yeomanry*, <https://www.fany.org.uk/history>.

We lunched at Mpulungu, and it was even hotter than in the Kalambo valley. In the afternoon we walked a mile or two along the shore to see prehistoric sites that Desmond had excavated some years before, and the heat was intense. The shore was on cobbles, the water was clear and inviting, and we decided on a quick bathe, ignoring crocodiles. We stripped off and went in. We had only been in a few minutes when we heard a motorboat, and round the point came a small dinghy with an outboard engine, containing both Miss Gamwells. We hastily submerged up to the neck and waved and smiled until they were out of sight. It seemed to take an age.

I have always been afraid of crocodiles, and this fear had been reinforced earlier that very year. In July, I had camped right beside the Zambezi at Binga's village, near the present administrative centre of the same name. It was a good campsite, under the shade of some really big musikili trees (*Trichilia emetica*), and is now under the water of Lake Kariba. These trees seem only to grow close to water. They have massive trunks, up to three feet in diameter, but they branch out quickly a few feet from the ground. They are evergreens with a very dark leaf, and a clearing under a clump of such trees makes an ideal site for a camp.

Those at Binga's were among the finest I know. The riverbank was high and steep, and the top was a flat plain about twenty feet above water level with the trees at the edge. In July, the evenings were quite cool even at this low altitude, about 1,500 feet above sea level, and, as there were few mosquitoes, it was pleasant after dinner to sit by a good log fire on the edge of the bank, with the pressure lamp hanging in the trees.

On this trip, instead of being alone, I had with me a young student, Neil Black. He had never been in the Valley before and enjoyed the newness of everything we did, even if it was hot and uncomfortable. Before we had reached the river, I had warned him most solemnly of the dangers of crocodiles, but I was not sure how seriously he had taken it. They are ugly brutes and have sly habits. They are generally shy, and often a stretch of river can look harmless enough yet be full of them. It was possible to live beside the Zambezi for days without seeing a sign of a croc. Then, one day, one would approach more quietly than usual, in a place where there was a sandbank some way off the shore, and there might be a dozen or more lying motionless in the sun. They lay so still that they looked dead. With binoculars, they still looked dead, but things that at first sight looked like logs left by the floods could also be seen to be crocs. In fact, the beautiful and harmless looking river could then be appreciated in its proper light.

They were worth watching. Although crocs will apparently eat anything they can catch – fish, animal, bird or human – there was one bird that was safe. It was about the size and shape of an English starling. Where there was a sandbank on which crocs were sunbathing, there were sure to be some of these little birds. Presently a croc would yawn. Instead of closing his jaws, however, he kept them open, and soon one of these privileged birds would hop into his mouth, or on to his snout, and start picking his teeth. Whether a croc was ever ungrateful enough to pay for this service by eating his dentist, I do not know, but it was a common enough sight to see these birds feeding on what they could find in those jaws. Ornithologists doubt this behaviour.[5] 'Dead' crocs on a sandbank came to life remarkably quickly when a shot was fired, and the speed with which they disappeared into the water left no doubt about their agility.

To return now to the camp at Binga's with Neil. We had been there together for about a fortnight and had enjoyed our evenings on the bank. During this time we had seen no sign of Europeans but, at about 9.00 p.m. one night, I thought I heard an engine, faintly, in the distance. Looking inland we could see no lights of Land-Rovers. A visitor by day would have been an event, but at night it was unthinkable. One just didn't travel on those tracks by night except in an emergency. Aeroplanes sometimes passed over, but the sound had died, and I thought I must have been mistaken. A few minutes later, we both thought we heard it again, but again it died. The third time, however, there was no mistake – it was an engine. But there were still no lights from the direction in which the track reached our camp. Of course, neither of us thought of the river, but soon we realized that the sound came from the Zambezi.

Now what was a motorboat doing on the middle Zambezi in the dark in the middle of the evening? We stood watching the river in the starlight, wondering. The noise came steadily upstream, and then a light, sweeping backwards and forwards along the bank on our side. The boat drew level with our camp, the motor was cut off, and the bow ran on to the bank. We ran down the path to the edge of the river as four men climbed ashore; two were European and two were Africans, and we realized that they were croc hunters. After introductions all round, we asked them up to our camp for a drink. They accepted and we climbed the bank to the fire, and the bright circle of light from the pressure lamp.

[5] An article by D.A Ewbank, 'Further records of birds associating with crocodiles', *Honeyguide* (2020) 66(2): 88, is followed by this note from the Editor, Brian Marshall: 'Associating with crocodiles is not without risk, even for a bird. Dinets *et al.* (2013, *Ethology, Ecology and Evolution* 27: 74–78) described instances where mugger crocodiles (*C. palustris*) in India and alligators (*Alligator mississippiensis*) in Florida waited near egret colonies with sticks balanced on their noses, and seized the birds when they tried to collect the sticks for their nests.'

They were all dressed more like Arctic explorers than people living in the tropics. One was even wearing a balaclava helmet. They had thick woolly sweaters with high necks, and wore thick trousers. Neil and I wore shirts, shorts and shoes without socks. We drank whisky out of enamel mugs while Million, my cook, stirred up the kitchen fire to make tea. Over drinks, our visitors talked of crocs and croc-hunting, and I became increasingly curious to see this business at first hand. After all, here was a golden opportunity that might never come again. Young Neil was obviously equally keen.

Cecil Manning and Doug Herder talked freely. Cecil had been hunting for three seasons and therefore had a good deal of experience. He was the natural leader, since Doug had been with him only half a season by then. He was of medium height, lean and hard-looking, dark haired with a broad, dark moustache. His eyes were very dark and very steady. Altogether, he looked the kind of man that I would choose to be with in an awkward moment. I was glad, because from what I had read of croc-hunting it seemed that such moments might not be uncommon. I was even more determined to go with them that night. Our luck was in, for Cecil soon brought up the matter and asked us both to go along, adding that we should need all our warmest clothes and would be back at about 2.00 or 3.00a.m. We dressed up in all we could find and followed Cecil and Doug down to the boat. There was no moon, and we pushed off the bank into the darkness of the Zambezi.

I suppose all croc hunters have their own ideas about boats and other equipment. This boat was home-made. It was sixteen feet long and had a beam of six feet six inches. It was flat-bottomed, and the sloping sides were about two feet high. The construction was simple enough: the frame was steel, angle-welded together; the bottom was 16-gauge steel plate, riveted on to the angle iron. Under the bows was an extra steel guard. The sides were made of lighter (20-gauge) steel plate. The flat bottom had substantial duckboards. There were no fixed seats.

The Zambezi had a strong current, probably more than five knots on average in this section, and I was curious to see what motor was used. The main motor in this case was an American outboard mounted on the stern, which developed about thirty horsepower. It was hinged on the transom so that it could be swung up if the water shallowed suddenly. This was very important, as a damaged screw could be serious. I was relieved to find that a spare fifteen-horsepower outboard was clamped on the one side, propeller inboard, in case of failure. This was reassuring. The prospect of drifting helplessly down the Zambezi in pitch darkness was not pleasant.

This boat was remarkably stable. All of us could sit on one side with perfect safety, and that immediately instilled a feeling of confidence. It drew very little water and was easy to manhandle if run on to an unexpected sandbank.

It was brought to the river on a trailer pulled by a vanette, but once there it could carry stores, petrol and camp equipment along the river from one hunting camp to the next. Once we were clear of the bank, Doug, in the stern, started the motor and steered out into mid-stream. It was quite surprising how soon one's eyes became accustomed to the starlight, and it was possible to see quite well.

The organization of this form of hunting was quite a team job. Cecil stood in the bows with the two rifles, a .303, which was used only for crocs over six feet long, and a .22 for the smaller ones. Behind him, one on each side, were the Africans with gaffs. These were large hooks of round steel about three-eighths of an inch in diameter, the hook itself being about six inches in diameter, sharpened, but with no barb. They were mounted in Malacca-cane handles about eight feet long and, as well as helping to haul the victims inboard, they were useful for poling the boat along over shallows when the propeller fouled in the sand.

Apart from the noise of the engine, the whole operation was conducted in complete silence; apparently, human voices disturbed the crocs, but the steady purr of the outboard did not worry them. Hunting had to be done at night, as the croc was too wary to be approached in daylight. The man in the bows was completely in charge. He had a powerful electric light, which threw a narrow beam, mounted on his helmet, rather like a miner's lamp. He swung this around and picked up the reflection from the croc's eyes. At night these beasts tended to lie up in shallows and among dense reed beds, and the red glow from their eyes could be seen for more than a hundred yards on a dark night. The motor man kept his eyes on the hunter in the bows and was guided entirely by hand signals, without any spoken words. If all went well, the procedure was simple enough. The hunter spotted the eyes and guided the boat, fast, towards them. With experience he could tell the approximate size from the eyes.

The run-in was surprisingly fast, and the motor was not cut until the eyes were right against the bow, slightly to one side. By this time, the appropriate rifle had been selected and the actual shooting was done directly over the side, almost vertically downwards at a range of about two feet. The shot was aimed directly between the eyes into the small brain. As the hunter aimed, the gaffs-man on that side was already standing up alert. As soon as the shot was fired there was a violent commotion as the victim tried to dive, but it was generally too late; the gaff was under his front leg and he was bundled over the side into the bottom of the boat, generally still with a good deal of life in him.

The crocs were surprisingly tough, and their reflexes kept them moving long after, by all the rules, they ought to have been stone dead. By far the best weapon for dealing with the crocs in the bottom of the boat was the ordinary geologist's or prospector's pick with a sharp point, which I had taken with me. This could be driven into the skull and soon put an end to the struggles in which the croc used both his tail and his jaws. In the darkness and confined space of the floor of the boat, this was quite exciting.

Cecil and Doug had been hunting downstream from my camp for some nights and had done well. They were just as ignorant of the existence of my camp as I had been of theirs. We were on opposite banks, separated by about two miles of river, including a large bend. In this part of the Zambezi the course of the river was very sinuous.

I knew its shape well from the bank and was immediately made navigator for the night's hunting upstream. We hunted, with fair success, for about two hours, covering some seven miles against the current and bagging fifteen crocs of various sizes, the largest about six feet. Even two- and three-foot crocs were worth getting, but under three feet Cecil varied his technique. Instead of shooting, he merely bent down and grabbed the croc behind the head and, with a smart flick, landed it, fully alive, inside the boat, where it was promptly brained with the pick. The first time, he did this without any warning and I got a considerable surprise. I had a further one when Doug and an African jumped into the river to grapple with a croc that had eluded the gaff. Neil, a keen photographer, took two or three flashlight photographs of this effort. If his pictures came out, they must be almost unique, but I have never seen him since the end of our trip.

After two hours of hunting, the silence was broken by Doug with the suggestion that it was time for tea and sandwiches, that the motor was hot and running short of fuel, that he was cold, and that he could see a nice sandbank just ahead. Without more ado, the boat was grounded and we all got out on to an island, mainly covered with reeds. Thermos flasks and sandwiches appeared, and we ate, drank, smoked and chatted for half an hour. It was a pleasant break: it had been surprisingly cold on the river.

After that, we turned downstream and I headed the boat for the south bank to find the long, narrow creek where the Logola river ran into the Zambezi. It was only about fifty yards wide, and about half a mile was navigable. I had a feeling that it would be a good place for crocs. It was. There must have been dozens of pairs of eyes, but overhanging trees made hunting difficult. In less than an hour we had picked up a dozen more, making nearly thirty in all, and the boat was by now heavily loaded. We turned for home and disembarked at about 2.00a.m. The faithful Million appeared and made more tea. After several seasons in the Valley with me, nothing I did seemed to surprise him. Doug and Cecil finally departed to offload the night's spoils at their own camp. There the crocs were handed over to specially trained skinners, who worked through the day, skinning, salting and packing the skins for market.

I had a further night with Doug and Cecil later in the season, and afterwards I thought about it a great deal. I was not concerned with the rights and wrongs of shooting such large numbers of these reptiles. I did not want to see any animal exterminated, nor was I a friend of the croc in particular. There seemed little fear of complete extinction, as hunting could only be effective under certain river conditions. There were plenty of crocs in streams that did not lend themselves to any kind of hunting that would be economic, and they would survive easily enough. A reduction in their numbers on the Zambezi could not fail to be a good thing from the human point of view. The whole picture has subsequently changed with the introduction of crocodile farms and strict protection on Lake Kariba. The species is in no danger at present.

I think the things that struck me most about hunting as done by Cecil and Doug were the apparent lack of risk and the perfect teamwork. It seemed by no means as hazardous as I had expected. I was probably lucky to go out with a pair of such experienced hunters. The main element of risk came not from the crocs themselves but from the river and the hippo. The river was dangerous, with half-submerged rocks capable of holing even their boat. Submerged stumps seemed no hazard at all: we grounded on them several times without

28 8.30 PM

Messrs Manning & Headen
Croc hunters arrived —
Stayed till 10.15 PM.
Gave me cigarettes!

Night
Went Croc hunting some miles
down river with Cecil Manning
& Doug Headen. Not too good 12
But Boat (16 x 8 ft) superb. Even
paddled when boat grounded. 3/4 moon

damage. Sandbanks were no worry. If the momentum was insufficient to carry the boat over into deep water, poling was tried first. If this failed, all but the motor man took off their shoes and socks, rolled up their trousers, got into the river, and pushed. We all did this more than once, and it was great fun. Underwater sandbanks are like desert dunes, with a gentle slope towards the current and a steep face downstream, and it was necessary to be careful not to fall into the deep water. Another very real danger came from overhanging branches on the bank. When one's eyes were straining to watch for the croc as the boat made a fast run-in, these branches could hit you hard and spikily.

Hippo were a special danger. An angry hippo could upset even a sixteen-foot boat with a load. Normally they should be on dry land, feeding during the dark, but even so, if they were near the bank when the boat was run in, their first impulse would be to crash back into the river. One might land on the boat, which would be calamitous. Those in the water were generally spotted by their eyes, which reflected in the lamp's beam. Cecil said he could generally distinguish their gleam from that of a croc, but mistakes might occur. It didn't happen while I was in the boat, so I didn't see the signal, but apparently it consisted of a good deal of waving and shouting with the general aim of telling the motor man to 'get into reverse and get the hell backwards' as fast as possible.

I am not disclosing how many crocs my friends collected in their three months' season, or the average price per skin: that would be breaking my promise to them. Taking their average expenses, the results sounded good, but there was a heavy capital outlay and good hunting areas were getting scarce. I doubt if any licensed hunting still survives.[6]

Crocs were getting scarce in the middle Zambezi even in 1956, and I noticed that the local people were taking far more liberties with the river. Women were going in waist deep to wash and draw water; children were going even further just to bathe. I mentioned this to Cecil, who remarked that they were being extremely foolish. The total croc population was certainly much reduced, but those that escaped were the particularly wary ones, and he forecast that there would be tragedies as a result of such casual behaviour.

[6] Crocodiles in Zimbabwe are now on CITES Appendix II. Crocodile management is the responsibility of the National Parks and Wildlife Management Authority, which is also the CITES Management Authority and Scientific Authority for Zimbabwe. During the late 1980s National Parks encouraged the effective management of crocodile ranching by the private sector. The Crocodile Farmers Association of Zimbabwe (CFAZ) is the producer representative body that works closely with National Parks in regulating the industry, and in monitoring the wild population.

I don't know how many tragedies there have been, but two of my European friends had bad accidents before the year was out. One took place on the Kafue and resulted in the death of a Fisheries officer. The other happened near the junction of the Sanyati and Zambezi, not far from Kariba Dam. If I did not know the man in question and could not vouch for the story, I wouldn't include it, for no one could - even in a schoolboy adventure story - invent anything so improbable as his escape. He would never allow the facts to be published in his lifetime, but he died in 1977.

At the time of his escape, Ronnie Rankine was a tobacco farmer, a bachelor of about forty-five. He was an authority on birds, and one of his relaxations was going on expeditions to out-of-the-way places organized by the Ornithological section of the National Museum of Southern Rhodesia, as it was then.

Late in the dry season of 1956, one of these trips was organized to go to the Kariba end of the middle Zambezi Valley, and Ronnie was invited to join it. On the journey to join the party at Kariba he had mechanical trouble with his Land-Rover and arrived several days late. When the main party packed up to leave, Ronnie, always a self-contained unit, announced his decision to stay on for an extra day or two, and he remained behind when they left the next morning.

Bird people work odd hours. They start at dawn, when the bush is full of birds, but they knock off at about 10.00a.m. because the birds very sensibly disappear for some hours in the heat of the day. Ronnie was also a keen fisherman, so it was only to be expected that, when the binoculars and shotgun were put away, the fishing rod appeared to take their place. This was especially to be expected when camp was on such a river as the Zambezi, where fishing was excellent. So it came about that, in the natural sequence of events on this day, as on so many like it that had gone before, Ronnie went fishing after an early lunch.

Zambezi fish, like all others, are capricious feeders, and on this day they were 'off'. Ronnie had a comfortable perch in the shade about six feet from the water's edge, with deep water in front of him. Apparently he dozed. He woke up at the bottom of the river, his left foot held in the jaws of a large crocodile. Several factors combined here to save his life. He was an experienced underwater swimmer, and subconsciously he must have taken a deep breath before hitting the water. He didn't panic, as nearly everyone else would have done, and he had a bush knife in the back of his belt. He drew this knife and systematically stabbed the eyes of this attacker until he drove it off. He struggled to the surface almost exhausted, swam to the bank, and hauled himself on to dry land in a state of collapse.

He did not know how long he lay there, but it cannot have been for very long. While he was still semi-conscious, he was seized by a second croc, this time by the left hand, and again dragged to the bottom of the river. For a second time, he had to repeat this incredible performance of fighting off his attacker with his knife by stabbing the eyes. Somehow he succeeded before his lungs gave out, and again he reached the bank and crawled out.

The damage was terrible. The first croc had torn out the ligaments behind one heel and some of the flesh of the calf; as a leg, it was useless. The second had practically bitten off his left hand, and blood was pouring from the severed artery at the wrist. As if this was not enough, he was entirely alone, and his Land-Rover was about 300 yards away in the bush. Holding the artery of the left wrist with his right hand, he set off to crawl to the vehicle, and succeeded somehow in reaching it. Luckily, his first-aid kit was there and not in his camp. He first put a tourniquet on his left arm, and did what he could for his injured foot. Then he put vein forceps on the cut artery and removed the tourniquet.

The position even then must have seemed hopeless; he could use only one hand and one leg. He was ten miles from any kind of help and the wrong side of the river, which was difficult to cross even in a Land-Rover driven by a man with experience and full use of his limbs. The drift over the Sanyati was 300 yards wide, the banks were high and steep, and the crossing was, at best, a hazardous one. His only hope, however, was to get to Kariba as quickly as possible, before pain and loss of blood could destroy his one chance of survival. To anyone who knows the local conditions, it sounds quite an impossible undertaking, yet he did it.

With one hand and one foot, he drove his Land-Rover over the Sanyati drift. If he had stuck even mildly in the river nothing could have saved him. Soon after crossing the river, he had his only stroke of luck. He met a man driving a lorry. Within minutes of this he was at the Kariba airstrip. A doctor was brought at once, and an aircraft, which was already waiting to take off, was pressed into service. One and a half hours after reaching the airstrip he was in Salisbury hospital. His hand was beyond all hope, but the foot was saved, although so badly damaged that to make it usable at all would take months of surgery and treatment.[7]

For most croc victims there is no such miraculous escape. The tragedy is swift and silent. Death is usually by drowning, and the chances are all with the attacker, the element of surprise being exploited to the full. There are very few comedies in such circumstances, but I have come across two. One was reported in the press and happened outside the Valley. An African woman crossing a shallow river was seized by the leg, probably by a croc of smallish

[7] An obituary to him says that 'his encounter with the crocodile ... cost him his left hand and a year later his right leg.' R. H. N. S[mithers], 'Obituary: Ronald William Rankine', *Honeyguide* (1977), 91: 9.

size. She beat it hard with her fists, and it let go of her leg but managed to get a grip on one hand. She is then said to have bitten its nose so hard that it let go and made off. She recovered in hospital.

The other involved a friend of mine, who really deserves a whole book to himself; for many years he lived a remarkable life. He is a namesake, Harry Bond, and I wish I could claim him as a relation, but it seems unlikely. He was a tsetse-fly ranger and lived in the Sebungwe for more than twenty years. He had naturally become the uncrowned king of the district. To appreciate his unique position, it is necessary to describe his life in a little more detail. Without understanding this, the point of the story would be lost.

When he first moved into the area to organize the fight against the tsetse fly, the journey to his camp from the main road involved walking a distance of about ninety miles. The take-off point was his nearest white neighbour. The track was impassable to any other means of travel and, once the rains came and the two main rivers beyond which he lived came down in flood, he was cut off. He built his own house, organized his life and the destruction of game, and stayed there.

He was first sent in more than forty years ago. He had no radio, no motor vehicle, and during the long periods of isolation he could have died several times over before help could arrive. He lived like this for about twenty years. Being a remarkable character, he survived it all. Towards the end of his stay, he was linked by two-way radio with the outside world, had a Land-Rover and a reasonable road in, and seemed to regard all of them as rather a joke. He knew his district inside out and the habits of the larger and smaller game animals equally well. A visit to his camp was an event to the few people who made it, and it came about that when, a certain friend visited him, the following incident occurred.

For some reason, perhaps to shoot something for the pot, the two of them went off a long way into the bush, apparently with only one of Harry's hunters. It was a hot day, as it generally was in his rather low-lying district, and on the way home they came across a rocky pool in the otherwise dry bed of the Mobi river. It was quite a small pool, perhaps the size of a living room of ordinary size, but it must have looked cool and inviting to the two hot and foot-weary men. I have no doubt that the temptation to bathe was very strong indeed, and I can imagine that they sat down beside the pool and considered the problems involved very carefully.

Harry took no chances in the bush: he knew it too well to treat it lightly. The problems in this case were two in number: bilharzia and crocs – the first a 'long-term' risk, the second a short-term one. Bilharzia is an insidious complaint involving an internal parasitic worm with a complex life cycle that starts and finishes in man and water. It is a disease picked up from streams in inhabited land; Harry's Sebungwe was practically uninhabited. It requires a snail as an intermediate host, and snails of this kind need permanent water and vegetation on which to survive.

Obviously the chances of bilharzia were remote in this case. That left only crocs. They are quite capable of long journeys over land, but they do not often get into small isolated pools devoid of fish, which are their main diet. Again, the chances of there being a croc at the bottom of this pool must have been small. However, to make even more certain that his friend should come to no harm while in his kingdom, Harry fired several shots into the water from his rifle. Nothing happened. Having taken all reasonable precautions for their safety, they stripped and climbed into the water.

But crocodiles are not, by human standards, reasonable creatures, and before they could enjoy the coolness of the water, Harry's friend was seized by the leg. The combined efforts of the three of them soon drove off the croc, and Harry's friend was safely deposited on dry land, without any very serious damage, although even small punctures from croc teeth are to be treated as serious because their bite is so filthy that wounds will go septic unless carefully treated.

I think most people would have been thankful that the material damage was so slight, taken the victim home and patched him up, and that would be the end of the matter. But not Harry. He was himself unharmed physically, but that a friend of his should have such an attack made on him in his presence and in his district was more than he could stand. His reaction was completely in keeping with his personality. After repairs to the damaged leg, he sent to the village for all the hunters he could raise and all the buckets they could find. They then set off for the pool, baled it dry and Harry shot the croc.[8]

In 1959, the opportunity came along to do a major consulting job in Malawi, then Nyasaland and one of the three territories comprising the Federation. I knew very little of this territory, though I had been on a fairly quick journey through parts of it in June 1955 with Ronnie Tyndale-Biscoe, a senior member of the Geological Survey in Salisbury. On that journey, we had been on both the Nyika and Viphya plateaux, in bitterly cold winter conditions, and gone as far north as Panda Hill in the Rukwa Rift of southern Tanganyika before returning through parts of the Luangwa Valley to Salisbury. But I had had no

[8] Harry Bond is credited with the 'altogether unexpected discovery' of *Glossina pallidipes* in the Sebungwe: J.K. Chorley, *Tsetse Fly Operations, 1942: Short Survey of the Operations by Districts for the Year Ending December, 1942* (Salisbury: Ministry of Agriculture and Lands, Bulletin No. 1232, 1943).

experience of the floor of the Rift Valley in Nyasaland, and had never seen the Lake at close quarters.[9]

Blantyre and Limbe were gradually outgrowing their power station, and a hydroelectric scheme on the Shire river was proposed to supply a new and more ample power supply. The site chosen was at Nkula Falls, only some thirty-five miles from Blantyre.

The Shire is Malawi's main river and it is the outlet from the Lake, but, hydrologically, it has had a peculiar recent history. For instance, for some twenty years in the early years of the twentieth century it is recorded that there was no outflow from the Lake, and consequently no flow in the Shire.

The reasons for this peculiar behaviour are not really understood but had been linked by an eminent civil engineer to sunspot patterns. Many things on earth, including fertility cycles in rabbits, have been so linked, and my own work on the glacial climate of the Zambezi valley 270 million years ago had shown up what seemed to be a fossil record of sunspots that had somehow affected the weather at that time. Leaving aside the cause, the fact remained that if the Shire did it again after an expensive hydroelectric plant had been built, the result would not only be farcical from an engineering angle but Blantyre and Limbe's power problem would remain. Some method of controlling the level of Lake Malawi had to be found, and the obvious answer was to build a barrage across the Shire to replace the ancient ferry at, or near, Liwonde. Two schemes were therefore necessary, and I was asked to do the foundation geology for both.

I was due to start the fieldwork in October. For most of August I had been in Leopoldville and Brazzaville for an important congress,[10] and soon after getting back to Bulawayo I went down with infective hepatitis, no doubt acquired from water in the Congo. It was a mild infection and needed only a week in bed, but I had no time for convalescence before leaving for Blantyre to do a month at Nkula Falls.

The journey was interesting. The first leg from Bulawayo to Salisbury was by Viscount turbo-prop airliner, very modern then. At Salisbury, a Dakota was waiting to do the next leg to Blantyre. At Chileka airport, I changed into bush clothes for the next stage, about thirty miles by Land-Rover, which took about two hours. This track finished on the east bank of the Shire. A canoe was

[9] Known as both Lake Nyasa and Lake Malawi, it is the third-largest lake in Africa. 'Nyasa (meaning "lake") is fed by 14 perennial rivers, the largest being the Ruhuhu; the sole outlet is the Shire River, a tributary of the Zambezi.' 'Lake Nyasa', *Britannica*, <https://www.britannica.com/place/Lake-Nyasa>.

[10] He was Chairman of Section I (Geology and Palaeontology) at the Fourth Pan-African Congress on Prehistory held in Leopoldville and opened on 20 August 1959. He reported on the Section's proceedings, sixteen papers and a two-hour discussion of African quaternary problems in B.D. Malan, 'A Report on the Fourth Pan-African Congress on Prehistory', *The South African Archaeological Bulletin* (1960), 15(57): 5–8:

waiting there to ferry me, complete with kit, to the west bank. Porters were already laid on to carry my bags the last two miles to camp at the Falls. This was a memorable mixture of transport systems all in one day.

The camp had been established by my only companion, a young engineer. Jeremy Burden had been there a week before I arrived, and his task was the initial survey on which the scheme would be based. His only method of cooking was by frying; my liver took a week to acclimatize.[11]

The Shire valley at Nkula Falls was beautiful, unspoilt and uninhabited. Although it was in the Rift Valley, the topography was quite rugged, the bush was thick, and in October it was hot. Being a bit lower than my middle Zambezi Valley, and a few degrees nearer the Equator, this was no surprise.

The name Nkula Falls was a misnomer. There was no single major waterfall. At this point, the river, after being a placid stream, suddenly bends at right angles to the left and enters a great half-circle bend to the right, which consists of an almost continuous series of rapids. The end of the half circle is the end of the rapids, and the river resumes its placid character, having fallen nearly 250 feet. A straight line from the start of the rapids to their end is about three quarters of a mile.

The hydroscheme involved no dam, but a nearly level tunnel was planned under the hills. At the outlet end of the tunnel the water was to be dropped down a steep slope by penstocks to the power station at river level, so making use of nearly the full available head for generation of electricity. I liked the scheme, for there would be remarkably little interference with this beautiful valley by creating it. Practically everything would be underground.

Fieldwork on mapping, and the subsequent drilling programme, involved many journeys between Nkula Falls and Bulawayo in the next few months. Some of them were extended to Liwonde to investigate the foundations for the barrage.

[11] Jeremy Burden is now retired in Australia: 'I graduated from Cape Town University in 1958 and commenced working for F.E. Kanthack and Partners in their Salisbury office in January 1959. I was on site at Nkula Falls from July to October of that year and again at the Liwonde Barrage site in 1960. I carried out all the survey work and supervised the drilling programmes. I remember Geoffrey very well. He was engaged by the firm to prepare a geological map of the tunnel route and also report on any other geological matters arising from the drilling programmes. I didn't have all that many visitors to site, so it was a pleasure to have his company. Another aspect I remember very clearly was the instruction of my boss – David Piésold, the local partner – to be sure that there was Black Label [Scotch whisky] available for Geoffrey for sundowners at the end of the day!

'The camp was certainly primitive – no tent, no gas cookers, no means of refrigeration, long-drop toilet and rapids for bathing, as there were crocodiles in the river. The main problem was not having a vehicle on site handy for use in emergencies. I eventually opened up a track on the west bank of the Shire from the closest bridge upstream, which was some 25 km away at Matope. Having a vehicle handy felt much better, but it became quite a long drive to Blantyre. It was originally estimated that I would be on site for six to eight weeks during the dry period, but as it turned out to be four months, it ran into the start of the wet season.'

Arriving and setting up a camp

Photos courtesy of Jeremy Burden

This site provided plenty of problems. The left bank was on solid gneisses, but there was a hot spring nearby that suggested faulting, and the drilling confirmed it. The Shire was running against a fault scarp buried in its own recent alluvium. We drilled in the river and on the north bank without bottoming this unconsolidated fill even at 250 feet. As a foundation for a heavy concrete barrage, it could hardly have been worse. It was porous and compressible. The final design consisted of sections with sliding joints sealed with bitumen to allow settlement if it occurred.

Left: The Nkula Falls. Right: the base of the Nkula Falls

Photos courtesy of Jeremy Burden

The official opening of the hydro-electric scheme, 1966

Photos courtesy of Jeremy Burden

No sooner was all the fieldwork, laboratory and design work for both schemes complete than political decisions postponed actual construction for several years.[12] Both were eventually built successfully, and I made many visits during construction. There is something rather satisfying in seeing such development completed.[13] Blantyre-Limbe no longer depend on a power station like a museum collection of ancient boilers, diesel engines and generators, running permanently at several per cent overload. Nor does one wait interminably at Liwonde for a ferry that is inevitably at the other side of the river, while the crew has lunch.

[12] From Hansard, House of Commons Debates, 8 February 1962, Vol. 653, cols. 604–5.

Mr Biggs-Davison asked the Secretary of State for the Colonies whether the Nyasaland Government will cooperate with the Federal Government's hydro-electric scheme at Nkula Falls; and what benefits the scheme is expected to bring in Nyasaland.

Mr Maudling Discussions have been taking place between the Federal and Nyasaland Governments and final decisions on the project have yet to be reached. Apart from the increased power supply for development schemes that the hydro-electric station would provide, the barrage upstream would regulate the level of Lake Nyasa with benefit to navigation, tourism and lakeshore dwellers. It would also facilitate irrigation projects and assist riverside dwellers in the Shire Valley.

Mr Biggs-Davison I thank my right hon. Friend for that reply. Is he aware that Dr. Banda, the Minister of Local Government and Natural Resources, has said that he would rather the Nyasaland people starved to death than had the benefit of Federal funds for the purpose of constructing this dam?

Mr Maudling I do not think that I am called upon to comment on statements of that character by individuals. I very much regret that there should be a dispute about this very valuable project. As I said, discussions are going on between the Federal Government and the Nyasaland Government and I hope that they will produce results.

[13] A write-up on the significance of this project appeared in John Amer and A. MacGregor Hutcheson, 'A Key to Development in Southern Malawi – The Nkula Falls Hydro-Electric Scheme', *The Society of Malawi Journal* (1966), 19(2): 29–51. Recently, 'due to its age, the hydropower plant experienced numerous forced outages over the years, which made a major overhaul an absolute necessity for a stable electrical power supply in the country.' Rehabilitation, upgrading and uprating of Nkula A was required, and the official inauguration took place on 31 October 2018: 'Latest news on HPP Nkula A, Malawi: All Units Are Put into Operation', 'Hydro News', *Andritz*, <https://www.andritz.com/hydro-en/hydronews/hydro-news-29/hy-news-hn29-05-nkula-abundance-hydro/nkula-a-news>.

– 15 –

Prelude to the Flood

There were signs from about the mid-1950s that two major changes were to take place in this country. The Kariba Dam was beginning to look like a certainty. This would drown much of 'My Valley', and I hated the scheme. A gang of wild young men were let loose in it to do the ground control for the maps to be drawn from air photography. They shot indiscriminately, leaving wounded buffalo and elephant, which were a menace. Previously, game was undisturbed, placid and plentiful, but now there was the menace of somebody else's wounded animals constantly to be faced.

The second major sign of change was the move to establish a university, or a university college, in Salisbury. I found this possibility disturbing and was opposed to it on several grounds, though I never voiced them publicly. Both changes threatened me.

The Kariba threat was easily defined. The privacy of My Valley would be destroyed and much of my favourite country lost under water. Selfishly, I wanted to preserve it for my own use, but the extent of the threat was measurable. The proposal to set up a university was of a different character, and far less definable. For one thing, it was to be in Salisbury and, as a staunch Matabele, I was instinctively against it.

Visions were conjured up of hordes of expatriate scientists, including geologists, descending on this country and trying to teach our local scientists their business. I had become aware, over the years, of a strong tendency to ignore or minimize the scientific expertise that had been built up in this country. It matched the general public's attitude towards anything manufactured here. If something was not imported, it was at least second grade.

This seemed particularly to be the attitude of government when any kind of special investigation was needed. Whenever the need arose, the first action was to import a high-powered 'specialist' from overseas. Almost anywhere overseas would do. Sometimes it would be more than one, in which case it became a Commission of Inquiry. Whichever it was, my suspicion was (and still is) that the specialists came for a limited time, picked the brains of our long-suffering resident scientists, wrote their report, collected a fat fee and departed. This seemed grossly unfair to our local scientists who had devoted years of patient work to our problems and whose knowledge of the background situation could never be equalled by any imported expert.

From my very first contact with the science in this country, the quality of the people and of the research that had been done by them had impressed me. It was based on my acquaintance with only two institutions: the National Museum in Bulawayo and the Geological Survey in Salisbury. Both contained scientists of world stature in their respective disciplines. Extrapolating from this sample of the country's scientific manpower convinced me that, for its size, it was served by a remarkably high quality of scientists. The broader spectrum with which I came in contact as a member of staff of the National Museum merely convinced me that my early assessment was in no way at fault.

The advent of the University College of Rhodesia and Nyasaland made me wonder whether the newly imported scientific staff would be open-minded enough to appreciate the local quality, or whether they would be typically arrogant academics who would look down their noses at our existing institutions from the lofty isolation of a small hill in Mount Pleasant, pronouncing the doctrine that only university staff were competent to undertake real research.

The association of the new College with London University did nothing to inspire confidence.[1] Here my own academic arrogance came into play. As a graduate of Imperial College of Science and Technology, I regarded the standards of the rest of the sprawling monster, the University of London, with some suspicion. On financial grounds I was unconvinced of the need for such an expensive institution, feeling that it was premature and that the money would be better employed by creating scholarship funds so that selected students could be sent overseas for higher education.

When I discovered that Geology was not among the subjects to be included in the new Faculty of Science, my mind was made up. It was obviously a bad lapse in planning to omit a science so basic to the mining industry. The new Faculty was to include only Chemistry, Physics, Mathematics, Botany and Zoology. It would be interesting to see how relations between the University College and existing institutions would develop. Bulawayo would be a good vantage point from which to observe the scene. If and when the College decided to establish a department of Geology, it would be even more interesting, though I had no intention of trying to join it, perhaps because of my traumatic experience in Hull in 1948.

In 1958, I heard that a department of Geology was to be established in 1960 and, like the other embryo science departments, that it would be headed by a Professor from the beginning. At least this indicated a serious attitude towards my discipline; this time there would be no nonsense about its status right from the start. I wondered who the new Professor would be, and whether his

[1] 'University teaching in Rhodesia can be said to have begun on the morning of 4 August 1952, with the course of lectures in accounting arranged by Commander J.S. Craig, in temporary quarters lent by the Education Department in the old Polytechnic building. ... The London link [was] responsible for its debut as the University College of Rhodesia and Nyasaland, which received its first students in 1957.' Michael Gelfand, *A Non-racial Island of Learning: A History of the University College of Rhodesia from its Inception to 1966* (Gwelo: Mambo Press, 1978), 13, 9.

research interests would clash with mine. It would also be interesting to watch how his relations with the Geological Survey would develop. As far as I was concerned, the Survey, established in 1910, was the official organ of Geology in the country and was the senior institution. I never saw the advertisement for the Chair of Geology and had no intention of applying for it. However, I reckoned without the influence of the Director of the Geological Survey, the late J.C. Ferguson.[2]

In about October 1959, a brief letter came from Fergie that virtually instructed me to apply for the new Chair. It left me with little option but to comply. My application was baldly factual and brief. I have seen dozens of applications from aspiring Professors since I became one, but none of them have been so patently and blatantly an attempt to show that they really did not want the job that was being offered. My letter was posted and I went off on long leave to brush up on recent advances in geology by working in my old London haunts at the Natural History Museum, Imperial College, and the Geological Society Library in Burlington House.

By December, to my surprise, I had been summoned to an interview for the Chair, and I began to wonder who else was in the running. The Selection Board consisted of six Professors, all of whom I knew. No one else was in the waiting room, which seemed odd. The other candidates must be from other parts of the world. That interview, like my application, was a matter of little concern to me, satisfied as I was with my position in Bulawayo. It lasted about three hours, most of which was spent in discussing and analysing the research potential of the central African region. I enjoyed myself, went back to work, and forgot about it.

I think it was in April, just before the end of my leave, that an official letter arrived from the Registrar of the University College of Rhodesia and Nyasaland with the formal offer of the position of foundation Professor and head of the proposed department of Geology in the College. The blow had fallen, and now a decision of great significance for my whole future would have to be made. By then it had also become a decision which would not only affect me but also a young lady in Bulawayo, Marina Levitsky, to whom I had become more or less engaged before she went off to America for a year's postgraduate study at Columbia University in New York. I sent her this cable:

'Furniture stock increased by offer 1 chair.'

[2] James Crighton Ferguson had joined the Survey as geologist in1930, became Director in 1948 until he retired in March 1960. He died on 23 August 1965. 'During the Second World War he served in the East African Engineers. He was deeply interested in the history of mining and in preserving machinery typical of the early days of mining in Rhodesia.' 'Obituary Notices', *Transactions of the Newcomen Society* (1964), 37(1): 147–8.

– 16 –

Salisbury and the University

The offer of the Chair of Geology at the University College of Rhodesia and Nyasaland presented me with a number of problems but potentially with a number of opportunities. The problems were headed by my previous unfortunate experience of starting a new university department in Hull. This time the geographical situation would be within my adopted country, with its beautiful climate and opportunities for research, but it would be in Salisbury. I regarded myself as a Matabele and Bulawayo as my home. Like so many of my friends, I regarded Salisbury with suspicion, for it seemed to grab everything - the *bamba zonke* syndrome.[1] I liked the harsher climate of Matabeleland, and the harsher bush, with its acacia thornveld, mopane woodland and teak forests, appealed to me more than the milder park-like msasa country around Salisbury.

I was still unconvinced that a university was needed at that stage, and I had already experienced, in Hull, the rather restrictive situation of being in a college attached to London University. I had heard a good deal about Dr Walter Adams, much of it indicating that he was a dictator, and difficult to work with.[2] I did not entirely believe these reports, for I had met him frequently and found him likeable and easy to talk to. He seemed to be a man with whom it should be possible to work, so long as he was kept fully informed and was in broad agreement with any proposed development.

The salary for a Professor was a distinct improvement on what the Museums were able to pay their scientific staff, and this was certainly a factor not to be discounted. It was plain that, if I accepted, there would be so much work involved in starting up a new venture that there would be little, if any, time for fieldwork and research. I decided to go to Salisbury and spend a few hours talking to Walter Adams to clear up as many doubts as possible before making a decision.

[1] 'A term meaning "grab the lot" or, more loosely, "skim the cream". Applied in jest in later years to Salisbury which, being the capital, was sometimes suspected of feathering its nest at the expense of others.' J. McAdam, 'Pat Judson: First Rhodesian-born Airman', *Rhodesiana* (July 1967), 16: 15, n.8.

[2] Adams was the principal of the College from 1955 to 1966.

I came out of that meeting feeling that Adams, and therefore the College, was serious in its intentions towards Geology. There would be money for staff and for equipment. The Department would have to be born in temporary quarters, but I was promised that within two years there would be custom-built accommodation on a scale ample enough to cater for several years of sustained growth. If I accepted, I would have a major say in the design of the Geology Department's own building.[3]

All this was on the credit side, but there was another side that I did not like. In answer to questions about the College's attitude towards consulting work, the answer was categorically 'no consulting', and certainly no question of accepting fees. This attitude contradicted the prevailing trend at such places as Imperial College, my yardstick in so many things. This was a serious snag. I was in the middle of several large consulting jobs, including the Nkula Falls Hydroelectric Scheme and the associated Liwonde barrage.

I told Adams bluntly that I would finish these schemes and get paid for them, and would reserve my position for the future. He merely shrugged his shoulders. I argued that it verged on criminal to lock up high-powered scientists on 474 acres in Mount Pleasant when the development of the country could use their skills to augment the scientific manpower already here.

This argument made no impact. Academics should teach and indulge in research, but should remain pure and unsullied by practical or applied work, except perhaps in agriculture. If this attitude was to be perpetuated, the embryo University's contribution to the country would, I felt, be lopsided and limited. If members of the Museum staff could do consulting work and be paid for it, our University should not only graciously permit its staff to do likewise, it should actively encourage them.

It took some years of patient behind-the-scenes pressure to change this position and bring the University into line with practice elsewhere. Adams was never reconciled to this part of the University's role, and the major breakthrough only came after his resignation in 1966.

The other point that I made abundantly clear to Walter Adams at that interview was that, if Marina declined to marry me on her return from America, I would stay in Bulawayo as Keeper of Geology in the National Museum and he would have to look for another Professor. To make the transition from Museum to University and from Bulawayo to Salisbury, I needed her support; without it the whole thing was unthinkable. We left it at that, with a clear understanding on both sides, and I went back to Bulawayo to think it over.

Marina duly returned and said 'yes'. I put in my three months' notice,

[3] See John Morgan's description in Appendix 1.

told Walter Adams that we would be married on 1 November 1960, and that after fourteen days' honeymoon on Paradise Island,[4] I would report for duty. His response was typical: my appointment (and therefore my salary) would date from 1 November, and he observed that it was probably unique in university history to acquire a wife and a Chair on the same day.

After the fiasco in Hull, I had very definite ideas on how a new department should be started, and what its objectives should be in the context of this country. Its growth would, of course, depend on factors beyond my control, and partly on the future of the Federation, which the College had been created to serve and which was beginning to look distinctly fragile. Walter Adams's opinion was that, if Federation collapsed, the future of the College would be bleak. Financially, it would be cut off from Copperbelt money. If it survived at all, it would be starved of students, and ultimately it would be little more than a second-rate teacher-training college. If this happened, my own future would be bleak, for this time the break with

[4] Now named Santa Carolina, the Island is part of the Bazaruto Archipelago National Park, off the coast of Vilanculos in southern Mozambique. In the 1950s, Joaquim Alves, a flamboyant Portuguese businessman, had constructed the Hotel Santa Carolina, a 250-room hotel that was very popular with Rhodesian and South African honeymooners. It was abandoned after Mozambique's independence and civil war. The Zimbabwean author, Douglas Rogers, visited the island in 2015 and described the hotel that his own parents had also honeymooned in:

> I feel as though I've stumbled on a lost city.
>
> Following a crumbling sidewalk, grass shooting through its edges, I make my way to the sprawling concrete shell, which fronts the island's eastern shore. Finding a flight of stairs, I climb up one flight, turn right, and step into a room. Hundreds of tiny pink tiles lie scattered across the floor. I pick a few up as mementos, then look out from the balcony toward a chapel, its roof swooping like some Palm Springs modern masterpiece, perched on nearby rocks.
>
> I continue to the top floor, and into the ballroom. Its walls and windows long since shattered by waves and cyclones, the space is open to the elements. But in its middle is a spectacular sight: a pillar inlaid with thousands of tiny blue tiles painstakingly pieced together by artisans half a century ago. Even more unbelievable to me is a piano, leaning against the back wall, the ivories long since pilfered.

Douglas Rogers, 'Upon the Beach in Sunny Mozambique', *National Geographic*, 12 January 2016.

Museums would be final and irreversible. But it was a chance that I had taken and would have to live with. We (it was no longer I) could not turn back.

I reported for duty on 17 November 1960 with the coveted title and status of Professor - without any students, staff, buildings or equipment, nor even an office that I could call my own. The existing six science departments were at the end of their first three-year degree cycle. Their Professors had been in post for four or five years. They were about to graduate their first crop of students. The staff were all expatriates, and I was the first local to be appointed to professorial rank. My total experience of university teaching consisted of two terms of one day a week at Queen Mary College, London, and my year at Hull. I expected to be looked down on as an inexperienced local by the old hands.

Walter Adams asked me when I would be ready to start teaching. The next intake of students would be in March 1961, at the beginning of the new academic year. My answer surprised him. I would be ready to take in my first students by the start of the next session. He pointed out politely that this was only four months away, and that all the other science departments had taken a minimum of eighteen months to reach that state of preparedness.

I set about ordering the basics and looking for staff. After a few days, I moved from a temporary office in the Chemistry department to one of the sections in the building that had given birth to all the science departments. I designed laboratory benches, cupboards, stools and rock storage, left my new baby in my new secretary's hands, and set off to collect the basic rock and mineral collections for first-year teaching. Many things could be collected within the country and I knew where to go to get them. Many others could not, and SOSs were sent to my friends overseas for duplicates to fill out the essentials. Soon parcels and cases began to arrive and, by 1 March, I was ready, except that I was still the only academic, the lead ox in a one-ox span.

Since I expected my products to go mainly into the expanding mineral industry both at home and abroad, my resolve was to bias the degree in that direction without neglecting the pure side of the science. Fieldwork was to be an important ingredient, and students would have to sacrifice large chunks of their vacation to take part in compulsory field courses.

Staff would have to be of my choosing, though selection boards would be composed of colleagues with the fixed notion that lecturers must all be people from other universities. I had other ideas. I did not want staff who had served several years in other universities. They might have done ten or fifteen years of lecturing, but this might have consisted of repeating the same thing ten or fifteen times over. I wanted people from the mining industry and geological surveys who had real experience of what geology was all about, and who had, in addition, a respectable published research record.

This was a difficult combination to find. Furthermore, I wanted people with as much local experience in Africa as possible. Actual lecturing experience was a secondary consideration. If they had the right personality, they would make a good lecturer, and, like Professor Read at Imperial College, they would draw on their field experience to embroider their lectures. Local people would make up a stable staff. University lecturers tend to migrate from one college to another, generally on promotion. They have an average tenure of about five or six years in one job before moving.

I was worried from the start by the constraints of the London affiliation. The degree pattern was three subjects in first year, drop one and carry two for two further years, rather like a knitting pattern. The student then graduated with two major subjects of equal weight. This differed from the pattern to which the mining industry here was accustomed, which was based on the South African university pattern of three years to a Pass degree in two majors, followed by a fourth year Honours in one subject.

My expatriate colleagues affected to despise the South African pattern because it began at 'Matric', one year short of our entry level of two A-level passes. It was claimed that the extra year at school allowed an Honours standard to be reached in two subjects after three years. I had the temerity to question this, and pointed out that the South African pattern was based on Scottish university practice, and that they admired the standards in Scotland. I wanted a fourth, Honours, year for my students, in which they did nothing but Geology on top of their previous three-year London degree. The Geological Survey told me quite bluntly that they would employ only four-year graduates, and would regard a three-year London degree as no more than a qualification for a technician. Unless I was to produce nothing more than school teachers, which seemed to be the main outlet for all the other sciences, I would be stuck.

Geology is not normally a teaching subject in schools in English-speaking countries, although at that time both Milton in Bulawayo and Churchill in Salisbury taught Geology at A level. Something had to be done if my graduates were to find proper employment.

A search of London regulations produced a possible answer. In certain subjects, and Geology was one of them, a fourth year in one subject was allowed and, at the end of it, the student, already a three-year B.Sc. graduate, could become a B.Sc. Special Honours in – in this case – Geology.

This was plainly the solution at which to aim, but no other department in the College seemed to think along the same lines, and I could see a battle in Faculty Board and Senate looming up. Why was what was good enough for them not good enough for me? The aim in other departments was for a Master's degree, generally by course work; why should Geology be different?

The answer, of course, was that Geology *was* different, and that neither the mining industry nor the Geological Survey would recognize their type of degree.

By the time we were into our third and 'final' year for the first time, my battles with my colleagues on Faculty and Senate were won, largely through strong support from friends in Imperial College, who backed my scheme wholeheartedly and saw it through at the London end.

The fourth-year Special Honours degree was designed from the start to fit the product for a career in geology and to be tough academically, psychologically and physically. Numbers would be low, probably no more than five, or at most six, in any one year. Geology is a small profession, and my aim was never to saturate the employment market. It is so much better to have jobs chasing graduates than graduates chasing jobs. There is nothing more soul-destroying for a new graduate after four years of hard slogging than unemployment in his field of speciality.

This advanced year was designed to start at the beginning of March and to go through to the end of November without any vacations. It was to include about ten weeks at the end on a research project made up of fieldwork and laboratory work, written up as a thesis.

The first products of this very testing programme emerged at the end of 1964, having come in as part of the very first intake of new students in March 1961. They had had the misfortune of being the guinea pigs of each year as we came to it for the first time.

Teaching at this very high level proved to be a great attraction for the staff, in spite of the heavy load it imposed on them. The success of these fourth-year graduates, both in the mining field and in academic positions, has amply justified that early decision, and has been my main source of satisfaction as a Professor and Head of Department.[5]

While this had been the main reason for setting up a new department in the first place, there is no doubt that its existence provided me with a far wider spectrum of opportunities than would ever have been possible if I had stayed in Bulawayo. However, there were certain directions in which it was necessary to move very carefully, and relations between the new department and the Geological Survey was one of them.

The main task of a Geological Survey in any country is to map the geology in as much detail as possible, and to catalogue its mineral resources. It is the

[5] See the tributes from former students in Appendix 1.

bridge between pure geological research, the accumulation of knowledge for its own sake, and the search for mineral deposits on which the mining industry survives.

This country has been wonderfully served by its official Survey ever since its inception in 1910. There was some knowledge of the rocks and minerals, and even of the fossils, before that date. The earliest reference I have been able to find to geology in this country goes back to 1868. In that year, Chapman's *Travels in the Interior of South Africa* was published and included a reference to 'a vertical stratum of sandy schist' near the headwaters of the Deka river.[6] This prompted an English geographer, G.W. Lamplugh, to deviate from his 1907 traverse starting at Wankie and aiming primarily to explore the gorges below the Victoria Falls.[7] He failed to find Chapman's locality, and recorded instead a horizontal sandstone.

Before the government took over Nantwich farm near the headwaters of the Deka,[8] to add to Wankie National Park, I decided to map the area, which included both Chapman's and Lamplugh's observations. As might have been expected, they were both right. Both kinds of rock exist there but, in their respective times, position-finding was a bit vague.

By the time I walked over this ground in 1950, there was a reasonable map on which to plot the geology. Chapman's 'vertical sandy schist' proved to be a remnant of a range of hills which, at the time the Wankie coal seam was being formed, about 270 million years ago, must have been quite an imposing feature and was probably covered by an ice cap. Sweating around Nantwich on foot, I could visualize the difference now, but it meant a small addition and correction to the map of the country.

Other early pre-Survey work had been done by Mennell, and by Molyneux, who pioneered coal fields in the Sebungwe on the Zambezi side of the watershed, and in the Tuli and Umzingwane areas in the Limpopo Valley. In 1901, he had published brief accounts of fossils from both areas which he had found in rocks associated with coal seams.[9] Fletcher and Espin had also produced a

[6] James Chapman, *Travels in the Interior of South Africa, Vol, II* (London: Bell & Daldy, 1868), 158.

[7] G.W. Lamplugh, 'The Gorge and Basin of the Zambezi below the Victoria Falls, Rhodesia', *The Geographical Journal* (February 1908), 31(2): 133–52.

[8] Nantwich had been run by Percy Durban Crewe, 'one of the old Rhodesians – a member of the entrepreneurs, in the words of one early pioneer, the "company of keen glorious adventurers", who preceded colonial government into Matabeleland. ... He called his main farm "Nantwich", a kind of geographic joke – Crewe and Nantwich are towns located next to each other in England.' Gary Haynes, *Hwange National Park: The Forest with a Desert Heart* (The author, 2014), 71–3.

[9] A.J.C. Molyneux, 'Notes on some fossils from the Sengwe Coalfields (*Proceedings, Rhodesia Scientific Association* (1901), 2: 88–92, 126–7.

rough map of the gold fields of Matabeleland before the turn of the century.[10] The founding of the Survey was, however, the start of systematic mapping and description of mineral deposits.

Their work was soon summarized in the publication of the first 'provisional' geological map of the whole country. At intervals of a few years ever since, they have continued to produce 'provisional' maps on the same scale (1:1 000 000), and the evolution of the geological knowledge of this country is clearly shown by comparing successive editions. At the time of writing, the last one was produced in 1977.[11]

For a country so large, and served by a Survey staffed for so long by so few, it is a remarkable map. It is based on the quite brilliant work done by the dedicated officers who, for many years, worked under the harshest of conditions. They built well-deserved reputations in the geological world, and the excellence of the work of our Survey is universally respected and acknowledged. Much of their work is of universal appeal and was published as papers in the international scientific journals.

Perhaps a good example of this was the description, by Dr A.M. Macgregor in 1941, of very ancient fossil algae which he found in a limestone near Turk Mine just north of Bulawayo.[12] At the time he published his paper, absolute dates were in their infancy, the world's oldest rocks were thought to date from about 1,760 million years, and the oldest fossils might perhaps be 600 million years old. Macgregor's 'fossils' were regarded as inorganic and were anyway in rocks that form part of the Bulawayo gold belt and were known to be very old, certainly too old for remains of life.

Perhaps because of this, and because of World War II, his paper was almost totally neglected until the early 1950s. It then suddenly became famous, and I conducted a series of distinguished American and European scientists to collect specimens, which were subjected to sophisticated techniques unknown in 1941. These proved beyond doubt that Macgregor was right. They were certainly fossils of calcareous algae, and their age was put at about 2,700

[10] 'The Goldfields of Matabeleland Ltd was a British Company that owned "development estates" within Rhodesia. In 1896, with the assistance of Fletcher and Espin, Bulawayo, and Government Surveyors, they published the *Fletcher and Espin's Map of Matabeleland*. In 1897, they published a second map, *Fletcher and Espin's geological sketch map of Matabeleland*.' 'The Goldfields of Matabeleland Limited', *Rhodesian Study Circle*, <http://www.rhodesianstudycircle.org.uk/the-goldfields-of-matabeleland-limited>. The map can be viewed here: <https://gallica.bnf.fr/ark:/12148/btv1b530292622>.

[11] 'The 1:1 million geological map of Zimbabwe, a very important mining investment decision tool, had its last major update in 1977 and new geological knowledge acquired since then is compiled but is yet to be captured on that map.' Zimbabwe Economic Policy and Research Unit (ZEPARU), *Reconfiguration of the Zimbabwe Geological Survey* (Harare: December 2016).

[12] A.M. Macgregor, 'A Precambrian algal limestone in Southern Rhodesia', *Transactions of the Geological Society of South Africa* (1940) 43: 9–15.

million years. The Huntsman limestone quarry is now a National Monument, but even older fossils of calcareous algae have recently been found near Belingwe. They are 3,500 million years old, and surely that locality deserved to be proclaimed as a National Monument.[13]

It was against this distinguished background that my embryo university department was founded. I was very careful to make it plain that it would be a very junior partner in the geological scene in this country, and would have to prove itself over the years. Its status would depend on results and not on its mere existence in the university.

When I finally decided to throw in my lot with UCRN, as it was then, I was fully aware that I was joining an institution that was almost universally disliked and widely distrusted, views to which I had to a large extent subscribed myself. I realized, however, that the future of the whole region depended ultimately on a non-racial approach. This was clearly stated in Clause 4 of the Royal Charter under which the College, and later the independent University, operated.[14] The early image of the College was not improved by the antics of some of the early members of the academic staff, who seemed to try to use an institution of higher learning and research as a political bludgeon.

Politics seem to have permeated every aspect of life, including sport and universities, and since World War II there has been a pandemic of politics. I have never taken any active part in politics and, if man is a political animal, then I am not human. Politicians are perhaps necessary largely to protect the public from the civil service, and maybe vice versa. I have tried to remain aloof both from the politics of the country and from the internal politics of academics.

I have doubts about the validity of democracy in its modern expression. It may have worked in Britain in Queen Victoria's day, when only those of independent means stood for Parliament; independent means may then have meant independent minds. Voting against the party might then have led to loss of seat at the next general election, but loss of livelihood did not follow. Nowadays, loss of seat means loss of job, and who wants to employ an

[13] Stromatolite Deposit, Huntsman Quarry, was proclaimed Monument No. 128 on 24 October 1969. The Huntsman quarry is now overgrown and uncared for. In 2015, Museums were considering declassifying the site as a national monument. Geologists have proposed that the Cheshire stromatolites south of Zvishavane ('near Belingwe') take their place. Neither has yet happened.

[14] Clause 4 of the Charter reads: 'No test of religious belief or profession or of race, nationality or class shall be imposed upon or required of any person in order to entitle him to be admitted as a member, professor, teacher or student of the University College or to hold office therein or any advantage or privilege thereof.'

unsuccessful MP? Such a fate can be avoided only by strict adherence to the party line. There is no room, at least in British politics, for the conscientious objector in the modern political machine. Survival entails slavish obedience to the three-line whip.[15] This, to me, is not democracy in the real sense.

Politics and even the breakup of the Federation notwithstanding, the College began to grow and gain in credibility and acceptance, and the platform from which I was operating became more secure. My range of opportunities began to expand in a way it could never have done had I remained as Keeper of Geology in Bulawayo. The process has accelerated ever since. In only one area has the move been a handicap, and this was in my own research output.

I had realized that starting up a new department would cramp my research opportunities for the first three or four years, after which I had hoped not only to be able to expand my own activity but to have postgraduate students working under my direction as well. This may be the case in old established universities, but under our local conditions it certainly was not. There seemed to be a succession of crises that were unsettling in themselves but also absorbed a great deal of time of senior staff.

The traditional method of selecting professors has always been to lay great emphasis on their research records. Their teaching and administrative abilities are of secondary importance. In my case, administration took up so much time that personal research inevitably suffered. Additionally, political and financial constraints limited growth, and research students were less plentiful than I had hoped. The output from the Department steadily increased in spite of these difficulties, but my own dwindled.

The first major outside activity with which I became involved was with the Museums: within eighteen months of leaving, I was appointed to the Board of Trustees.[16] I wondered how my erstwhile colleagues – and particularly Director Reay Smithers – would react to this. They seemed quite glad and, as one put it, it was good to have a Trustee who really knew what went on in a museum.

It was a period of major expansion in the Museum service. Reay Smithers was adept at raising money and, in a brief period of a few years, had found the means to build the impressive building in Bulawayo's Centenary Park and

[15] 'In the UK a three-line whip is an instruction given to Members of Parliament by the leaders of their party telling them they must vote in the way that the party wants them to on a particular subject.' 'Three-line whip', *Cambridge Dictionary*, <https://dictionary.cambridge.org/dictionary/english/three-line-whip>.

[16] The Board of Trustees was established by the National Museums and Monuments of Rhodesia Act [*Chapter 313*], 1 October 1972. See also Chapter 4.

the new Queen Victoria Museum in Salisbury. This was soon followed by a new building in Umtali. The momentum had to slow down, but subsequently the organization has blossomed with the Midlands Museum in Gwelo. By 1970, I was Chairman of their Board, and it was decided to institute proper identity cards for members of staff and for Trustees. As Chairman, I was told that mine would be numbered 001. I suggested that, with my name, 007 would be more appropriate, and that is the number with which I was issued.

The Museums and the National and Historical Monuments Commission had always been very closely linked. One collected and cared for material objects that could be housed and displayed indoors; the other was responsible for those material objects that had to be maintained in situ, such as ruins, memorials and painted caves. Neville Jones had been the first Secretary of the Commission when it was established in 1936 and had combined this with being Keeper of Antiquities in the National Museum. Roger Summers had carried on the same tradition. Both had been active research workers, and had done much pioneer work.

By 1970 it was becoming clear that the work of the Commission and the Museums was so closely linked that it would make good sense to amalgamate the two into a single organization. In 1972 this came into effect. Museum directors then became responsible for all monuments work within their designated regions. After some initial growing pains, the system has worked well.

A major difficulty arose over Zimbabwe Ruins (Great Zimbabwe), our premier national monument and major tourist attraction. Somehow it had become not only a proclaimed Monument but a National Park as well. There was, therefore, dual control, which was not entirely satisfactory. Neither organization was prepared to pay for development, which badly needed to be done. The Museums maintained the fabric of the Ruins and were responsible for the control of all archaeological research. The role of National Parks was far less clearly defined, but included tourist accommodation and its booking. Entry to the Ruins was free.

The area of the Park took in a slice of surrounding country. This had been defined as long ago as 1896 when someone had the foresight to have this area designated as a 'Scientific Reserve', a land category long since forgotten, and probably a unique one in the history of this country. Persuading government to de-proclaim this area as a National Park and vest total authority in the Museums and Monuments Commission was a tedious operation, but it was done. Tourism was booming and it was decided to impose an entry fee to Zimbabwe Ruins.

This caused a major stir in Fort Victoria, but it was done, just in time for the almost total collapse in the number of tourists due to the escalation of the Bush War. Instead of becoming a major money-spinner for the organization, the place became a major liability. However, the Board pressed on with its plan to become the legal owners, and the deed of transfer was duly signed. Zimbabwe Ruins became the property of the National Museums and Monuments Commission. Its future development will be an interesting exercise.

Before I joined the staff of UCRN, the first Professors of Zoology and Botany, Eric Edney and Arthur Boughey, had realized the scientific significance of the creation of Lake Kariba. They shared my view that this was not just a hydro-electric scheme but could also be regarded as a gigantic scientific experiment, the creation of a vast lake where there had previously been dry land in a major river valley. The biological impact would be immense. Money was needed to tap this important research resource. They applied to the Nuffield Foundation for £35 000 to set up a research station on the lake shore, and run it for at least three years. This was granted.[17]

They were neither of them familiar with the Valley. Eric Edney had been on the Museum staff in Bulawayo before World War II, working as an entomologist under Dr Arnold, but had been away from the country for more than fifteen years before he was made first Professor of Zoology. Arthur Boughey was new to this country. They invited me to come to Salisbury to discuss the siting of the proposed research station, using my detailed experience of the Valley.

The meeting took place in early 1960. The aims of the station were to be partly terrestrial, to monitor the effects of environmental change on the larger mammals, and partly hydrobiological on the new lake. The whole project was very exciting and I put a lot of thought into it, as well as a study of maps and aerial photographs on to which I had plotted full storage shorelines for the new lake.

There were many fundamental considerations. It was plain that the area around Kariba township would soon become developed and that this development would quite rapidly begin to expand around the eastern end of the basin. Tourist centres would be established, and boats would begin to operate from Kariba itself. Additionally, Binga, 120 miles up-lake, was already taking shape as an administrative centre and would inevitably attract visitors. The upper end of the lake at Mlibisi was also too accessible.

These considerations demanded as remote a site as possible, with game in abundance and as undisturbed as possible. Some part of the Chete Safari Area was, therefore, indicated as the prime target. Some spot on the coast of this section of lake shore that stretched from Chete Gorge in the south-west to the

[17] The Nuffield Foundation is a charitable trust, established in 1943 by William Morris, Lord Nuffield, the founder of Morris Motors. *Nuffield Foundation*, <https://www.nuffieldfoundation.org/about>.

Mwenda river in the north-east offered the best chances.[18] My own preference would be for a place that had no road access so that casual visitors could not disturb our game by driving through the area. To arrive by boat would cause no disturbance.

There were two major river estuaries in this section: the Ruzuruhuru (Luizikukulu) and the Mwenda. Between them was a rather diffuse area of what would be shallow-water coastline, subject to major changes as the engineers manipulated the lake level for their own purposes. A sheltered, reasonably deep-water harbour was an essential. The Mwenda formed the eastern boundary of the Chete Safari Area, and was an obvious possibility. The Ruzuruhuru was another. Whichever site was chosen, I knew that the staff living there would be exposed to very hot conditions in summer, but it might be possible to find a site where the extremes could be avoided, making life just bearable. Winter conditions would be pleasant, but research must go on throughout the year.

The Ruzuruhuru looked as though it would provide completely sheltered harbour conditions, but the only reasonable site for buildings faced west and was virtually cut off from cooling breezes from the Lake. In the Mwenda estuary there was a promising-looking promontory behind which there would be a fairly sheltered harbour. Behind the promontory there was a steep sandstone scarp that would cut off the afternoon sun quite early from the buildings on the crest of the promontory. Buildings so sited would be exposed to breezes from the north, south and east blowing across the water, which would cool them somewhat. Perhaps this would be a possible site, but an inspection would be needed so that direct comparisons could be made before deciding on either. It was eighteen months before this took place.

The Nuffield Lake Kariba Research project was to be a joint venture between UCRN, the University of the Witwatersrand ('Wits') and the universities of Oxford, Manchester and Glasgow, a rather grandiose concept, which depended on scientists who had expressed a personal interest. There was no financial commitment from the UK participants. I had doubts about the real value of their contribution from the start.

Eventually, in December 1961, the inspection party assembled at Binga. It consisted of Prof. Harper from the Chemistry Department at UCRN,[19]

[18] Chete is a controlled hunting area. It was proclaimed a Game Reserve in 1963 and re-designated a Safari Area under the 1975 Parks and Wildlife Act [*Chapter 20:14*], Fourth Schedule (Section 36), Item 2.

[19] Prof. Stanley H. Harper had been interviewed in London in January 1955, when he was Reader in Organic Chemistry at King's College London, before being appointed Chair of Chemistry. He was president of the first committee of the Association of University Teachers at the College, formed in March 1958, and was appointed Vice-Principal when the 1965 session opened. Michael Gelfand, *A Non-racial Island of Learning* (Gwelo: Mambo Press, 1978), 119, 139, 201.

The *Lord Malvern* and its sister ship, the *Sir Patrick Fletcher*, at Musango, 1962

the Professor of Zoology (Edney),[20] Balinsky from Wits,[21] with one or two of his research assistants, and myself. We were to be ferried down to do our inspection by Ivor Cockcroft, now DC Binga, on the *Lord Malvern*. This was a 45-foot twin diesel steel hull built in Salisbury for work on the Lake. It was my first journey on the new lake, which by then was only twenty-five feet from full storage level.

The whole thing got off to a difficult start. The Professor from Wits was adamant that he wanted to walk to the area from the road, thirty miles inland, straight across country. This was sheer madness. It was the hot season, the country was unknown, probably waterless, full of big game, and he was nearly sixty-five. Neither Ivor Cockcroft nor I would have attempted it except in an emergency, but the old man persisted. It took about two hours to convince him that what he proposed was sheer folly, and that if he got into difficulties it would be next to impossible to find him. Finally, the DC used

[20] Eric B. Edney was born in England, moved to Southern Rhodesia at the age of thirteen and was educated at Plumtree School (Gelfand, *A Non-racial Island of Learning*, 120). He passed away suddenly on 28 May 2000 in Vancouver, BC; he was 86 years old. 'In Memoriam: Eric B. Edney, Professor of Zoology UC Los Angeles, 1913–2000', *UCLA Ecology & Evolutionary Biology*, <https://www.eeb.ucla.edu/dr-eric-berry-edney>.

[21] Boris Ivan Balinsky (1905–1997) was Professor and Head of the Department of Zoology at the University of the Witwatersrand, South Africa. He went to Wits in 1949 from Kiev, where he was Professor of Embryology. He died in September 1997, just short of his 92nd birthday See B. Fabian, 'Balinsky's Darwinian roots', *South African Journal of Science* (2009), 105: 410–14.

his authority and threatened to call the whole trip off unless he came with us by boat. Grudgingly and grumblingly he capitulated, and the trip was on. It lasted three days. The Ruzuruhuru estuary was ruled out and Sinamwenda was chosen.

Before anything got started, Edney and Boughey left UCRN and went to America.[22] I found myself landed with the Nuffield Kariba project based on their estimates, almost at the start of the year 1962. How the figure of £35,000 was made up is still a mystery. It was meant to build and sustain a lake-shore station but seemed to have no provision for boats, radio communication or even setting-up time. Perhaps research stations were expected to grow overnight like mushrooms. The whole thing looked impossible, but luck stepped in as usual.

By chance, I heard that a vessel built for the Game Preservation and Hunting Association of Northern Rhodesia, an amateur equivalent of Operation Noah, was lying idle in Andora harbour and was for sale. A quick visit to Kariba revealed the ideal boat, the MV *Erica*. She was the same-sized hull as the *Lord Malvern*, built in the same yard, and differed only in superstructure. She had a forward compartment with PK (toilet) and washbasin, a forward cabin with two bunks, a galley with gas stove and small refrigerator, a wheelhouse of half the beam, and a large well deck aft of the wheelhouse. She was forty-five feet in length at the waterline, and was powered by two 65 hp Perkins P6 diesel engines with marine gear boxes. She had a cruising speed of about nine knots and the clocks showed that she had done only 700 hours running. The price was £5,000. Such a bargain was too good to miss.[23]

Erica was bought and the research station began to take shape. A retired Royal Marine officer, Captain Robson, was made the station manager, and he set about building the station, complete with electric light, pumping facilities, living quarters and a laboratory. It took nine months, during which the Robsons lived on *Erica* and explored the Lake.

[22] Boughey went on to become the Founding Chairman (1965) of the Department of Population and Environmental Biology at the University of California, Irvine. See 'Interview with Arthur S. Boughey', 14 December 1973, *UCI Libraries*, <http://ucispace.lib.uci.edu/handle/10575/11331>.

[23] The *Erica* was launched on 3 January 1960 by Lady Hone, wife of the Governor of Northern Rhodesia, Sir Evelyn Hone. 'Erica Critchley was of old settler stock, being the daughter of F.J. 'Mopane' Clarke – the nickname given to him by Lobengula and by which he was known throughout his life.' R. Critchley, 'Operation Noah', *Oryx* (1959), 5(3), 100–7. The editor of *Oryx* added this footnote:

> On the point of publication, the following details were received of the new vessel, the first since the Ark to be built especially for animal rescue work. She will be named *Erica* as a compliment to Mrs Erica Critchley, than whom nobody has done more to arouse public opinion to the desperate plight of wild animals in Northern Rhodesia. *Erica* is being built by C.B.R. Bentall Ltd., in Salisbury, to the design of Commander Collumbell, managing director of the firm. ... There are four sleeping bunks, but the vessel can be operated by one man. Her chief function will be that of a parent ship from which boats can operate, but 180 sq. feet aft, which can also be used for stores, has been specially prepared for rescued animals. A searchlight with half a mile range has been fitted.

The *Erica* as it was in the 1960s

University College of Rhodesia and University of Witwatersrand, Nuffield Lake Kariba Research Station, Sinamwenda, *Report, 1962–1968* (Salisbury: University College of Rhodesia, 1969)

The *Erica* as it was in 2020 in the grounds of the Lake Kariba Research Station

Jonathan Waters, *Kariba: Legacy of a Vision* (Harare: New Zanj Publishing House, 2020)

Although I felt satisfied, I immediately had problems with the Nuffield Foundation. Why did I need such a large boat, the cost of which had eaten up so much of the grant? Explanations by letter got me nowhere. Sitting snugly in London, it was clear that no one there could visualize the size of Lake Kariba nor the size of the waves generated by winds during storms. The engineers had forecast that waves up to ten feet from crest to trough would be generated, and our early experience soon proved them correct. It was no lake for cockle-shells; even *Erica* could not be turned in heavy weather. She had to be kept bows on to the waves and throttled back; even then, waves came right over the wheelhouse.

Only a demonstration would convince my doubters of the wisdom of our decision to buy such a boat. It was eventually agreed that a high official of the Nuffield Foundation, Dr John McAnuff, should visit Sinamwenda and that Dr Cain from Manchester University should come with him.[24] I went up-lake with them. For the whole eighty miles from Kariba to Sinamwenda we had perfect weather; the Lake was like glass. As a demonstration of the dangers of the sheet of water, it was useless, and I could see the doubts in their minds growing into the conviction that *Erica* was far too big and expensive. The

[24] 'Arthur Cain was a polymath, and one of Britain's leading evolutionary biologists. A strong believer in, and advocate for, the importance of natural selection, he greatly influenced evolutionary thought during the period from the 1950s to the 1980s. ... he was chosen, in 1961, to represent Oxford University on the Inter-Universities Committee for research at Lake Kariba. He acted as the scientific secretary to this committee in the UK, and during this period he made several trips to Kariba and Johannesburg.' Arthur James Cain, 25 July 1921 – 20 August 1999, *Biographical Memoirs of Fellows of the Royal Society* (2008), 54: 47–57.

The Research Station at Sinamwenda in the 1960s

fact that she had carried every bag of cement and sheet of roofing material, a complete prefabricated wooden house, diesel engines, pumps, lighting plant, and, indeed, every nail and nut and bolt that had gone into the building of the station, was overlooked. She had even transported a concrete mixer.

We were afloat at dawn the next day for the return journey to Kariba. The harbour was sheltered, but I could see white horses on the Lake and there was a stiff northerly breeze. Once clear of the shelter of the estuary we hit the weather. It was not really severe, but spray flew everywhere. Robbie kept the throttles open and *Erica* performed. The kettle would not stay on the stove and we could not cook breakfast. Portholes were battened down, and my visitors hung on to anything they could to stay upright. The blow lasted only a couple of hours and then gradually subsided.

No word has been heard since of any criticism of the need for a large boat on the Lake. She is now twenty-one years old and still one of the most sea-worthy boats on the lake. Research has continued ever since. *Erica* has done thousands of Lake miles, and even carried out several rescue operations. She is a proud old lady, and, like a certain even older gentleman, is still going strong.

In 1966 the opportunity came to become a director of two mining companies - parts of one group and, therefore, closely related.[25] This was a tempting offer, but it posed a major problem in terms of the prevailing climate towards outside work and the retention of fees. It was also a novelty, as no such offer had previously been made to any member of staff. To me it signified more than just a personal matter. It seemed to indicate that the outside attitude towards the College was beginning to soften, and it might even be a turning point of significance in public relations.

I consulted with the Chairman of Council, a banker of distinction. His reaction was interesting and, after some initial hesitation, he agreed that I should accept 'providing it did not interfere with my university duties'. A relaxation in the general situation on consulting soon followed, always with the same proviso. This was difficult to interpret, as it could have a number of meanings. In my case, for instance, eleven board meetings each year, with a

[25] This was the Rio Tinto group, now RioZim. 'Rio Tinto Southern Rhodesia Ltd. was incorporated in 1956 and was set up initially to develop and mine the Empress Nickel deposit in Gatooma and was the first mining operation to be set up outside Europe by Rio Tinto plc in March, 1957. By 1959 it acquired the Cam and Motor Mines at Eiffel Flats. ... From 1965–1979, Rio Tinto (Rhodesia) Ltd detached itself from the parent company, grew into the single largest money-earner and sanctions-buster, exporting millions of pounds worth of nickel, emeralds and tin, as well as becoming one of Zimbabwe Rhodesia's most successful manufacturers.' 'Rio Tinto Southern Rhodesia Ltd', *Rhodesian Study Circle*, <http://www.rhodesianstudycircle.org.uk/rio-tinto-southern-rhodesia-ltd>.

Graduation ceremony, 15 May 1976

few visits to mines, would not constitute any great reduction in the time spent on pure university business, but my direct association with the individual mining house might prejudice my relations with others in the same field of activity. The chance seemed to me to be worth taking and I accepted.

In the result, I think the university lost nothing, because my relations with other mining companies seemed, if anything, to be improved. There is a considerable camaraderie among the very top echelons of those involved in our mining industry, and this was enhanced by the difficulties created by sanctions and the war situation.

In another direction, I became a little less ignorant about finance, and this soon helped me considerably as Vice-Principal to comprehend the complexities of university budgets and general financial management.[26] Professor Robert Craig, as Principal, professed ignorance of financial matters and left most of it to me. Since he was a Scot, I found his alleged ignorance about such matters genetically improbable.

[26] He was part-time Vice-Principal from July 1974 and retired as Professor of Geology in 1977 to become full-time Vice-Principal and Deputy Vice-Chancellor. He was Vice-Principal from June 1981 until his final retirement in June 1982.

– 17 –

The Natural Resources Board

In 1941, the Southern Rhodesian legislature passed an Act that made provision for the setting up of the Natural Resources Board.[1] Its grassroots support was to be a network of voluntary farmers' organizations to be called Intensive Conservation Areas.[2] Because of World War II, the start of the ICA movement was delayed, and the first committee was formed in 1946. But it quickly gained strength among the European farming community, and in a few years grew into a powerful influence for good conservation of soil and water.

I was soon aware of the remarkable degree of conservation consciousness not only among the farming community but also among the general public even in the towns and cities. Our white farmers have, of necessity and even by choice, been frequent visitors to their nearest centre of business and industry, and this may have helped to spread the conservation message, but the media gave much assistance. The movement has been of incalculable value to this country, which has certainly been the leader in soil and water conservation in Africa. In my wanderings in other African territories, I have been able to compare conservation and farming skills in many parts, and none came anywhere near the level attained in this country.

If the same foresight had been applied in so many territories in which the threat of desertification has now become a grim reality, much land would have been saved for food production. Even Australia, so much of which is desert or semi-desert, seemed to shrug off accelerated erosion as no particular threat when I did a quick tour a few years ago. The eastern strip of Australia is the really fertile area, but flying over it I could see donga and sheet erosion of an alarming extent. When it was brought tactfully into conversation, the answer nearly always was along the lines that if land was being degraded it

[1] Natural Resources Act [*Chapter 21: 13*], Part II, sections 3–21.

[2] Natural Resources Act, Part V, sections 33–44. 'If the holders of land in any area wish on their own initiative to undertake the construction of works and other measures for the conservation or improvement of natural resources in such area they may petition the Minister in writing to declare such area to be an intensive conservation area.' Section 34(1).

did not matter greatly. By cutting down a few more million gum trees, plenty of new land could be opened up for agriculture. After the conservation-conscious attitude in this country, such a reply was horrifying.

Very naturally and very properly, the Natural Resources Board came to have a most enviable reputation as a power in the land. I respected it greatly, and I admired the quiet and efficient way it went about teaching the community as a whole that soil and water were the most precious of the country's resources.

At the same time, the message came across loud and clear that soil in a tropical and sub-tropical region is fragile and in a very delicate equilibrium. If this is upset, damage that cannot be repaired in the time span available to man can easily occur. Dongas must be detected early and steps taken to remove their cause so that healing can take place. Sponges in vleis must be protected so that rivers can run longer in the dry season. Sheet erosion must be prevented by a good cover of vegetation so that river courses are not choked with sand. All this, and much more, was brought constantly to the attention of farmers and the general public, and backed by sound technical advice based on research.

I was surprised, in about June 1977, to open an official-looking letter and find myself invited to become a member of that prestigious body. I was to replace, for the last three months of that year, a member from Matabeleland, a well-known farmer who had died after several years of service on the Board. In addition, from 1 January 1978, I was to take on a three-year spell as a member and to be Vice-Chairman under Lance Smith. I joined, and began to realize the complexity of the Board's activities. Before long I concluded that to

What are the Natural Resources?

They are: **THE SOIL, WATER and MINERALS**

THE ANIMAL, BIRD and FISH LIFE

THE TREES GRASSES and other VEGETATION

LANDSCAPES, SCENERY and other NATURAL BEAUTY

Who are their Trustees?

EVERYONE OF US . . . . AND PARTICULARLY

The Natural Resources Board

Salisbury, Southern Rhodesia

NRB

be a fully functional member three qualifications were required: one should be young, retired and rich. I possessed none of these qualifications, but it was some comfort to observe that none of my fellow Board members conformed to this ideal either. We all did the best we could, busy though we all were with our normal daily lives.

There was a certain amount of muttering in Matabeleland on my appointment. They had lost a member; he was also a farmer. I was based in Salisbury and had never been a farmer. The mutterers were mostly old friends from Bulawayo days; I met and talked to as many as possible. My role on the Board would be, I hoped, in line with the broader concept of the Board's responsibilities that the Chairman had in mind. Soil and water would not be neglected, but the Board would take an active interest in all the resources of the country, not forgetting its human resources. The mutterings from Matabeleland gradually seemed to subside. I was fully in sympathy with the way the Board viewed its overall responsibility as the conscience of the nation to see that none of its resources was squandered or mismanaged, if this could be avoided.

Lance Smith revived one of the major committees of the Board, which had been dormant for a few years. The Environmental Conservation Committee became my special responsibility as its Chairman. Terms of reference were agreed, and they could be roughly summarized as an investigation of environmental pollution of all kinds and on all scales, stimulation of public awareness, and an examination of the law relating to all kinds of pollution. The membership was small, highly specialized and basically scientific. It was decided to work in two ways, firstly by inviting experts to give evidence, and secondly by making visits to potential or actual centres of pollution.

I was pleasantly surprised how easy it was to get the real experts to come and discuss their specialities, and to put their evidence in writing to build up the data from which the Committee hoped eventually to compile a report on the state of pollution in this young and relatively unspoilt country. Such a report could be a baseline from which trends could be measured, or at least seen, in the future. Without people there would be no pollution, but people need industry, and most industries - and by this term one embraces farming and mining as well as the urban-based forms of industrial activity - are potential polluters of our environment.

After two years of monthly meetings, the Committee had gathered an impressive body of evidence on water, air, pesticide and noise pollution. The distinctions between global, regional and local kinds of pollution were becoming clear in our minds, as were the differences between blatant and cryptic forms of environmental hazard. The Midlands of England, the coalfields of South Wales, and many other areas of heavy industry, provided us with awful warnings of blatant pollution, rampant, out of control, and callously inflicted on an innocent countryside by unplanned growth and total disregard of even elementary scientific considerations.

There were terrible lessons to be learnt from the old mistakes in the developed countries. In the long run, prevention of pollution is cheaper than its cure, and the technology exists now to make industry of nearly all kinds virtually pollution-free, but at a price in initial capital outlay. Surely, in an unspoilt country, that price is worth paying. The attitude of top management was most encouraging, and almost without exception their determination to limit all forms of pollution to an acceptable level was evident. Total prevention is often impossible.

Cryptic pollution is a much more difficult problem to define and to control. It is mostly concerned with the long-term effects of certain pesticides on the ecology; without them, economic agriculture has become virtually impossible. It is also concerned with trace quantities of certain substances, mostly heavy metals, which escape into waterways. There have been great tragedies involving the deaths of hundreds of innocent human beings caused by the escape of minute traces of mercury from industry. Over long periods, as much as twenty-five years, air minutely contaminated by certain metals can result in cancers of specific organs. Modern knowledge can practically eliminate such hazards - if proper vigilance is maintained. Even the entertainment industry is thought by some to provide a hazard to hearing, to which millions of young people eagerly submit themselves.

– 18 –

Broadcasting

My association with broadcasting came about through a distant relative of my mother, Thea Alcock, who was a BBC producer in the late 1940s and used to spend holidays in our village. I met Miss Alcock on a long leave in 1948, and was persuaded to do a fifteen-minute programme on BBC overseas channels on 'Bushman Paintings in Rhodesia'. This was done completely solo from a prepared script, from Broadcasting House in London – *live*. This was really throwing the novice in at the deep end, but it went off tolerably well and was reprinted in *The Listener*. Before this effort, I had always heard that broadcasting was a nerve-shattering experience. After it, I knew it really was, and even now I can have some disconcerting attacks of microphone nerves, especially if someone I'm interviewing seems to be feeling jittery; perhaps it's infectious.

While I was living in Bulawayo, I did very occasional short broadcasts, but had quite a lot of contact through Charles Hodgson, who for a long time ran a programme in *Young Rhodesia* (or its equivalent) called *Information Desk*. This frequently had scientific questions, of which two in particular caused problems. One was: 'How does a cat purr?', asked by some innocent ten-year-old; the other was about hippo that run out of oxygen under water, rise quietly to the surface, take a huge lung full of new air, and quietly sink out of sight. Try it in your pool and you will see.

The first of these questions eventually had me writing to Professor Bullough at Birkbeck College in London, who replied very promptly.[1] His letter was a beautiful exposition of *why* a cat purrs but completely dodged the

Note: This chapter was originally published as an article in *Zimbabwe Science News*: G. Bond, 'Science broadcasting in Zimbabwe: The lighter side', *Zimbabwe Science News* (1980), 14(6): 131–2, 134.

[1] William S. Bullough was Professor and Head of the Zoology Department at Birkbeck for twenty-nine years. 'Professor Bullough was also a gifted teacher: he took his work and enthusiasm to a wider audience with television programmes.' 'Obituary: William S. Bullough', *Birkbeck, University of London*, <https://www.bbk.ac.uk/about-us/obituaries/obituary-william-s-bullough>.

issue of *how*.[2] We eventually got some kind of answer, but the one on hippo has defied and worried me ever since.[3]

Then Bruce Baker, who was in charge of studios in Bulawayo in the old Post Office building next to the Museum, asked me to do a series of weekly five-minute talks on geology, also in *Young Rhodesia*, so I came to have my own programmes.[4] I called these *Stones and Bones* and kept all the scripts. Again, I found it a bit nerve-racking because I had to talk to a microphone like a person, all alone in the studio, and I could see Bruce frowning at his stop-watch in the control room. I still envy the professionals their ability to talk to a microphone in so personal a manner. Monica Marsden used my scripts again many years later.[5]

When we moved to Salisbury in 1960, I thought my broadcasting connections would be cut off, and so they were for a short time. Then the Rhodesia Scientific Association (RSA) decided to persuade the Rhodesia Broadcasting Corporation to run a weekly *Science Half-hour* programme. Leslie McKenzie presented the first series.[6] I took part in one or two and got to know Leslie,

[2] A 1991 study concluded that 'the primary mechanism for both the sound and the vibration of purring is a centrally driven periodic laryngeal modulation of respiratory flow.' Dawn E. Frazer Sissom, D.A. Rice and G. Peters, 'How cats purr', *Journal of Zoology* (1991), 223: 67–78.

[3] 'Hippos can even sleep underwater, using a reflex that allows them to bob up, take a breath, and sink back down without waking up. Yet despite all these adaptations for life in the water, hippos can't swim – they can't even float! Their bodies are far too dense to float, so they move around by pushing off from the bottom of the river or simply walking along the riverbed in a slow-motion gallop, lightly touching the bottom with their toes, which are slightly webbed, like aquatic ballet dancers.' 'Hippo', *San Diego Zoo Wildlife Alliance*, <https://animals.sandiegozoo.org/animals/hippo>.

[4] Bruce Baker was Deputy Director of the RBC when he retired in 1978. He had joined the Federal Broadcasting Corporation in 1955 and was Regional Manager for Matabeleland in Bulawayo until he returned to Salisbury in 1965 as Head of Programmes in the General Service. 'Broadcasting official retires', *The Herald*, 30 September 1978.

[5] Monica Marsden was the producer of drama shows put on by the Runyararo Drama Club in Mbare (then Harari). She 'endeavoured to make township life a little more interesting. She also fought the colour bar through entertainment by organising and staging drama shows in the city centre. The white community soon realised that for educating the community on health, social and political issues.' Joyce Jenje Makwenda, *Zimbabwe Township Music* (Harare: The author, 2005), 27–8.

She was the author of more than thirty children's adventure tales and mysteries, including *A Matter of Clues: A Light-hearted Rhodesian Adventure for Girls* (London: Collins, 1962). 'British & Irish Women Writers of Fiction 1910–1960 (Ma)', *Furrowed Middlebrow*, <http://furrowedmiddlebrow.blogspot.com/2013/01/british-women-writers-of-fiction-1910_39.html>.

[6] Leslie McKenzie started his broadcasting career with the SABC in 1959. He went to Rhodesia in 1968, where he ended as the head of the RBC's commercial radio and television division, before returning to South Africa in 1982, where he joined Springbok Radio and became head of station until it closed down in 1985. Thinus Ferreira, 'Legendary radio broadcaster Leslie McKenzie (80) dies', *Channel 24*, 30 March 2019, <https://www.news24.com/channel/The-Juice/News/legendary-radio-broadcaster-leslie-mckenzie-80-dies-20190330>.

who by then was our next-door neighbour but one in Pomona, quite well. The second series was presented by Professor Bursell,[7] and a pattern of alternate quarters with a wildlife programme began to emerge. I also took part in some of Professor Bursell's programmes. They were all done in the old broadcasting centre in Manica Road. This old building had never been intended for studios, but it had a remarkable atmosphere and served its purpose surprisingly well.[8]

There was an occasion when Leslie McKenzie was recording during his series, and I was in the middle of talking about something or other, when a programme assistant came in and said to him: 'Mr McKenzie, it's four minutes to six and there is no one to read the six o'clock news.' Les stopped the recording and disappeared. He came back at a quarter past six having read the news and weather forecast 'cold', picked up the discussion, and went on as though nothing had happened.

One series of thirteen programmes was enough for Einar Bursell, and the lot fell on me to take it over. For a time, the RSA produced all the ideas and the panels and I presented, but gradually this fell away, and I have done it ever since as part of the normal RBC/ZBC schedule, alternating every quarter by quarter, with *Wild Life Forum*.[9] The programme has had three different titles: *Looking at Science* was the original one; then *Scientifically Speaking* (a title borrowed from the BBC); and then *Science Forum*. In total, I have now done 251 of them, which has taken about ten years, and that, in brief, is its history.

I have had quite definite aims in mind in shaping these programmes. They are aimed at the general public, not at the specialist, and broadcasting must entertain as well as inform. Zimbabwean problems must loom large, but visiting scientists are grabbed and made to talk whenever possible. I have tried to make them informal and unscripted, except for a brief introduction which must be written in 'spoken English', a very different language from written English. They must, as far as possible, be a private discussion on which the microphone just happens to be eavesdropping.

Just about every staff member of the ZBC has, at one time or another, been the producer, and each of these professionals has taught me something new

[7] Einar Bursell had joined the Department of Zoology at UCRN in 1959. J.P. Loveridge, 'Einar Bursell FRSSAf 1925–1994', *Transactions of the Royal Society of South Africa* (1995), 50(2): 169–72.

[8] This was in the Founders Building, 75–77 Manica Road (now Robert Mugabe Road). It had become the Main Post Office in 1911. Part of the building was used by the GPO as a broadcasting studio and the Radio Branch took over the whole building in the 1940s. Peter Jackson, *Historic Buildings of Harare* (Harare: Quest Publishing, 1986), 46. The Corporation (now the Zimbabwe Broadcasting Corporation) was established in 1958 and moved to its present premises in Pocket's Hill in the 1970s.

[9] *Wild Life Forum* was presented by Peter Rollason. 'The first edition went out on the air in October 1969. It ran on a system of three months on and three months off air, until 2004, a total of some 35 years. We broadcast some 850 editions. It was one of the most popular programmes ever broadcast.' Peter Rollason, *One Man in His Time* (Bulawayo: The author, 2008), 92.

about the business of interviewing. Discussion programmes are essentially interviews, and there is more than one style of interviewing. I have regarded the programme as belonging to the interviewee, the interviewer's role being merely to start it off, to keep it going with occasional questions and, finally, to wind it up. But much depends on the nerves of the interviewee. For instance, the topic of one programme was 'Explosives', and I had a preliminary discussion with the two experts in my office. They were real extroverts; both had given up the 'filthy habit' of smoking and tried to persuade me to do likewise. The omens were set for a lively and easy recording session. However, when faced with a microphone in the studio, they were both so nervous that they could hardly say more than 'yes' or 'no', and within five minutes were begging for cigarettes. I did more talking than they did. I learnt another useful tip from this difficult session: don't let people answer the questions fully in the preliminary discussion or they will invariably say something else when the time comes during the recording.

When dealing with unfamiliar topics, and even with familiar ones, I like to get a few questions from the interviewee beforehand. These starters don't preclude going off at a tangent during the programme, but it is important to avoid 'throwing' the interviewee by slipping in something totally unexpected. This produces long awkward silences that have to be tightened up afterwards in an extra editing session; it also upsets the timing. Thirty minutes of air time means 28 minutes and 30 seconds of talk time, to which are added the opening and closing 'billboards' which are pre-recorded. To sit down with two or three people, wait for the cue after the opening billboard, go through the necessary talk time, and wind up so that the closing billboard finishes a few seconds before the full thirty minutes is rather like doing a hole in one at golf, except that finger signals from the producer through the glass panel are a great help. Two contrasting situations frequently occur: one is that the panel dries up two minutes early, and I make a frantic effort to 'stretch' the discussion; the other is when one of the panel gets into his stride two minutes before time and I have problems stopping him and winding it all up.

Controversial subjects crop up from time to time. The proposal to ban DDT, on the grounds that certain predatory birds may become locally extinct because of their extreme susceptibility to DDT poisoning, has been given a good airing, both from the angle of the environmentalists and that of the agriculturalists. I was surprised how little mail this generated, even though it is a highly emotional topic.

I looked forward to a good fight between a water diviner and a geophysicist, but that turned out to be rather a damp squib. They agreed more than they disagreed on the question of the behaviour of ground water, though they differed on the nature of the 'force' that acts on the rods.

While two American biologists were here, I discovered that they were out-and-out Fundamentalists, and firmly believed the literal truth of Genesis Chapter One. They were fluent talkers and had a wealth of evidence at their fingertips. However, when they tried to persuade me that there was a record of a trilobite on the same slab of rock as a print of a human foot wearing a sandal, and that dinosaur and human spoor occurred on the same bedding plane, this was too much for my geological upbringing to swallow. For once, I abandoned my neutral role, and climbed into them boots (sandals?) and all. But I failed miserably to make a dent in their convictions.

Not all programmed material is collected in the studio, and many people react much better in the informal atmosphere of their office or mine, or even out of doors. There are hazards in doing recordings outside the studio. Out of doors, aeroplanes fly past unexpectedly; in offices, telephones ring and people walk in without knocking; and there is always the risk of poor quality. I've done recordings in out-of-the-way corners of Lake Kariba on board the MV *Erica* with loud night noises as background. We recorded a programme on solar energy on the flat roof of the Engineering Faculty on a hot January afternoon, and we have done 'OBs' in laboratories. With a panel of three people, we try to get well-differentiated voices - an Australian man and wife plus an American, for instance, though this is an extreme case.

Just a few people can fill half an hour entirely by themselves. Mike Raath from the Museums and Monuments and David Mitchell could both do this, but they are in Johannesburg and Australia, respectively, though I got a full programme from Dr Mitchell while he was in this country on holiday.[10]

[10] Mike Raath was a former Director of the National Museums and Monuments of Rhodesia. From 1978 to 1987 he was Director of the Bernard Price Institute for Palaeontological Research at the University of the Witwatersrand, and at the same time was Head of the newly established Department of Palaeontology and Palaeoenvironmental Studies. Bruce S. Rubidge, 'The BPI: 50 Years of Palaeontological Activity', *Palaeontologica Africana* (1997), 33, 1–9. See also his tribute in Appendix 1, and his obituary to Geoffrey Bond in Appendix 2.

David Mitchell was 'a friendly and inspiring lecturer in the Botany Department at the University. He taught the ferns and other 'lower plants' to undergraduates but his particular interest was in aquatic plants and their ecology and he taught this section of the M.Sc. (Tropical Resource Ecology) programme at UR/UZ. He later moved to Australia and became Chief of the CSIRO Division of Irrigation Research and is now retired. 'Mitchell, David Searle', *Flora of Zimbabwe*, <https://www.zimbabweflora.co.zw/speciesdata/person-display.php?person_id=41>.

He received many awards, including the Hilary Jolly Award for 'service to Australian Limnology'; the Aquatic Plant Management Society, USA, President's Award in recognition of his career devoted to scientific discovery and contributions to aquatic plant biology and control; and was the 2003 International Fellow of the Society for Wetland Scientists for 'a lifetime of contributions to wetland science and management'. He was named a Member in the general division of the Order of Australia in the 2014 Queen's Birthday Honours list. 'David Mitchell', *Institute for Land, Water and Society, Charles Sturt University*, <https://www.csu.edu.au/research/ilws/news/news-and-achievements/2014/david-mitchell>.

Mike neatly turned the tables on me when he was to present a series of thirteen programmes, called, I think, *Nature's Note Book*. He had his first panel lined up for recording on a Monday evening, but by lunch time all had backed out. He rang me up, and we met at Pocket's Hill at five o'clock. We eventually did thirteen programmes together, with hardly a preliminary note between us.

A certain amount of mail is generated, and I have a thick file, but not many useful suggestions come from listeners. Mail has come from surprising distances: the old Katanga Province of Zaïre, Zambia, Malawi, Republic of South Africa, and even South West Africa, have all been represented in the mail. But listeners' ideas and questions would never keep the series running, and it has been quite a battle to think up topics either for full-length programmes or shorter items for the magazine type of editions. The idea is one thing, finding the right person to deal with it is quite another, especially for the January quarter of the year.

In all broadcasting there are behind-the-scenes hassles of which the listener is blissfully unaware, and *Science Forum* is no exception. Dr Phelps did a lovely twelve minutes with me, on my own cassette recorder, on the unlikely topic of 'Sausage Flies'.[11] I took this in for dubbing on to an open-reel tape so that editing would be easy. Somehow the cassette machine in the studio got my tape in a tangle. It was late evening but I saw one of the engineers, who opened the cassette, took one horrified look, and handed the bits back to me on my clipboard.

Next day, I took the pieces to a good friend with neat fingers, in our own Educational Technology Television studio. He promised to try, and got a long way before he ran out of time. For safety, he left the job in his control room with the loose end hanging off the bench, so that this gossamer-thin strand should be free. No one is allowed into that control room but, in his absence, an over-enthusiastic minion went in with a vacuum cleaner, which swallowed the entire tape. The machine made a good recovery, but Dr Phelps had to do his recording all over again.

[11] Robert John 'Fido' Phelps was a glossinologist – a term the Rhodesian government gave to biologists employed by the Tsetse and Trypanosomiasis Control Branch of the Agriculture Research Council, where he worked. 'Phelps, Robert John "Fido"', *Flora of Zimbabwe*, <https://www.zimbabweflora.co.zw/speciesdata/person-display.php?person_id=224>.

– 19 –

Near the End

This ramble through a lifetime has covered very nearly the traditional three score years and ten. Chronologically, it falls into three sections: early life in England, the Museum period, and the University period. They have not been given equal space for what seem to me to be good reasons. The first period was the most formative, and it has been treated self-analytically, largely so that I could find out what sort of a person I thought I was. This may be quite different from other people's view of me.

The second period was, geologically, the most productive, and by far the most interesting as far as I was concerned. I have often wondered whether I came to this country fifty years too early, fifty years too late, or approximately at the right time to suit my temperament and particular inclinations. Since no one has any choice of arrival date on planet Earth, it is rather meaningless speculation, but, if I had the choice, and with the benefit of hindsight, I think I would have chosen just that period. The Valley that was to be my main area of activity in the field was still wonderfully wild and known to only a few, privileged, people. Nature gave me a reasonably good physique to traverse it, and the kind of temperament that enjoyed the solitude that can be found in such an environment.

The nature of my employment gave me the chance to capitalize on the opportunities for research. I saw something of the way in which the pioneers of geology laid the foundations of geological science in this country, and I came to appreciate the immense value of their early work. At a later time, I had the benefit of working under modern conditions, with topographic maps, controlled vertical air photographs, and instant radio communication with my base from even the remotest spot - so long as the battery of my Land-Rover was in good order. This lent an air of great security, for, in the event of serious mechanical breakdown, illness or injury, I could count on a rapid rescue operation from my own department. All I had to do was to give an accurate map reference and help would arrive in hours, or at most in a couple of days.

This was in contrast to my solitary wanderings while I was based in the Museum in Bulawayo. Then no one knew precisely where I was, nor when

I would reappear. A serious breakdown could have involved me in a walk of perhaps 150 miles to the nearest telephone or radio. I drove very carefully, knowing full well the inconvenience of a broken half-shaft, a hole in the sump, or even a miscalculation of my petrol position, and such a situation never arose. This was in great contrast to the slap-happy treatment meted out by some university postgraduate students to the transport with which they were provided. One even managed to break not only a back half-shaft but a front one as well - on the same journey. Being equipped with a radio transceiver, all he needed to do was call for help and wait for it to arrive.

Healthwise, I was equally lucky. Never did I fall into a hole and twist my ankle, or get bitten by a snake, though I have been stung by scorpions several times. I have never had malaria; some people seem to possess a natural resistance and, beyond the old-fashioned Paludrine, I have never taken any other anti-malarial drug. Minor tummy upsets have occurred, probably due more to Epsom salts (magnesium sulphate) in many natural waters in the Valley than to their bacterial content. Some of the waters have been of very doubtful purity, but water is water, and one drinks what one can find. Boiling was usually sufficient precaution, though if it was got by digging neat holes in sandy river beds, I usually regarded it as well filtered. Elephants find water under sand cover and drill for it with their trunks. I drank such water frequently. It often tasted strongly of elephant, but it was welcome just the same.

I spent three uncomfortable days under a tree in the Ume valley with a pain in my guts, thinking I had got it from some particularly unsavoury water, but it passed off. Back in Bulawayo I described the symptoms to my doctor. He assured me that I had given a textbook description of appendicitis, so he took it out. I think that the Museum geologist probably gave his guardian angel a good deal of extra trouble.

I told this story to a group of second-year geology students on a field trip while we were camped at approximately the same place, and suggested to them that to be without one's appendix was a distinct advantage for a field geologist. Several days later, while we were camped near Selukwe, one of them developed all the symptoms of appendicitis and was duly deposited in Gwelo hospital. I contacted the surgeon next morning to find out how the operation had gone, but I was told that no operation had taken place. The surgeon quizzed me about the student and how our trip had gone. I told him that I had recounted my own experience of appendicitis in the Valley. This very astute surgeon had suspected something like this and explained to me that I had virtually talked the student into believing that he had pain in his abdomen, that he was being given harmless pills and being made a fuss of in hospital, and that he would probably never have a recurrence. He never has to this day.

This was yet one more example of the powerful influence that a professor can exert on impressionable young minds. I have tried to avoid doing things that were obviously stupid, and I treated Africa with the respect she deserves. In return, she has been exceptionally generous. Probably the silliest thing I did was to stalk a sleeping male warthog with a spear borrowed from my Batonka assistant. Luckily he woke up and made off at speed when I was about two yards away.

Baboons were a menace to the villagers in the Valley, and for good public relations I shot quite a few. I hit one old male in the shoulder and found that I had used my last round. It would have been wrong to leave him to suffer, so I followed the blood spoor, caught up with him in a boulder riverbed, and finished him off by stoning. In the process he was understandably aggressive, and my respect for our Stone Age ancestors was greatly increased.

I soon learned to stop still when I, or my Batonka assistant, spotted an elephant and look around 360 degrees. This generally disclosed many others, and prevented me from blundering right in among the herd. Rhino I feared most of all for their unpredictable nature and perpetual bad temper. On a particularly hot day, I was traversing on a rhino path with the freshest of spoor, when I came to the edge of a vlei filled with tall dry grass, into which the spoor vanished. The choice was either to brave the grass and go through the vlei, perhaps a hundred yards wide, or make a detour of a couple of miles round its head. I took a chance. In the middle of the vlei, there was a sudden tremendous commotion and I braced myself to dodge the charge. Out popped a beautiful bushbuck, but in my state of nerves it was hard to realize what had happened. The noise made by that small antelope was out of proportion to his size, and it was greatly magnified by my nerves.

The Museum period was full of adventure, mostly self-inflicted. By contrast, the University period was dull, though the things that happened were probably of much more lasting significance. The students turned out by the department of Geology have dispersed to many corners of the world, and many of them have been very successful in their careers. The University itself has grown in size and in its importance as a provider of skilled manpower in many disciplines, and its influence on the development of the country has likewise increased. But in many ways it is like so many other universities, which is just as it should be. However, in consequence, life in one university can be very like life in another, so I have given less space to this period. Some historian will write it all up one day.

I have one period left: retirement. I think I am coming to terms with old age but, hopefully, have not yet taken refuge in senility. I like to think that in both the Museum period and in my time at the University of Zimbabwe, I have put something back into the country that has given me so much.

I suppose that, in remembering those people
who shaped my way in the world,
and to whom I have tried to express my belated thanks,
I ought to include Hitler.

Without his War,
I would never have had anything to write about.

Appendix 1

Geoffrey Bond
by some who knew him

The Colleagues

Terence Coffin-Grey

Terence Coffin-Grey was born on 19 July 1926 in Port Elizabeth, South Africa. He married Yvonne in 1948 and they moved to Bulawayo in 1950. Terence was soon mounting birds and mammals in their small dining room. He responded to an advertisement in the *Bulawayo Chronicle* for a museum taxidermist/technician and was appointed by Reay Smithers, the Director of the National Museums of Rhodesia.

He and Yvonne started Taxidermy Enterprises in the 1960s, and he worked at the National Museum until 1978 and then co-founded the Taxidermy Association of Southern Africa (TASA). They moved back to South Africa in the early 1980s. On 24 July 2021, Terence was officially inducted into the Taxidermy Hall of Fame. His daughter, Cathy Peacock, asked him about his memories of Geoffrey Bond.

Dad recalls that Geoff was a very friendly, welcoming man. As a young man, aged twenty-four, it was heartening for him to be treated as an equal by a man of learning, a scientist. He says that the rest of the staff, except for Reay Smithers, were a little stand-offish. They both loved the veld.

Dad remembers him as an absolute gentleman and that it was 'quite a blow to him when Geoff left the Museum to become a University lecturer'.

He says that one could track Geoff in the bush by two things: the tracks of his short-wheelbase Land-Rover and a trail of Guineafowl feathers. He and his companion/co-driver/cook Million lived on Guineafowl. He remembers Geoff's rifle, a .22 Hornet; he never used a shotgun.

Dad recollects that Geoff walked miles. Million would drive ahead and make camp. Dad inherited Million when Geoff left the Museum, and remembers him as a very trustworthy man.

Although Geoff was employed by the Museum, he was permitted to do private jobs. He was involved with the Sandawana emerald mine and received a very big piece of rock containing a sizeable emerald, which became part of a Museum display. His office was never empty. There were always prospectors bearing samples for examination.

Mike Raath

Mike Raath was born in South Africa, grew up in Rhodesia, and obtained a B.Sc.(Hons) degree and teaching diploma from Rhodes University. He began his working career as a high-school biology teacher in Bulawayo, but resigned after two years and joined the National Museums of Rhodesia as zoologist at the Queen Victoria Museum, now the Zimbabwe Museum of Human Science, in Harare.

During this time he completed a Ph.D. through Rhodes, with Geoff Bond as co-supervisor of his thesis. They had met through the Museum, as Geoff was Chairman of the Board of Trustees, and their shared interest in palaeontology cemented the bond. In 1978 he emigrated to South Africa to become Professor of Palaeontology at the University of the Witwatersrand.

A brief interval of eight years followed at what was then the Port Elizabeth Museum complex (now Bay World, Gqeberha), followed by a return to Wits to see out the rest of his working career until his retirement in 2006. He now lives in Simon's Town, South Africa.

I first came into contact with Geoff Bond when I joined the National Museums in 1965 as Zoologist at the Queen Victoria Museum. I had had a wish to follow a career in active research ever since my Honours year at university, encouraged by my academic mentor there, Dr R.F. 'Griff' Ewer FRS. So, as soon as I got an opportunity to leave teaching on the collapse of the Federation of Rhodesia and Nyasaland, I eagerly took it and luckily got a job at the Museum.

Professor Bond, legendary Head of the Geology department at the University College of Rhodesia and Nyasaland, was a member of the Board of Trustees of the National Museums and Monuments that employed me. As I had an interest in palaeontology, Bond seemed to show some interest in my work. I was flattered by this attention, as it came from one who inhabited the exalted uppermost levels of the organization of which I was but a new-come green junior, still very wet behind the ears. But as my time at the Museum passed, I was fortunate enough to gain promotions that brought me into ever closer direct association with Professor Bond through the meetings of the Board of Trustees, which I attended *ex officio*. Geoff was by then Chairman of the Board, a position he held for a record thirteen years until he retired, not long before his death.

While I had been a high-school teacher at Northlea school in Bulawayo, I had been involved in the excavation and study of a small vertebrate fossil that had been found by boys in the school's Archaeology Club in the Nyamandhlovu District, Matabeleland, in 1963, the year before I joined the school from university. My studies had revealed that this little fossil was the almost

Geoff Bond in the excavation pit next to the sacrum and pelvis of a specimen of the giant sauropod dinosaur, *Brachiosaurus* (now *Giraffatitan*), Kadzi river, Zambezi Valley

Photo: Mike Raath, 1974

complete skeleton of a new small theropod (meat-eating) dinosaur, which it was my privilege to prepare, describe and name as a species new to science.[1]

[1] Tim Broderick reports: 'This was named *Syntarsus rhodesiensis*, later arbitrarily renamed *Megapnosaurus rhodesiensis*, as an entomologist had found a bug with the name *Syntarsus*. It was Mike Raath's identification of an early dinosaur (sauropodomorph) femur from the Dande in 1994 that eventually led to the naming of *Mbiresaurus raathi*, a small but much earlier dinosaur.' This recently made the news following the publication of a study in *Nature*: Shingai Nyoka and Oliver Slow, 'Africa's oldest dinosaur found in Zimbabwe', *BBC News*, 1 September 2022. They note that: 'The near-complete skeleton of the *Mbiresaurus raathi* is stored in a room in a museum in Zimbabwe's southern city of Bulawayo.' For more details, see Michael Greshko, 'Fabulous 230-million-year-old fossil is Africa's oldest known dinosaur', *National Geographic*, 31 August 2022.

Geoff Bond on Island 126/127 off Bumi, Lake Kariba, where the remains of a new basal sauropod dinosaur, *Vulcanodon karibaensis*, were discovered

Photo: Mike Raath, 1974

My interest in this little dinosaur shaped my research direction, so I focused on dinosaur remains in Rhodesia. This had also been a long-standing interest of Bond's, who had published extensively on his own research, so our paths crossed on several levels. We accompanied each other on several fossil-hunting field trips, and this initiated and cemented a mentor-mentee relationship between us, whose memory I cherish to this day.

As time moved on, I eventually had the good fortune to succeed my original chief, the equally legendary Dr Reay H.N. Smithers, as head of the National Museums and Monuments of Rhodesia when he retired. This also meant that the bond between Geoff, as Chairman of the Board of Trustees, and me, as Executive Director of the National Museums and Monuments, became even closer.

The times we spent on field expeditions together, especially into the Zambezi Valley, which both of us loved, were times to be cherished, usually in search of dinosaur fossils, but anything else counted as well. Geoff was fun to be with in the field. He had a wealth of amusing anecdotes, he was an engaging conversationalist, his geological knowledge was encyclopaedic, like me he was a fan of Gilbert and Sullivan operettas, and like me he was a chain-smoker!

He was a very effective popularizer of science, and he did a lot of popular radio and television work. He was a favourite among the production teams at the Rhodesia Broadcasting Corporation because of his easy-going, friendly demeanour, his attractive, well-modulated, beautifully enunciated speaking voice, his photogenic good looks, and his relaxed attitude even when things got frantic.

He had an endearing quirk when the sound engineers needed to take microphone sound levels before a recording, either on radio or television – he would recite his own little variant of *Mary Had a Little Lamb*:

Mary had a little lamb.
He had a sooty foot.
And into Mary's bread and jam
His sooty foot 'e put!

It invariably raised a giggle – and got the audio levels right for the recording!

Geoff was respected – indeed loved – by his staff, his students, his TV and radio audiences, and by essentially everyone with whom he came into contact. He was what some today would approvingly term 'a Mensch'. Those of us who knew him personally were privileged to be a peripheral part of his life, however small. Certainly that's how I feel. Marina, Ivan and Nina lost the most, but the rest of us also feel his loss. He is deeply missed and his memory is treasured.

Craig Gibson

Born in New Zealand in 1927, Craig Gibson moved to Scotland in 1936. On leaving school in 1945, he volunteered for the Royal Navy and after demob in 1948 went to Edinburgh University, where he obtained a First Class Honours degree in Geology.

He was recruited by Anglo American to work in South Africa before moving to Southern Rhodesia in 1955 as assistant to Dr A.M. Macgregor, who was then Consulting Geologist for Anglo. He subsequently joined Mineral Search of Africa, Rio Tinto's exploration arm, and was based in Salisbury. After he left Rhodesia, his later positions included CEO of Rossing Uranium in Namibia, Managing Director of Rio Tinto Metals in England, and, ultimately, Mining Director of Rio Tinto, London.

I first met Geoff Bond when he was Keeper of the Bulawayo Museum and I was Assistant Resident Geologist in the Salisbury, Rhodesia, office of Anglo American. It was 1955 and I had just been transferred to Rhodesia from exploration work in Namaqualand and the Eastern Transvaal lowveld.

As my first task, I had been asked to do a survey of the fire-clay resources around the Wankie coal mine. My first boss in Salisbury was Dr A.M. Macgregor, recently retired Director of the Geological Survey, and he arranged for me to travel to the colliery in company with one of the survey geologists who was also heading Wankie way. So we set off from Salisbury, with much of the way still being strip road.

Arriving in Bulawayo, the Survey geologist, Larry Watson, said that he wanted to stop off at the Bulawayo Museum and speak to the Keeper, a geologist and palaeontologist. I was invited to join him, so into the Museum we went and there I met none other than Dr Geoffrey Bond. I sat and listened while the two talked sediments and fossils, but as we drove on up to Wankie my Survey passenger filled me in on Geoff and his background.

My next contact with Geoff was in 1958 at the First Annual Congress of the Geological Society of South Africa held in Johannesburg from 11 to 14 February 1958. By that time I was working in Johannesburg for Mineral Search of Africa, Rio Tinto's exploration company in southern Africa. We chatted, but Geoff was mostly in contact with the collection of leading Society geologists.

Subsequently I was transferred to Salisbury and spent time on general exploration before being sent, in 1960, to the Sandawana emerald mine in the Belingwe TTL as mine manager. I saw little of anyone in those days, but returned to the office in Salisbury in 1963 and, among other things, joined up with a group of geologists, members of the Geological Society of South Africa (GSSA), and we worked at establishing a Rhodesian branch of the Society.

Geoff was a leading light in that group, and his background as a Council Member of the GSSA was an enormous help in achieving establishment of the branch. Other participants in forming the branch included John Lenz and Chick Böhmke.

The branch was a great success and survives to this day as the Geological Society of Zimbabwe. Our outstanding success as a branch, of which Geoff became the Chairman, was the organisation of Granite '71, an international conference attended by geologists from all over the world.

Geoff had been elected President of the GSSA in that year and served as Conference Chairman. The Post Office issued a set of commemorative stamps.[2] His service to the GSSA, which he joined in 1952, was great, serving on the Council from 1959 to 1973 and becoming President in 1971. His work was recognized well beyond Africa's shores and he was awarded the Murchison Award by the Geological Society of London.[3]

Geoff at that time, apart from his many other posts, was a Director on the Board of Rio Tinto Rhodesia. In due course I was appointed to the Board as an Executive Director, and Geoff and I became the sort of geological sounding-board of what became Rhodesia's leading mining company, with gold mines, chrome mines, copper and nickel mines, as well as a gold refinery, a copper/nickel smelter and a refinery complex.

With his intimate knowledge of the Mafungabusi/Zambezi Valley area and its rocks - and indeed of its fossils, about which he had written so much - he was a great aid in Rio's assessment of the Sengwa coal field. In the event Rio did not start mining there, but I believe that interest is now being shown in the possible development of the deposit.

We became good friends, saw a deal more of each other, and before long started playing regular weekend tennis at the University courts. Joined by Marina and Margaret, we had many a happy weekend, but Geoff was clearly the most competent player. Our children played together and from time to time we shared an outing to the Rio Tinto 'chalet' at Inyanga. They were halcyon days.

[2] See 'GRANITE 71: Issued 30th August, 1971', in Rhodesian Study Circle, *Encyclopaedia of Rhodesian and Zimbabwean Postage Stamps since 1964*, <http://www.rhodesianstudycircle.org.uk/wordpress/wp-content/uploads/2016/12/028-1971-Granite-71.pdf>.

[3] Geoffrey Bond is listed as a Past Winner in 1958: 'This Fund is awarded to early career geoscientists who have made excellent contributions to geoscience research and its application, in the UK and internationally. Recipients must be within ten years (full-time equivalent) of the award of their first degree in geoscience or a cognate subject.' 'Murchison Fund', *The Geological Society*, <https://www.geolsoc.org.uk/About/awards-grants-and-bursaries/society-awards/murchison-fund>.

Granite '71

When I was young, I worked with Read,
Who wrote a monumental screed
On granites, and the mystery
Surrounding their long history.
And then, much later in my life,
I listened to the Learned Fyfe
Disclose by graphs and subtle mirth
The sordid story of their birth.
It seems they rise like blobs of oil
When crustal heating makes them boil.
But whatever the truth its been lots of fun
So roll on Tonalite 81.

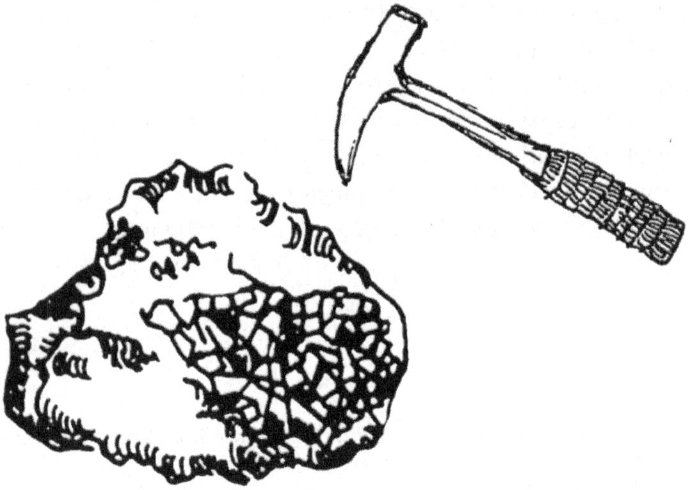

Geoff Bond, *Verse & Worse: From Bad to Worse*

David Cumming

Dr David Cumming graduated in Zoology and Entomology from Rhodes University in South Africa in 1961. He joined the Department of National Parks and Wildlife Management in 1964. After twelve years at the Sengwa Wildlife Research Institute, he became Chief Ecologist and headed the Branch of Terrestrial Ecology. He retired as Deputy Director in 1988 to set up the WWF Multispecies Animal Production Systems Project, which grew into the WWF Southern African Regional Program. He was Program Director until early 2001, and is now an independent consultant.

I was born in Pietermaritzburg, Natal, in 1939, and my family emigrated to Southern Rhodesia in 1947. We lived in Tjolotjo, some 60 miles west of Bulawayo, with its experimental farm and agricultural training school, for the next thirteen years.

My school days were spent as a boarder at Milton Junior and Senior Schools. School holidays were spent happily in the bush, in what was then the Gwaai Reserve, collecting birds for museums, ringing birds, and engaging in taxidermy (under the tutelage of Terence Coffin-Grey), falconry and photography. Whilst at senior school I had a special relationship with the National Museum in Bulawayo, and many Monday afternoons were spent there, helping with skinning specimens collected the previous weekend or going over the road to the National Free Library. All of this laid the foundations for a lifelong interest in biology and wildlife.

After graduation, I worked as a Fisheries Research Officer at Henderson Research Station for eighteen months and then tried my hand at teaching before being appointed a Wildlife Research Officer in the then new department of National Parks and Wild Life Management in 1964. I was immediately posted to Sengwa Gorge to set up a temporary field station to examine the impacts of tsetse-control hunting on large mammals. The intended eighteen-month posting ended up being twelve years. The temporary research area of 360km^2 became part of the Chirisa Game Reserve in 1968, and the Hostes Nicolle Institute of Wildlife Research was formally opened by Ian Smith in September 1972.

Early days at Sengwa were focused on game–tsetse research. Warthogs were the primary host of tsetse fly, and their ecology and behaviour formed the subject of my doctoral thesis, which was submitted in 1970. At the end of 1976, I moved to Head Office as Chief Research Officer (Wildlife).

It was during the period that I was at Sengwa, and later at National Parks HQ in Harare, that I interacted with Geoff Bond. Shortly after the inauguration of the Hostes Nicolle Institute, Roelf Attwell established a committee to guide wildlife and ecological research at Sengwa with Geoff Bond, who was Professor of Geology at the University of Rhodesia, as Chairman. With no plans for geological research in the area, this seemed, at first sight, an unusual choice. However, as a well-known host of a Sunday evening radio programme on matters relating to science and conservation, and in his role on the Natural Resources Board, apart from his standing as a scientist, it was soon apparent that this was an excellent choice.

Geoff enjoyed his bi-annual visits, in a light aircraft, to Sengwa, and wisely mediated over sometimes contentious issues, such as those relating to the management of elephant in the research area and problems that occasionally arose in dealing with visiting researchers and graduate students. On one occasion, Geoff invited the Principal of the University, Professor L.J. Lewis, to join him on a visit to Sengwa and sit in on a committee meeting. He also arranged for both the Hostes Nicolle Institute and the Lake Kariba Fisheries Research Institute (also managed by the Department of National Parks) to be affiliated to the University of Zimbabwe.

The committee included Prof. Einar Bursell, and David Mitchell from Biological Sciences at University of Rhodesia, Roelf Attwell, and Graham Child, Director of the Department of National Parks, with myself as head of the Institute and secretary to the committee. After Bursell and Mitchell left Zimbabwe, they were replaced by Prof. Kit Cottrell of Biological Sciences and Prof. Malcolm Blackie of Faculty of Agriculture. In the early 1980s, the name of the Institute was changed to Sengwa Wildlife Research Institute.

The committee generally met twice a year at Sengwa, and between December 1973 and 1981 Geoff chaired sixteen meetings; he and his committee made an invaluable contribution to guiding the Institute's research. During the sixteenth meeting in 1981 he indicated that he would soon be retiring.

One year, while we were still at Sengwa, he invited my wife Meg and me to 'house sit' his home in Harare while the Bond family were away on holiday. A key responsibility was to take care of the dormouse residing in the postbox at their garden gate. We were given strict instructions from Geoff to not overfeed his special dormouse.

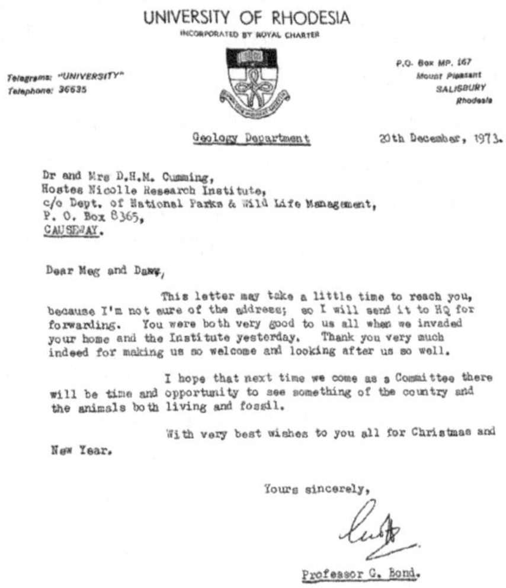
UNIVERSITY OF RHODESIA
INCORPORATED BY ROYAL CHARTER

Telegrams: "UNIVERSITY"
Telephone: 36635

P.O. Box MP. 167
Mount Pleasant
SALISBURY
Rhodesia

Geology Department

20th December, 1973.

Dr and Mrs D.H.M. Cumming,
Hostes Nicolle Research Institute,
c/o Dept. of National Parks & Wild Life Management,
P. O. Box 8365,
CAUSEWAY.

Dear Meg and Dave,

This letter may take a little time to reach you, because I'm not sure of the address; so I will send it to HQ for forwarding. You were both very good to us all when we invaded your home and the Institute yesterday. Thank you very much indeed for making us so welcome and looking after us so well.

I hope that next time we come as a Committee there will be time and opportunity to see something of the country and the animals both living and fossil.

With very best wishes to you all for Christmas and New Year.

Yours sincerely,

Professor G. Bond.

The Broadcasters

Jill Baker

Jill Baker spent her childhood years in Matabeleland before moving to Mashonaland at the age of fourteen. Marriage to a tobacco farmer meant that their lives on the farm ended shortly after UDI and Jill and her family moved to Salisbury, where she worked for the next fifteen years as a journalist, news and documentary presenter and producer in radio and television. She also opened her own business, Jill Baker Associates, in public relations and marketing.

On moving to Australia in 1983, Jill managed a radio station for five years, before opening her own business in consultancy to the tourism industry.

I remember being absolutely horrified when I was asked to stand in on an interview that should have been handled by John Pank to talk, of all people, to that most erudite scholar, Professor Geoffrey Bond![1]

I had always enjoyed his regular science programmes and found him to be an academic who spoke with a refreshingly common touch without in any way diminishing his eloquence or knowledge. But to interview him ... that was very different.

From what I can remember, Geoffrey Bond had just returned from a conference in Europe in the mid-1960s, having recently headed up the department of Geology at the newly formed University College of Rhodesia and Nyasaland. I was to talk on both aspects, and I had been given a twenty-minute slot. I knew absolutely nothing worthwhile about either.

I met him in the foyer and walked nervously up to the studios, where we sat for a precious half hour or so before going live – as things were in those days! I hadn't met him before, he didn't know me, and I confessed that I knew ridiculously little about what we were to discuss except that there had been some problems with student rioting in the dining hall. He seemed delighted!

[1] John Pank was born in Plymouth in 1939 and settled in Southern Rhodesia at the age of nineteen. He become private secretary to the Prime Minister, Sir Roy Welensky, qualified as a Chartered Secretary, and after the break-up of the Federation joined Anglo American Corporation in 1964. He was transferred to Johannesburg in 1976 and emigrated to South Africa. In his spare time, he became a 'highly professional and immensely popular television presenter with his own weekly quiz show, for which he won the coveted South African national Artes Award.' 'John Pank (24 October 1939 – 27 June 2019)', *Old King's Club*, <https://www.oldkingsclub.org.uk/obituaries/pank-john>.

It was only much later that I grew to appreciate and know his wonderful gentle humour and wit.

'Just ask me a few common-sense questions about evolution and minerals, don't worry about the students, and let me take it from there!'

Well, of course, I was absolutely held in thrall by what he was talking about and found my questions coming thick and fast. It seemed only minutes later that I received a wind-up wave from the floor manager and brought the interview to a halt. He was an interviewer's dream: his specialized knowledge delivered with expertise, humour and clarity. I sighed in relief as the cameras closed down and dear Professor Bond came over, put his arm round my shoulder and said a meaningful and genuine 'well done'.

Suffice it to say, that first live television discussion – with anyone – set my own path into a television and radio career of interviewing about a wide range of subjects. It was important for another reason, too. Women in those days were still battling to escape from the 'bake a cake and make a dress' imagery of television interviews. John Pank's sudden inability to be there gave me a chance to break through that perception, and I could not have had a gentler, more humorous, more informed or more fascinating interviewee to help me do so.

What an attractive man he was, with his slightly military bearing, his ability to seem to be as interested in you as you were in him, and his sometimes quite acerbic but always perfectly apt and very funny comments. He must have been a delight to his students, keeping them as intrigued and alive to what he was saying as I had been in that first tentative interview.

Sadly, I didn't get the chance to work with him very much again. I remember doing a couple of topical radio programmes largely on the worldwide recognition he was getting for his enormous range of publications, and then again following his appointment as Vice-Principal, when the mood changed into some very tricky years at the University.

I found the after-interview chats – freed from the sometimes dictatorial constraints of programming – the most memorable of all. Radio freed us up to be able to chat about a glorious compendium of thoughts from a tungsten-based mineral found in the Essexvale (Esigodini) area – which I remember because of the topicality of its location to me particularly, as it was where I was brought up – to the rigours of high office in the University as we approached a critical point in the country's political life.

I feel humbled to have been asked to contribute in this small way to a book about the life of a remarkable man who left an indelible imprint on a young and growing nation through his wide-ranging, important work and his discoveries, particularly those down in the Zambezi Valley. Thank goodness his world of writing is still available for students of today to study in awe at how much was achieved by one man. He will long be remembered.

The University Staff

John Morgan

John Morgan's geological 'career' started at the tender age of sixteen at the Geological Survey and Museum in Exhibition Road, London. He was an Assistant (Scientific), the lowest grade in the Scientific Civil Service. The 'highlight' was being appointed a chief laboratory technician at the University College of Rhodesia and Nyasaland in 1961, where he remained until 1982, when he returned to the UK. He now lives in Australia.

An offer of the position as Chief Laboratory Steward in the department of Geology, University College of Rhodesia and Nyasaland, required important decisions to be made. Pam claims I never asked her for her hand in marriage: I just asked if she would like to go to Africa. The opening of a Geology department is a rare occurrence, and I considered myself most fortunate to have been selected.

I believe that Professor Bond – henceforth known as, and referred to as, 'Prof' – appreciated our situation and kindly suggested that we should travel by sea rather than fly. After nearly two weeks at sea and a train journey from Cape Town to Salisbury, we were met by Prof. He was in his field-working clothes, complete with 'notebook', similar in size to a thirty-pack cigarette box. We often wonder what he thought of this young couple!

It all started in August 1961. At first we were all housed in two rooms in an area referred to as 'the stables', originally part of the Agriculture department. Eliminating a termite infestation by spraying had some undesirable affects on animal experiments in adjacent laboratories.

I joined Prof, his secretary, Joan Wahl, Tony Gifford, a lecturer recruited from Imperial College, Brian Honey, technician, Wellington, office cleaner and later the thin-section technician, and a Land-Rover. Clive Stowe, lecturer, arrived from the Rhodesian Geological Survey about a year later, as did Linley Lister – then Linley King, the daughter of Geology Professor Lester King of Natal University – who was recruited from Durban to teach geomorphology. At UCRN, Tony Gifford was involved in the Film Society as well as in the Exploration Society, and he organized a student expedition to the Rwenzori, the Mountains of the Moon, in Uganda.

Geoffrey Bond and Tony Gifford had begun teaching the first first-year Geology students in March 1961, and rock-cutting equipment and collections of rocks, minerals and fossils from Gregory Bottley had arrived from England.[1] Zeiss Junior Polarizing microscopes were used by the students. The Aldis epidiascope was used to project the glass 'lantern slides', 3¼ × 3¼ inch, black and white. The Zeiss Prado projector catered for the 'modern' 35 mm colour transparencies.

We had been introduced to these first students on a visit by Marina Bond to Sinoia (Chinhoyi) Caves, where Prof was conducting field studies. Pam was impressed by the one student who ate raw eggs and raw meat. Second-year field trips became memorable events: up through Gokwe, the view over the Zambezi Valley, and a cruise on the *Erica* to study beach formation on Lake Kariba, Prof's favourite research project. As a result of a common interest in amateur radio, all notes and memos ended with '73s, GB' – 'Best Wishes, Geoffrey Bond' – usually sent in Morse code!

The new Geology block was completed in mid-1963. The architects were Ross Mackenzie, van Heerden and Hartford, and the builders, Richard Costain Ltd. It was not at all popular with residents higher up the hill on the campus because of the reflection from the aluminium facing. The building came to be referred to as Bond's Biscuit Box, and the lecture theatre as a carbuncle. It was, however, very practical, with all services on the outside of the building, and the internal wooden wall panels could easily be rearranged. I believe the design won an architectural award.

The architects designed the metal-framed office furniture, slotted microscope benches, museum cabinets and display stands. Important features of the building were the air-conditioned lecture theatre and the 'dumb waiter' intended to carry specimens from the collections stored on the ground floor to the teaching laboratories on the top floor. I have been told that it could even convey students!

Another feature was the indoor garden on the ground floor. This was never successful, but a creeper did reach the top floor. Slate and marble were used in the stairway and the entrance. Some outside marble had to be replaced owing to weathering.

During the construction of the building, I could not resist making daily visits to the site, and I got to know the Clerk of Works quite well. I did manage to persuade the architects that we needed a larger photographic darkroom.

[1] 'James Reynolds Gregory ... one of the principal London mineral dealers in the last half of the 19th century, was born in London on December 29, 1832. The company he founded, in 1858 (today known as Gregory, Bottley & Lloyd) had a reputation as one of the best in the business, providing mineral samples for scientists as well as private collectors. ... Recently located in Deal, Kent, after various premises in London ... the firm has been run since 1982 by Brian and Mary Lloyd who took over the business from Mr and Mrs Bottley.', 'End of an Era for Gregory, Bottley & Lloyd', *The House Directory*, <https://www.thehousedirectory.com/end-of-an-era-for-gregory-bottley-lloyd>.

'Bond's Biscuit Box' under construction

Photo: Vernon Stocklmayer

The equipment and collections were transported to the new building by the University's maintenance staff. The technician at that time was Mrs Iris van der Poll. Vernon Stocklmayer remembers helping in the museum when a young lady (Margaret?) was employed to do the artwork; it cost us a fortune in Letraset! The specimens in the glass display cabinet in the entrance hall were provided by Tony Gifford, who sourced them from the Marungudzi ring complex.

Initially, the Geology department occupied one wing, which housed, on the top floor, the Kariba research laboratory, the map room, seminar room and first-year teaching laboratory. The second floor held the staff rooms and library and Prof's personal specimen collection. The ground floor comprised the museum, specimen collections, rock-cutting and -polishing labs, camp store, workshop and preparation room. Second-year and third-year laboratories and postgraduates were on the top floor of the 'right wing', the ground floor of this wing housing the embryo School of Medicine's Anatomy department, complete with cadavers.

After nearly forty years, we and some of these students still get together and remember Professor Bond and those early days. Successful past students are now scattered around the world, some now retired. We remember him as an approachable, supportive human being.

'If you seek his monument, look around you.'
(with apologies to Sir Christopher Wren)

The University Geology Students

Vernon Stocklmayer, 1961–1964

Vernon Stocklmayer was one of a small initial group of Geology students at UCRN who graduated with a B.Sc. (Special Hons.) in 1964. He immediately began work as an exploration geologist, initially in Zambia and then in Namibia, before joining the Rhodesian Geological Survey. His eleven years in the Survey, during which he published two *Bulletins* and completed his D.Phil., were followed by another short period as an exploration geologist before he emigrated to Australia to join BHP, a major Australian mining company.

Some twenty years later, he became a consultant, completing projects for numerous small mining companies in Australia and around the world. Having retired, he continues to work as part of the editorial group that produces the *Australian Journal of Mineralogy*.

Having enjoyed geography at school and having spent a day in the field at Umvuma with Neil Bliss of the Rhodesian Geological Survey, I was delighted to discover that the University College of Rhodesia and Nyasaland was offering Geology as a new science subject, commencing in 1961.

Geology at UCRN in 1961 was a very small affair - we had a couple of lecture rooms and a small laboratory in the old Agriculture building. There was one lecturer, Professor Geoffrey Bond, one technician, John Morgan, and about ten students, the majority male with two women among them.[1] We lost the two girls at the end of first year and, by a process of slow annual attrition, reduced the males to the final four doing Special Honours in 1964.

The staff complement gradually increased: first came Tony Gifford, who lectured us in petrology, then Clive Stowe (structural and economic geology), and Linley King (geomorphology). At the same time, the number of students

[1] As far as he can recall, the other first-year students were Nick Bell, Carol Davidson (?), Ken Ireland, Muhammad Patel, Ian Robertson, Nigel Rowlands, Craig Smith, Pat Stidolph, Nan Winter.

increased and our premises became too small, so we moved to a new building – known as Bond's Biscuit Box because of its external aluminium cladding – which we shared with the fledgling Medical faculty.

As the first Geology students to go right through to Special Honours, we were the guinea pigs and must have caused the teaching staff many headaches that we were unaware of. One particularly potentially awkward situation occurred in the Special Honours year when, just before final exams, Clive Stowe was chatting to us about pegmatites only to find that we knew nothing about the economic importance of some pegmatites; this section had somehow slipped through the teaching net and, as it happened, there was a question on it in the exams. A crash course on pegmatites followed!

Prof. Geoff Bond was the complete antithesis of the average university professor, who, apart from being exceptionally intelligent and knowledgeable, were mostly poor communicators and generally unapproachable. Geoff, on the other hand, was more like an older family friend who was willing to impart his great understanding and experience to his 'geological family' in his easy-going, relaxed style.

He considered all his Geology students and staff to be part of an ever-growing family and seemed to take pleasure in their achievements. I worked for him as a research assistant during the Christmas break between the third and fourth years, and he published two papers partly based on my results. I was delighted to see that he had attributed me as co-author of both, a great honour.[2] Even much later, when I returned to Rhodesia with a baby after five years in Zambia and Namibia, he told me how pleased he was 'to meet another departmental grandchild'.

He was very supportive of his students, as was well shown by his stand against his university peers during the sudden flare-up of racial unrest in 1963 that involved one of his Geology students.

Lectures were punctuated by his numerous anecdotes, and students of all years would be familiar with 'the one-armed geologist' and the 'lonely grave containing two people'. However, his love of palaeontology and the Karoo sediments were apparent, and we heard many stories of fieldwork, including such diverse topics as the use of a geo-pick for crocodile-hunting and the scavenging methods of vultures.

Field trips were a joy, especially the long second-year trip that ended with a day on *Erica*, the UCRN/Wits University boat on Lake Kariba. Prof. Bond appeared to know everybody – from mine managers through prospectors to local farmers – and often had a little anecdote about them. He took the invention of Fred Crynge in his stride and, I think, rather enjoyed telling irate

[2] See Appendix 3: 'Geoffrey Bond's Publications'.

On the second-year field trips: 1962 (left) and 1963 (right)

and perplexed mine managers not to be too concerned when Fred did not return from an underground visit.

In hindsight, I consider that our Geology student group, as guinea pigs, were extremely fortunate to have Geoff Bond as our first professor in 1961. I am sure that many of us would partly attribute our future successes to the excellent grounding we received from him.

Charles Castelin, 1963–1967

Charles Castelin graduated from UCR with a B.Sc. Special Honours degree in Geology. He later carried out research work at the University's Institute of Mining Research. He spent most of his working life at Risco/Zisco and Buchwa Iron Mining Company, where he rose from Geologist to General Manager, a post he held for twenty-three years. He was acting chief executive of Zisco for six months in 1994. After retiring in 2008, he spent nearly four years at RioZim as geological consultant, working mainly on gold and coal, and finally as General Manager, Business Development. He was vice-president of the Chamber of Mines of Zimbabwe for three years.

In the early 1960s, registration to read for a degree in the Faculty of Science at the then University College of Rhodesia and Nyasaland required acceptance by three departments. On registration day, their stalls were laid out in a line in the quadrangle between the Geology, Chemistry, Zoology and Botany departments. For me, going down the line was a depressing experience of reluctant acceptances or outright refusals - until I came to the Geology desk. There, a benevolent-looking, bespectacled professor by the name of Geoffrey Bond welcomed me with open arms, stating that the country needed home-grown geologists - rather like the First World War poster: 'Your Country Needs YOU'. That was my introduction to 'Prof'.

I soon discovered in my first year that Prof didn't only teach geology but, just as importantly, taught the *love* of geology. It worked. I was hooked, and by the end of 1963 I knew I wanted to be a geologist. Geoff says in his memoir that he learnt that attribute from his own mentor, Professor H.H. Read, at Imperial College.

Prof was particularly good to me, or so it seemed; perhaps he was kindly and caring to all those under his wing. In my case, the first occasion was when I predicted that I would fail second-year Chemistry, despite being told that 'no one ever fails second year'. I proved them wrong, and Prof, by now Dean of the Faculty of Science, changed the timetable so that I could study both Geology and Zoology. To help me keep up my interest in repeating second-year Geology, he employed me as a demonstrator in the practicals, a task I relished.

Around this time, he proposed the formation of a student geological society, and he encouraged me to stand for the first committee and then to chair the second year of its existence. Prof was very keen to name the society after his Rhodesian mentor, Dr Alexander Miers Macgregor. Somewhat to his disappointment, student power ruled, and the society was named after F.P. Mennell, an early pioneer in geology in Rhodesia. Geoff did agree that it was a good choice, and he was a strong supporter of the activities of the Mennell Society thereafter. He was made the first honorary member and Vice-President of the Society.

In June 1967, I was privileged, at the urging of Prof, to lead a six-day Mennell Society expedition to the Zambezi Valley to a dinosaur fossil site in the Kadzi river area. We were honoured to be accompanied by Mike Raath, the curator of the Queen Victoria Museum in Salisbury, and, in addition to having a most enjoyable trip, we mapped the location and successfully retrieved numerous petrified bones.

Prof suggested that the Mennell Society should produce its own journal and, after much deliberation between students and the Prof, they agreed upon the name *Detritus* – very appropriately, as Geoff was primarily a sedimentologist. The cover of *Detritus* (quartz crystals) was designed by Zelide, the wife of Gerard Siedner, a lecturer who had by then joined the Geology department. The journal survived for many years and published both a heavy fraction (erudite scientific papers) and a light fraction (items of geological humour). In the mid-1960s, the Society was the largest on campus and had a following from students in many faculties because of its interesting talks and outings.

Upon graduating and finding myself having exhausted my bursary, Prof found me a scholarship to undertake the Special Honours fourth year in Geology. During that time, he urged me to apply for a fellowship of the Geological Society, London, and was my sponsor. After the Honours year, he wanted me to continue my studies at Imperial College but it was not to be, as I had to return to Risco to pay back my bursary.

Some years later he persuaded me to join the committee of the Rhodesian branch of the Geological Society of South Africa, soon to become the Geological Society of Zimbabwe. There were many other acts of kindness, and in later years I wondered if it had something to do with my antipathy towards chemistry, a feeling he shared.

Everyone who attended Geoff's lectures enjoyed the jokes that he interspersed with the coursework, and it was rumoured that he had them written into his lecture notes. I was able to verify this when I repeated my second year and, sure enough, out they came at precisely the same point as they had in the previous year.

One of our favourites was the story about a father and son walking through a graveyard: The son, upon studying one of the gravestones, said, 'Dad, why are there two people buried in this grave?' The father replied, 'How so?', to which the boy retorted, 'It says on the gravestone: "Here lie the mortal remains of a geologist and a gentleman".'

One of the highlights of reading Geology in the 1960s was the second-year field trip led by Prof himself. Part of the trip was through his own stamping grounds among the Karoo strata in the Sebungwe area, north of Gokwe. Prof's company and fascinating explanations of the geology were a never-forgotten experience that I cherish to this day.

One of my greatest regrets is that, after finishing my studies and then working at Zisco, 230 kilometres from the university, I didn't see Geoff as much as I would have liked to. However, when I did, his door was always open, and an invitation to a cuppa in the staff tea room was always made. He had expressed an admiration for the steel 'splash' that was given to VIP guests at the steelworks, and it was my pleasure to present him with one in the 1970s.

Prof. Geoffrey Bond was a mentor to me in perhaps the same way as Alex Macgregor had been to him, and I owe my career in geology and mining to him in a way that can never fully be repaid. The greatest tragedy is that his life was cut short in 1983 at the age of seventy-one, when he still had so much to offer to the world of geology.

Adrian (Bert) Mann, 1963–1969

Bert graduated from UCRN with B.Sc. Special Honours, and then a Ph.D. after three years barefoot in the Sanyati valley, telling himself jokes he hadn't heard before. He joined Falcon Mines in 1970, thus returning to his first love, mining, but soon shifted to rock mechanics. He transferred to a consultancy in Johannesburg in 1980, completing an MBA at the University of the Witwatersrand in 1985.

He then moved to Calgary, Canada, where he hung his shingle as an international mining and oil and gas consultant. To take up any slack, he makes marionettes and writes fiction under the *nom de plume*, Arthur Gardiner. He retired in 2010, 2015 and 2020, and intends retiring again in 2025.

For every student, the second-year field trip defined the quintessential Geoff Bond. Indelible in my mind is the moment I realized that Geoff was no mere mortal. Our second-year field trip was returning from Nyamandhlovu Pan at sunset – sundowner time, the time when Prof's bottle of Limousin emerged from that wooden box, that one that housed his paraffin lamp – the one that Willie Oldham had reversed over! Between Prof and his brandy-and-water was a herd of elephants, doing elephant things. He lent on the horn – to no effect. He gave another blast – longer, more assertive. The elephants continued gossiping. Thrusting an arm from the Land-Rover, he hammered on the side of the door. One disdainful old cow dusted off a tuft of buffalo grass on her bum, then began munching it. Enough!

'Get out of the way, you bloody fool!' trumpeted the great man.

She did. They all did. Even elephants listened to him. I was in awe!

At noon in the Sengwa river gorge, a dozen fit twenty-year-olds struggled to keep up with the old man (in their eyes), then aged fifty-something. There was not a cloud anywhere, save above the man, who smoked a pack of fifty Matinée a day. With neither map nor compass, Geoff Bond out-walked the lot of us, veering neither left nor right as we returned – not just to the road, not just to the line of vehicles, but to the door of his Land-Rover. I'd swear the man was part homing pigeon!

Coming down from the Gokwe plateau, somewhere near Chief Nembudzia's village, the road was blocked by a sign that said, DEVIATION, and pointed left, to the west. The only discernible continuation was to the right. Geoff stopped, plugged another cigarette in his amber filter, lit it as he thought about things, then turned right. Within a few kilometres the track had deteriorated to a path that no self-respecting goat would follow. Prof persevered, until an un-navigable donga and an emaciated *mombe* confronted us. He lit another cigarette before he began to back up, setting off frantic honking from the truck behind us and from the Beasley-mobile behind that; rush-hour traffic gridlock in the mopane veld en route to Binga! There was a grunt from our driver, and then the immortal words: 'Why do these damned idiots leave roads where I don't want to go?' Aah, Prof, our Prof!

Susan Warner (now Stocklmayer), 1965–1968

Susan Stocklmayer (née Warner, also Anderson) graduated from UCR with a B.Sc. Hons. Geology in 1968, and was the first female to take the Honours course. She then joined the Rhodesian Geological Survey and held the position of mineralogist for ten years. During a break of three years, she accompanied her husband in the mapping season to the Pfungwe and Wedza areas, at the same time studying with the Gemmological Association of Great Britain by distance learning and qualified as an FGA.

Since 1980, she has worked as a gemmologist and jewellery valuer in the UK and Australia, and she continues to work as an examiner. With Vernon Stocklmayer, she was a contributing author to the publication *Gemstones of Western Australia*, written in collaboration with the Geological Survey in Perth, Western Australia.

In my final year at Roosevelt Girls High School, the A-level class attended an evening lecture presented by Prof. Geoffrey Bond in the Arts lecture theatre at UCRN. The talk on minerals and their unique properties was an important one to me. The venue was impressive compared with a school classroom, the subject matter was interesting, and the talk that came across - in what I was to learn was Prof's usual manner - was coherent and with a staged unhurriedness, the subject matter sounding uncomplicated. This event forged my career path, all decided in one evening by hearing an inspirational teacher.

I was a student at UCRN from 1965 to 1968. On registration day of my second year, which would have been March 1966, Prof was on duty in the Beit Hall at the Geology department's registration desk, making welcome all those who were putting their names down to continue in Geology. It was also - if my dates are correct - the time when his daughter, Nina, was born, because he was pleased to share this good news with everyone.

In the first three years of studying geology, it was obvious which subjects were Prof's specialties: palaeontology, sedimentology, and especially coal measures and the Karoo System in particular. However, in my first practical class on observing graptolites (carbonaceous fossils found in black shales - many varieties look like pencil lines to the uninitiated!), Prof appeared. He was not impressed by my straight-line diagrams of what I had observed of these animals. He took pains to point out that they actually had small anatomical structures, which I might have noticed if I cared to use some magnification.

This was typical of him: he always showed patience and took time to make sure that you understood what was important in each of his subjects. His fascination with palaeontology was rewarded by the discovery of a new dinosaur in the country, and we were kept up-to-date with his recounting of his discussions with staff at Bulawayo Museum, particularly Mike Raath.

Prof had a long history of bush work and encounters with prospectors and mining characters, and his lectures were always punctuated by anecdotes gathered from his wide experiences. One that easily comes to mind is the advantage of the one-armed geologist, who couldn't use the excuse that 'on the one hand, this could be the explanation and, on the other, it could be this one'.

His research projects were endless and ongoing; the large number of papers he produced evidenced this. He once did accuse me of flattery when I handed him an assignment that happened to include many references to papers that he had authored.

In our second year, we had a shortened field trip to Gokwe. We did collect some dinosaur bones and agates, among other things; this was a time when it was commonplace to take a piece of everything geological. Prof's bush attire always included long khaki shorts, a geological hammer hanging from his belt, and a cotton floppy hat. One of the stories he told us was about when he was walking in the forests of the Eastern Highlands with the entomologist George Arnold, and into a net he was carrying fell a strange insect, quite by chance. Although they hoped to have found a new species, after it had been studied it wasn't named for either of them. However, Prof did tell us that the only being or thing of the natural world to which he had been honoured to give his name was a fossil cockroach, *Rhodesiomylacris bondi*.[3]

Another accessory that he always had with him was the cigarette packet kept in the top pocket of his shirt. Apart from the smoking, which were unfiltered Matinée cigarettes, the back of the box was always used to make notes while out and about. Our group was also well aware of Prof's dependence on tea, especially when the day was hot, and we were informed that tea should be served from a brown china teapot.

[3] In fact, there were two: 'Bond's palaeontological work culminated in his having two fossils named after him, the graceful, gazelle-like extinct springbok, *Antidorcas bondi*, and a cockroach, *Rhodesiomylacris bondi*; characteristically, he was especially proud of the latter!' M.A. Raath, 'In pursuit of Paradise Lost: A call to rekindle palaeontology in Zimbabwe', *Palaeontologia Africana* (December 2009) 44: 175.

'The new species [of Bovidae] has therefore been provisionally referred to [as] *'Gazella' sensulata* as it is considered undesirable to create a new genus upon the material now available. It is proposed to give the specific name *bondi* in honour of Dr Geoffrey Bond who was chiefly responsible for undertaking the work at Chelmer and who discovered the type specimens.' H.B.S. Cooke and L.W. Wells, 'Fossil remains from Chelmer, near Bulawayo, Southern Rhodesia', *South African Journal of Science* (1951), 47(7): 207.

'The [*Rhodesiomylacris bondi*] species is dedicated to Dr Geoffrey Bond, in recognition of his fruitful activities in the geological field of Southern Rhodesia'. F.E. Zeuner, 'A fossil blattoid from the Permian of Rhodesia', *Annals and Magazine of Natural History* (1955), Series 12, 8: 685–8.

The Honours year class of 1968:
Susan Warner, Nick Lockett,
David Armstrong,
Craig Smith and David Edwards.
The photo was taken by John Morgan,
'for whom we caused much extra work'.

Following on the B.Sc. general degree, Prof considered that it would be a good idea for me to continue and take the fourth-year Honours course. I did, and became the first female to do so.

My project that year was based on the Kalahari sands in Mondoro TTL. Part of the heavy mineral concentrates from the area also contained 'chips' of carnelian agate, an orange-brown type of agate. Somewhat bewildered by these occurring within some metres of the present land surface, I talked to Prof about their significance. Immediately he had the answer: these were, in fact, flake tools – the shards of material chipped off a larger specimen when early man was making tools. Surprised by this identification, I corroborated the findings with the Queen Victoria Museum archaeologist and recounted this to Prof, who was amused by my initial scepticism.

Like many of his Honours students, Prof had small projects that needed students' input, and I was one among many who worked on some of these projects in vacations. Mine was a study by SEM (a scanning electron microscope) of surfaces on sand grains from various sediments. Such practical work helps to build up experience of geological applications outside lectures.

Prof was also very active in many radio broadcasts. One series was a Sunday evening science-based one in which he interviewed different people about their special subjects. I made one programme with Prof, talking about minerals and gemstones; I still have a copy of that tape.

In December 1971, Prof kindly accepted an invitation from my family to give a speech at my wedding, held at army headquarters in Salisbury. I don't remember much of his speech, but he did mention the amount of mascara that was always attached to the microscope eyepiece after I had used a department instrument!

My recollections of Prof have always been of someone who was very approachable and patient, someone of whom you never hesitated to ask questions for fear of being thought somewhat inadequate not to have worked it out for oneself. After all, Prof was a polymath.

How lucky we have been to have shared the years at UCRN in a well-run department in an almost brand-new building with well-equipped facilities and a friendly and competent staff headed by Prof. How can you not look back on these times other than with great affection?

Mennell Society groups, 1966 and 1967.
'I happen to be sitting next to Prof in both examples.'

Nick Lockett, Geology, 1965–1968

Nick Lockett graduated with a B.Sc. Hons Geology at UCR in 1968. In 1970 he joined the Geological Survey of Rhodesia and was assigned to map the Dett (now Dete) area in north-west Zimbabwe. The results were published in 1979 as *Bulletin* 85. In the same year he was awarded a D.Phil. from UCR for a thesis on the structure and metamorphism of the Dett–Kamativi area.

From 1976, he worked variously in coal and metals exploration in South Africa and Australia, before concentrating his efforts on geological image interpretation as applied to mineral exploration. From 1989 to 1998, he was employed by Ashton Mining Ltd in Perth, working mainly on remote-sensing aspects of Ashton's diamond- and gold-exploration projects in North and South America, Africa and Australia. In 1999 he founded his own in geological image interpretation consultancy based in Fremantle, Australia. He retired in 2018 due to the worsening effects of Parkinson's disease.

I first met Prof. Bond on registration day prior to the commencement of the 1965 academic year at UCRN. Earlier in the morning I had enrolled as a first-year student with the departments of Chemistry and Physics; however, my application to enrol in Mathematics had been rejected owing to poor A-level results. I was directed by the Maths department to 'go and talk to the people at Zoology or Botany, they will probably have you'.

I was feeling dejected by my failure to enrol in the science subjects on which I had set my heart when I wandered, somewhat aimlessly, past the Geology department desk staffed by Prof. Bond and Dr Tony Gifford. Prof. Bond greeted me with a cheery, 'Hello. You look a bit lost. Are you interested in geology?' I don't recall my halting reply, something along the lines of knowing nothing about the science. A brief explanation by Prof about studying the Earth's crust finished with the comment 'You look like an outdoor sort who might enjoy geology' was sufficient for me to enrol in the subject - which would lead to postgraduate study and eventually to a lifelong professional career as a geologist in industry and with the Rhodesian Geological Survey.

Throughout my undergraduate days I was infected by Prof's technical enthusiasm, and I enjoyed his lectures, which were always well delivered and entertaining. After he moved on from Geology to become Vice-Principal at UCR, I was well guided in my research work by his academic appointees, Prof. Jim Wilson and Dr Clive Stowe. It was said that Prof. Bond trained a generation of geological generalists. Many of his students had careers in mineral exploration and management, but there were also those who became academics, as well as consultants in such varied specialist fields as geo-statistics and remote sensing.

Beverley Wahl, Geology 1967–1971

Beverley Wahl, the daughter of Prof. Bond's first secretary, Mrs Joan Wahl, graduated from UCR with a B.Sc. Special Honours degree in Geology. She says that she then 'went right off the rails' and lived in Norway for forty-five years, acquiring a Cand.mag. degree (roughly equivalent to a BA) in languages at the University of Oslo. She is still working actively as a government-authorized translator of Norwegian into English, now based in the UK.

I think I may well have been the very first of Prof. Bond's students to meet him. Certainly my mother, Mrs Joan Wahl, was the first member of staff of the Geology department hired by Prof in 1960. The two of them presided over a couple of rooms in what later grew into the Faculty of Agriculture, and their neighbours inexplicably included a cage of monkeys and a young baboon named Charlie, to whom my mother grew very attached. We've puzzled over their presence in retrospect, and concluded that they must have belonged to the Faculty of Medicine, then still in its infancy.

Farewell Mrs Wahl

Dear Mrs Wahl, we miss you so.
We mourn the day you had to go,
But hope that chances there will be
For you to drink our health in tea –
Since you left, the tea leaves faded
And the Staff are somewhat jaded.

We wish you luck and health and strength
And Life of an enormous length.
From Linley, Prof. and Clive and John,
Tony, Iris and Wellington.
Jerry and Biset joined in, too,
In sending this from us to you.

Geoff Bond, *Verse & Worse: From Bad to Worse.*

As new members joined the staff of the Geology department, they were invited to play tennis at our house, so I knew them as Marina, Clive, Linley, Tony, John and Pam. But Prof was always Prof. He and I discovered that we had something in common: both lefties, he had been forced to learn to write with his right hand but wielded his racquet in his left; born in a more enlightened age, I wrote with my left but played tennis with my right.

Added to this was our shared passion for dinosaurs, about which I intended to write a book. Prof gave me every encouragement in this project but suggested I study geology first; and so it came to pass that, just under a decade later, on my eighteenth birthday, I signed up for Geology at UCRN. It was like coming home to family.

Under Prof's avuncular wing – not forgetting the wings of the other members of the tennis set – I progressed with some ups and downs to a rather ordinary B.Sc. The first three years were so enjoyable that I signed up for a fourth, which took me to Kariba to join the second-year field trip on Lake Kariba. Prof and I were flown there very competently by Dave Bowen, but it was a bit bouncy, especially our landing. I was not feeling my best when we disembarked to join the boat that was transporting the second-year students round various islands in Lake Kariba. Val Smith was also there, making us fabulous curries to round off our days when we camped for the night. Somewhere I have a picture of her in a zinc bath of the type we used to take washing out to the line, legs dangling over the side and with a hat tipped over her nose.

I was supposed to be collecting sandstone specimens for my project, but in point of fact Prof collected most of them. Always a perfect gentleman, he climbed the first outcrop to show me how it was done, and then we froze into a sort of pattern, where I was too embarrassed to say 'Er, should I take over now?' and he was too polite to suggest it himself.

It was extremely hot, especially on the last night, and some of us slept out on deck. Prof slept down in the cabin – not because of the mosquitoes, he assured us. He didn't react other than with a small, non-itchy puncture mark. On our last day it was windy and I disappointed myself by getting seasick travelling up the Zambezi. I mentally wrote off the round-the-world-on-a-yacht trip I had always assumed I would make at some stage.

Prof, Val and I travelled back by to Salisbury by Land-Rover. They were both heavy smokers, but I wangled a window seat and just about survived the trip back. I remember Val remarking that she smoked more when she was with Prof because he constantly had a cigarette going ... but often he wasn't actually smoking it. So she would light up when he did, with the difference that she smoked them. For myself, I hung out of the window as much as possible, opting for dust rather than the blue air in the cab.

The grand finale for my part was the following day, when my 200-odd mosquito bites came out in full bloom. I spent the next three days confined to the house, covered in calamine lotion, cursing Prof and his little non-itchy red marks.

Those were wonderful years, and I have very fond memories (even though my love of dinosaurs proved insufficient to compensate for my inadequate powers of observation, so essential to geologists). I definitely have the department of Geology, with Prof. Bond at its heart, to thank for them.

John Orpen, Geology, 1973–1978

After first obtaining a B.Sc. Special Hons., John Orpen graduated from UCR in 1978 with a D.Phil. in Geology. He lectured in the Geology department at the University from 1980 to 1990, and was Chairman of the department for the last three years until he was appointed Director of the Zimbabwe Geological Survey. In 1993 he was contracted by the Free University of Amsterdam to establish an M.Sc. in Exploration Geology back at UZ. In 1997 he started his own consultancy and also sold diamond-drilling rigs and equipment. He went to South Africa in 2001, initially working for Reeves Wireline before again doing consultancy and drilling/geotechnical assurance reviews, which work he continues to this day.

Geoff Bond's memorial service was packed. It was a warm day, and fortunately I sat by an open window in the University of Zimbabwe's chapel. A lady with a beautiful voice sang *Where'er You Walk* as I gazed out over the campus that Geoff had played no small part in building up - memorialized as Bond's Biscuit Box. The Geology department is an architectural marvel, constructed with meticulous attention to detail and designed to last, starting with its aluminium façade - hence the sobriquet.

That was Geoff. A man who could see the end from the beginning, and who was a chess master in navigating the difficult waters of the time with a firmness and kindness that few dared question, qualities that many benefited from - not just the students and staff that walked daily through the cool foyer of his 'Box', subliminally calmed by its exquisite Italian marble tiling.

There was even a serenity in the way he smoked. Blue-grey eyes glinted over the lectern as he charmed knowledge into young minds, with a relaxed resonance that I can hear now as he spiced his notes with jokes that made recall easy when exam time came.

Anyone who came to know him soon found that behind the grey tweed was a steely character that preferred the challenges of fieldwork - mapping the rocks in the wilderness of the mid-Zambezi Valley to prize out the secrets of the dinosaurs and unveil the landscapes through which they had walked. Men like that are a rarity, whose lives continue to enrich many if they leave their biography, and happily Geoff did that.

A day I will never forget came after Geoff's retirement as Vice-Principal, when he was an honorary research fellow. I had developed an interest in the Lower Zambezi Basin and was studying aerial photos when he walked past my office. 'Geoff!' I called. 'Is the Karoo folded anywhere in the Valley?' 'No, John,' he said. 'It's only block-faulted.' I showed him a photo I had been staring at. 'That folding is in the gneisses above the escar ...' He tailed off. I laughed, as he leaned over my desk in astonishment, placing his finger on the folding near the bridge over the Angwa river, well into the sediments below the escarpment.

Thus began my little foray into the Valley, but sadly Geoff passed away shortly after, and I often wonder what discussions we could have had.

The FJC Syndicate, 1966

'FJC Syndicate' was the 1966 Honours year, comprising Alan (Beastly) Beasley, Gerald (Mr Responsibility) Harper, Richard (Legs) Linnell, Adrian (Bert) Mann, and John (Wake up!) Stephenson.

Beasley joined the UCRN Geology department as a lecturer, and did his doctorate in mathematical geology, working on a hydrology problem on an aquifer in Matabeleland.

Linnell moved up the corporate ladder in South Africa, where he ended on top of the heap with Samancor.

Harper and Mann did their doctorates on adjoining areas in the Sanyati Valley. Harper then headed for Canada, eventually becoming president of various companies and CEO of PDAC (Prospectors and Developers Association of Canada, a very prestigious organization).

Stephenson moved to the University of Manitoba, where he also got doctored, then moved up the corporate ladder to become President of Kennecott, then of CRA/RTZ Australasia in Perth, WA.

Geoff's poem refers to the party that FJC held at the end of the year. FJC Syndicate actually did some contracting geological work during the year, some for Lonrho, some for Henry Martin, taking our pay in cases of beer.

At the end of the year, Linnell and Harper joined the University of Cape Town's oceanographic research, which fell apart for some reason; Beasley joined UCR; Stephenson married and headed overseas, and Mann started his research in the Sanyati.

Thanks to Crynge for a Bynge

FJC and Partners
Have had a busy year.
Consulting work was minimal
And income nil, we hear.

The partners, now, are parting –
The firm has lost its shape.
One will go to Canada,
Two others to the Cape.

Two remain in Salisbury
(The Beasley and the Mann).
We know that Gerald Legs and John
Will write us when they can.

We hope that some years later
A meeting there will be,
At which the partners re-unite,
Each one a Ph.D.

The AGM was marvellous,
With booze and super skoff.
We thank you all sincerely,
From me and Mrs Prof.

Geoff Bond, *Verse & Worse: From Bad to Worse*.

Frederick Jurassic Crynge, B.Sc. (failed), 1961–

Fred started at UCRN in first-year Chemistry in 1961, but had the grave misfortune to fail three times, through no fault of his own. By 1963, he started to transfer his allegiance to Geology at the same revered institution, joined there by his colleague and nonetheless friend, Angus Stürer. Both these stalwarts suffered no ill effects from question marks on their lungs, shown on their TB X-rays.

Despite being sent down several times for being drunk and disorderly, neither he, nor fellow students Andrew and Penelope Nuss, were ever permanently rusticated. The charge of theft of a Coca-Cola sign was dropped against his alibi that he was actually replacing a stolen Fanta sign.

The Winchester quart of absolute alcohol he signed out from the first-year chemistry laboratory was never recovered, although the bottle subsequently reappeared. His defence, when discovered on the roof of Swinton, was that he was retrieving the Hall bicycle, which was in his possession at the time. Although frequently abandoned underground at Kamativi tin mine, Dalny gold mine, Mufulira copper mine, and Wankie colliery, he survived into his fifty-eighth year social, second-year academic.

To edify my curious mind,
In leisure, relaxed and refined,
I, Frederick Crynge, Renaissance man,
Launched my career, as any can,
Who passed half way through Standard Ten:
That's Standard Five, if maths you ken.

My life of crime precluded me
From joining the BSAP.
And Gwebi told me I'm too dumb
A mielie farmer to become.
At TTC, where teaching's taught,
My application came to nought.

And so, to help improve my lot,
A 'mature student' pass I got;
And thus, to UCRs and N.
I dedicate, with gown and pen,
My studious life, which first began
In chemistry, in sixty-one.

But I had to do the year again,
For chemistry requires some brain,
Zoo was flies, and smelly bones,
Maths was numbers, planes and cones
And then I did the year once more,
And once again, that made it four.

Physics really was no fun,
Except the part with light and sun,
And botany was full of weed,
But not the kind that filled my need.
And then, behold, the Bond I found
The Bond that bound ev'ry rock-hound.

The Bond, who had humanity
An academic man was he,
A sophist, true, but no mere chaff,
An honest leader, who could laugh
As much with us, as at himself
And at me, the mysterious elf.

Reflect, ye mortals, how I've striven
A dedicated student, driven
For six decades, my craft to learn
Thus to avoid the need to earn
An honest crust, by breaking rocks,
Or dishonest one, by picking locks.

'Twas never me, I tell no whopper,
Who cast the pawpaw at yon copper;
Nor did my X-rays, just for larks,
With massive shadowed question marks
Besmirch my pure clean innocence
With baryte in a greasy fence.

A shy young fellow, not too bold,
I always did what I was told
Or nearly so; and seldom went
Away from there where I was meant,
But once or twice, I will admit
Perhaps I strayed a little bit.

'Twas not my fault, that every time
We headed down a deep, dark mine
I lost my way, fell down ore passes,
Took wrong turns, mislaid my glasses,
Blundered to 'madala side',
Or missed the bus, the cage, the ride.

It's dark down there, there is no light.
There's beasties there, and bugs that bite,
And creaking rocks, and scary sound,
And men who with a hammer pound
The stulls that are the stopes main liner
To stop the key of A flat minor.

I'm just a simple kind of chap,
Who swapped some samples with the clap
With those of students, nun and bride
To show they're on the pregnant side.
And just for sense of humour's proof
Put Swinton's hall bikes on the roof.

I'm not naughty; never tried
To make nitrogen tri-iodide.
Never! Would I lie to you?
Did I, like bunting, at the SU
Hang condoms filled with hydrogen
To celebrate November ten?

I didn't steal the Fanta sign,
The footprints on your car aren't mine,
And tiddlywinks in Cecil Square?
Not guilty, sir! I wasn't there.
The drapes from Meikles snooty bar?
What makes you think I'd go that far?

The farmer's daughter, from Karoi,
Who had that fair-haired little boy?
That lovely lass I've never met,
I'll swear it's true! I'll place a bet,
Those nine months past, I was below
Down some damn mine! The records show.

The Second-Year Field Trip, 1964

Join the shaky memories of a couple of geriatric geologists
on a virtual re-enactment of that nostalgic fortnight over fifty years ago.

Two University Land-Rovers and the B— mobile (a third Land-Rover), packed to the gunnels with a dozen students, lurched from the car park of the shiny new Geology department. Bulging with enough booze, food and a Professor to reach wherever they had to reach, plus a couple of tents (remember that number), they trundled west towards the great unknown.

They crossed the Great Dyke. What, no stop? After passing the Macheka-*waka-sunga-beta* mountains (I'm just proud to remember the name), they blunder on. At Que Que, turn right. Yonder lies your hinterland - and onward, through the Experimental Irrigation Scheme, down across the Um-something (-nyati, -fuli -tali or -sweswe?) river drift, and upward, to the Gokwe plateau.

On the edge of the Mafungabusi escarpment, the Prof stands atop a vehicle to explain the wonders we are about to behold in the Zambezi Valley beneath us.

With some difficulty, and an intermittent stream of *sotto voce* expletives, the convoy unravelled itself from an unscheduled detour, retracing its way back to the road, where the correct detour was lurking but a metre beyond the sign. Northward Ho! Ferricrete, dust, bone-jarring potholes, wind-blown sands, mopane scrub, mopane veld, mopane bees, all were piled one upon the other. Then, stop!

The world disappeared. As the dust cloud thinned, so the clouds of mopane bees thickened. A rusty iron pipe protrudes from the sandveld. What wonder is this? From it emanates a fetid stench and gushing, boiling water. Miracle of miracles, it is the Kavira artesian well, drilled by the Ministry of Agriculture. The pilgrims admire the pipe, swat the mopane bees, then light up cigarettes, the only way to keep the little blighters away.

Under interrogation by the Prof, M— explains that *kavira* is a corruption of *kuvidza*, which means boiling. That helps to explain why all hot springs seem to be called Kavira. Paying little attention to the discussion, L— plunges his hand in to collect some algae that thrive in the water.

'Sh**, that's hot!' he exclaims, nursing his scalded paw.

'Hence the name,' was Prof's instant rejoinder.

All aboard! The convoy stirs its dust clouds again, but the journey is now short. We follow the stench and water for a few kilometres, then stop to allow the mopane bees to catch up. This is our first campsite, on the banks of the Salami ... Sessame ... Sesami ... river.

No one died from the designated cooks' food, not least M—'s 'mushi mashed potatoes'. A case of Castle initiated some discussion about its provenance. Prof, as commissary officer, insisted that the beer provided by the department - beer provided by the department! - was not to be consumed until Binga. But said beer was not under commissariat jurisdiction, being a part of our emergency stash.

Tragedy strikes! No one has a bottle opener. Crisis averted. G— demonstrates that Series 2 Land-Rovers are mobile bottle openers.

Up betimes, and north, ever north, to the Hostes Nicolle Trail, where the convoy turned left near Samapakaruma fly gate ... aah, those names ...Matusadona ... Mavuradona ... no idea where, but their names are etched in memory.

There was geology, plenty of geology. We whack rocks, scribble illegibly in dog-eared notebooks, then hurry on after the maestro, to hear him expound on the next outcrop. Drive on, drive on. Stop! Another year is added to the Fred Crynge Rock. We pose beside it, then on, on to the next campsite.

A fine rock pool was cleared for our communal bath in the Sengwa river, where a small baobab now announces that 'David Livingstone slept here'. It was impossible to keep up with the Prof on the sand bed of the Sengwa river. Only years later did I learn the soft-sand shuffle necessary for gliding along in those sand beds. Kavira hot springs ... ripple-marked flagstones in the Siabuya ... mopane bees, goddamned mopane bees ... 'Beware low-flying aircraft' sign in the middle of nowhere ... Mcheni Gorge ...

'Hey, Prof. Stop! Stop!'

The Land-Rover lurches to a halt.

'What's up?'

'Could you back up a bit, please?'

'Back up? Why? What's up?'

'You missed one.'

'I missed one? One what? I never miss anything.'

'A pothole, Prof. You hit every other one on the road; we want you to maintain your perfect record.'

Chizarira plateau ... another Kavira hot springs ... then Binga, an ocean cruise on the DC's launch. Here was civilization, a real shower, salubrious accommodation: tin roof above, clean concrete beneath; luxury indeed! It was spoilt by a substandard *braai-meister*, whose culinary skills belonged in a crematorium.

Stratigraphy, sedimentary petrology, sedimentology, then reality ... igneous and metamorphic paragneisses, schists and ... pegmatite, Kamativi tin mine. A Don Juan narrowly avoided having his face slapped when he offered to hold M—'s accumulator while she 'got it on'. It was a battery pack, not what she thought he'd meant.

Heading into the shades of Hades, all twelve students (Fred included) gave voice to *O Sole Mio*. Binga was luxury? Kamativi was Utopia. Unlimited beer at the mine club, a catered buffet dinner, then drive back to camp. It's as well we're driving, for we're in no condition to walk, as demonstrated by M—, in the middle seat, who earns the nickname 'Blattie' thereby.

Onward, ever onward, to Wankie colliery, down Kamandama shaft, the site of that massive tragedy eight years later. We are treated like royalty, with white coveralls, white hard hats, and ill-fitting rubber boots; in return, we treat them to an underground reprise of *O Sole Mio*. From Utopia we have arrived in Paradise: hot showers, real beds, with bed linen and air-conditioning.

Life becomes a blur ... glossopteris at the clay pits ... new road cutting to see jointing in basalt ... old railway cutting, for jointing in basalt ... Matetsi river bridge, for jointing in basalt ... more sedimentary strata ... Prof sketches the stratigraphy (Escarpment Grit over the Madumabisa mudstones) on the back of a packet of 50 Matinée. (Amazingly, C— still has that sketch!) ... the Victoria Falls – 'S—, if you fall over, I'm not fetching you; and you can drive yourself to hospital' – down the funicular to the power station, to see jointing in basalt.

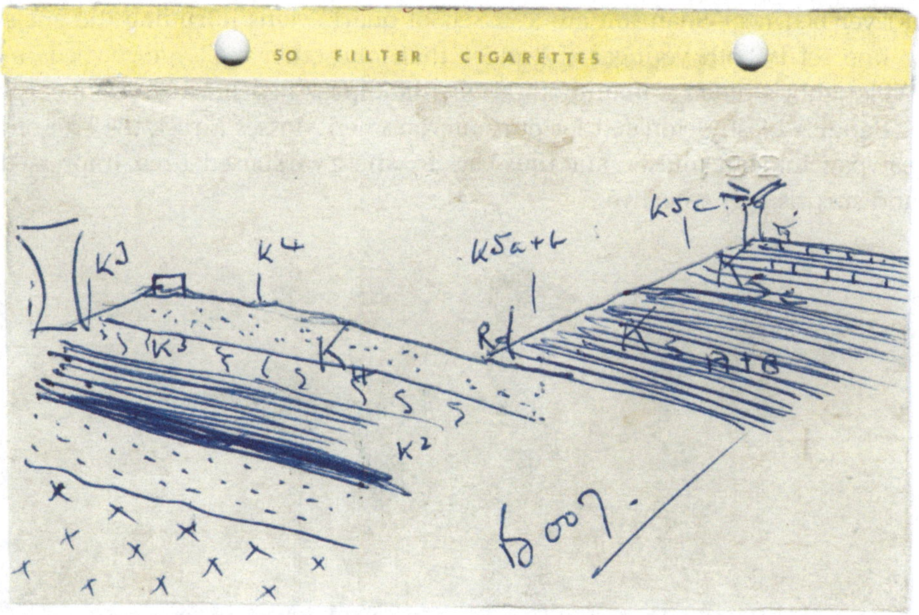

Lukose river near the Trappist monastery ... whoopee! ... Upper Wankie sandstone, not basalt! Roll on, turn right at the crossroads, and we go into Dett. Past Dett is Wankie Game Reserve and the rondavels of Main Camp. We examine perched water tables under pans ... Seif dunes, now fossilized and reduced to a couple of metres in height, running tens of miles west towards the Kalahari ... wind-blown sand ... mopane bees.

Oops! Prof has a hissy fit. Nothing is going well for him today; honking great jumbos blocking the road, our first near fatality (gash in the knee) from a week-old bread loaf in an impromptu game of rugby, one tent left behind at Sengwa, now his flashlight batteries are flat. He replaces them, tossing the old ones into the braai ... spectacular! Bang, bang! And bang again! Even the gods err! So the lost tent episode has to be forgiven.

Off again. Fossil ostracods at Commander Coombs's farm ... dropped in on Mr Mennell in Bulawayo ... awesome gold and mineral specimens at the National Museum in Bulawayo ... then out past the Ntabazinduna mesa to view stromatolites in the Huntsman quarry.

Prof: 'These are the oldest fossils yet discovered. The limestone is Bulawayan.'

R—: 'Wow! How old would that be, Prof?'

Prof: '2,900 million years.'

All (pre-programmed by R—): 'Happy birthday to you, happy birthday to you.'

Somabula flats, aka MAMBA (miles and miles of bugger all). Exposed in a river bed near Shangani are four sets of quartz veins intruding the gneiss ... one set is both younger and older than one other set ... gravel beds on Willoughby's ranch ... mammoth kimberlite pipe ... Redcliff iron-ore open pit ... Bang! A blast, detonated for our entertainment, tosses a rock the size of a pawpaw into our midst ... and thus back to where we started: tired, impressed, and surprisingly still alive.

The Family

Marina Bond: Geoff Bond, My Husband

To write about Geoff is one of the most difficult tasks I have been set. He was my friend, lover, husband and father of Ivan (born 1962) and Nina (born 1966). We were married in the Presbyterian Church, Bulawayo, on 1 November 1960 by Reverend Manod Williams. It was the simplest of ceremonies, attended only by the closest of friends and followed by a luncheon. We used to joke about it: how many men can say that they took an academic chair and a wife on the same day. What to say about the twenty-one years of marriage? That they passed in a flash.

We declined the offer of university accommodation, which would have cost us an exorbitant 20% of Geoff's salary. Instead, we rented half a house on the corner of West and Aberdeen roads while we looked for a place to buy. Tang was the first of several Siamese cats. Geoff did not want a dog, having had a surfeit of them in the foregoing years.

Geoff set about building a Geology department from scratch. Many of the specimens were collected by us. Perhaps it is correct to say that only now, with 20/20 hindsight, I can see the full extent of his achievement: he built a department as well as a family and a home from scratch – and was successful in both. I did not fully appreciate the enormity of this.

Despite all my efforts, Geoff could not be persuaded to diminish, let alone stop, his smoking habit. It had become a true addiction. It cost him his life. He died at home on a Sunday night in the presence of Ivan, Nina and myself. His last words were: 'Thank you for everything.'

Dad Stays Home, Aug. 1973

I really don't enjoy this life
When separated from my wife,
Who cares for me in many ways
To brighten up my nights and days.

The children, too, give me much pleasure
Although they gobble up my leisure –
Without them leisure seems to pall,
So hurry back, both one and all.

The chance to work seemed only golden
To my family not beholden!
A week without a single chore
Was in reality a bore.

I hope that while you were away
You found enjoyment every day.
Each day to me seemed almost endless
Each night was empty, cheerless, friendless.

Marina

Some presents are for use and some for show.
This may be neither for all I know.
The opium pipe is useless in this day and age,
Though centuries ago was all the rage.

I like its shape despite its lousy plate,
Which I'll replace at no far distant date.
Useless or showless as this pipe may be,
It comes with all my love to you from me.

Happy Christmas, Geoff.

Geoff Bond, *Verse & Worse: From Bad to Worse*.

The Geologist Sums Up

Man appeared so late in time
He's hardly worth the cynic's rhyme.
For ages before man arrived
This planet spun along and thrived.

Species came and species went,
None of them with ill intent.
Then came man and round his village
Started out to kill and pillage.

Man the technocrat and master,
Now, despoils his planet faster
Than nature's forces can restore
The damage that's been done before.

The world will wake one day to find
That man has jeopardised mankind.
There still is time for a solution
To soil erosion and pollution.

With thought we might still obviate
The ruin that man will create.
To overcome the self destruction
Man must limit reproduction.

Its plain that all this spoil and plunder
Sprang from creation's biggest blunder.
If nature did her work again
She'd give our species far less brain.

Appendix 2

Published obituaries for Geoffrey Bond

In Memoriam: Geoffrey Bond, 1912–1983

OLM, BSc(hons), PhD, DIC, ARCSc, C. Eng, MIMM, FGS, FGS(Am), FZSA

M.A. Raath

Professor Geoffrey Bond, a founding member of the editorial board of this journal, died at his home in Harare, Zimbabwe, on June 19th 1983 at the age of 71.

He was born in Brighton and grew up in Yorkshire. It was the limestone country of the Yorkshire dales that instilled in him his early love of geology, and it wasn't long before he turned to it in something more than an amateur fashion, firstly at the Chelsea Polytechnic and later as a Royal Scholarship student at the Imperial College of Science and Technology in London. The War interrupted his studies from September 1940 to December 1945, and also resulted in his initial posting to what was then Southern Rhodesia (now Zimbabwe). He returned to Imperial College early in 1946 to complete his degree, graduating with First Class Honours in July 1946. He received his PhD in 1949, and his Diploma of Membership of Imperial College (DIC) in 1953; the two latter theses dealt with themes that were to dominate much of his subsequent geological writing – the palaeoclimate of the Pleistocene of Matabeleland, and the Karoo Sequence of the Zambezi Valley. Palaeontology had been a particular interest of his at Imperial College, and it remained one of his abiding passions for the rest of his life. But he was no narrow specialist; his interests ranged widely across much of the spectrum of geology and the wider horizons of natural science generally.

He lived most of his working life in Zimbabwe (then called variously Southern Rhodesia, Rhodesia, Zimbabwe–Rhodesia, and finally simply Zimbabwe). His love affair with Africa, and particularly with Zimbabwe, began in 1941. He himself was fond of telling a story about his posting to Bulawayo by the Royal Air Force during the War as a result of a military signalling mix-up. According to him, the signal from the Commonwealth Air Training School at Heany Junction, near Bulawayo, ordered 'send 400 airscrews' but it was unscrambled in London as 'send 400 aircrew'; thus he and several other Air Force servicemen arrived at the air training station in 1941, and except for some minor breaks, he stayed in the country for the rest of his life.

When he returned to Zimbabwe after the War and after completing his degree, he accepted appointment to the post of Keeper of Geology at the National Museum of Southern Rhodesia, a post for which he was ideally equipped by virtue of his wide interests and knowledge, and his friendly approach to people. He served in this post for the next fourteen years, with a break of one year from June 1948 to June 1949 when he returned to England to found the Department of Geology at Hull University. This experience also paved the way for another very important chapter in his life – surely THE most important chapter: the establishment in 1960 of the Department of Geology at the newly founded University College of Rhodesia and Nyasaland during the days of the ill-fated Federation of Rhodesia and Nyasaland. He was appointed inaugural Professor and Head of the Department. He continued to serve the University in a variety of capacities until shortly before his death, particularly as Vice-Principal and Deputy Vice-Chancellor from July 1974 until the time of his official retirement in mid-1982. The latter years of his service with the University were at a time of major political upheaval, especially during the tense years of the so-called 'Rhodesian Bush War', and his duties required him to tread a very narrow diplomatic path in this multiracial institution where inter-racial feelings often ran high. In these situations his tact, diplomacy and fair-mindedness were unfailingly valuable to the University.

He retained for the rest of his life close ties with the quasi-governmental national museums organization of his adopted land, even after leaving to join the university staff in 1960. He was appointed to the board of trustees of the national museums organization in 1963, and he continued to serve on the board for the next twenty years, thirteen of them as its chairman under three vastly different political regimes. It was as an employee of the museums that the writer of this memorial first came into contact with him in 1965, when appointed to the post of zoologist at the Queen Victoria Museum in what is now Harare. Because of a mutual interest in palaeontology a friendship quickly developed that was to last until his death; it even survived a time when he was supervisor of the writer's PhD study! Geoff Bond had the most endearing and charming ability to convince all men that he was genuinely interested in them and what they were doing.

As a geologist he rendered valuable service in opening up his part of Africa to the scrutiny of science. He himself traversed much of the country on foot, meticulously recording his observations, many of which were to have substantial economic impact subsequently; for example, he is credited with first identifying the emeralds for which the Sandawana area is now so famous. Through his department at the university he was instrumental in attracting many geologists from overseas to visit the country to carry out meaningful geological studies. He was a particularly worthy successor to such geological pioneers in that part of Africa as A.J.C. Molyneux, A.M. Macgregor, F.P. Mennell and others. One notable achievement of his was to persuade ordinary people whose work took them outdoors into the largely virgin Rhodesian bush to keep their eyes open and to report to the appropriate authorities observations that seemed in any way unusual or significant to them. In this way many new and important fossil sites were added to the few that were known before Geoff Bond came on the scene. He will undoubtedly be remembered in Zimbabwe as a man who loved to teach and to popularize science. He was a lucid, eloquent and entertaining speaker who was constantly in demand to address audiences of all kinds; he also hosted a succession of radio and television science shows for many years.

Apart from being a busy academic, administrator, field and laboratory geologist, speaker and raconteur, Geoff Bond was a 'people's person' and family man. His was a particularly close and happy family and his home was always welcoming for any visitor. He was a keen sportsman, and he still enjoyed a regular game of tennis despite the arthritis that was beginning to trouble him right at the end of his life. Cricket – that enigmatic, incomprehensible and uniquely English pastime which is the elixir of life for any Yorkshireman worth his salt – naturally loomed large in his life! He was particularly proud of one cricketing achievement when, in a 'Fathers-and-sons' match at his son's school, Geoff made sixty not out at the age of 65. His other loves included an abiding love of wildlife and the great African outdoors, 'ham' radio operating, and being aboard his beloved university boat *Erica* on Lake Kariba. With one hand on the wheel, inevitably clutching the radio mike, the other lovingly coaxing the throttles of the twin diesel motors, and a chart of the lake spread out on the wheelhouse table in front of him, Geoff was in his Seventh Heaven!

With over sixty scientific publications to his name, he was a productive scientist. Most of his publications dealt with the two major themes identified earlier: Pleistocene palaeoclimates, and Karoo stratigraphy and palaeontology. But he also wrote on matters of economic geology, mining, engineering geology, gemmology, sedimentology, and the history of geology in Central Africa.

He was the recipient of several honours and awards, including the Rhodesian civil honour of the Order of the Legion of Merit, honorary Fellowship of the Geological Society of London, of America, and of the Zimbabwe Scientific Association, and the Draper Memorial Medal of the Geological Society of South Africa, which he received at a presentation ceremony in Johannesburg just a few weeks before his death. He is survived by his wife, Marina, who supported him so strongly in all he did, his son, Ivan, and his daughter, Nina.

Originally published as M.A. Raath, 'In Memoriam: Geoffrey Bond, 1912–1983', *Palaeogeography, Palaeoclimatology, Palaeoecology* (1984), 45(2): 101–4.

Obituary: Geoffrey Bond, 1912–1983

J. F. Wilson

Professor Geoffrey Bond, Fellow of the Zimbabwe Scientific Association and former Professor of Geology and Vice-Principal of the University of Zimbabwe, died suddenly at home in Harare on 19 June 1983 aged 71 years.

Born in Brighton, England on 27 May 1912 Geoffrey Bond spent much of his childhood and youth in Yorkshire. At the age of 18 he left school in Cumberland to face the frustrating task of job hunting in the depression years of the thirties. He eventually found work as a schoolteacher in London.

His formative years among the largely limestone country of the Yorkshire dales had instilled in him, from an early age, a love of geology and of the wide open spaces. As a schoolteacher he further pursued his hobby of geology in evening classes at the Chelsea Polytechnic. Friends eventually persuaded him to write the examinations for a Royal Scholarship in which, much to his astonishment, he found himself successful. Geology had suddenly become something more than a mere hobby and, thus financed; he embarked in 1937 on a 4-year B.Sc. course at the Imperial College of Science and Technology in London. World War II intervened and from September 1940 to December 1945 he served with the Royal Air Force. He returned to Imperial College in January 1946 and, after a supreme two-term effort, graduated in July of that year with First Class Honours.

Most of his wartime service was spent in what was then Southern Rhodesia. Geoffrey Bond always maintained that his wartime posting to Africa, and that of the several hundred others who accompanied him, was the result of a clerical error where '400 airscrews' became '400 aircrew'. This posting changed the whole course of his subsequent life; it instilled in him a love of the bush and a determination to return to this country after the war. So, in September 1946, he arrived in Bulawayo to take up the post of Keeper of Geology at the National Museum.

Though palaeontology was still his prime interest, Geoffrey Bond, with the late H.H. Read as one of his professors, was a good all-round geologist and therefore able to cope competently and efficiently with the numerous and diverse tasks which confronted him. He soon became immersed in his new job.

For the next 14 years, save for one (June 1948 - June 1949) when he returned to England to start a department of geology at Hull University College, he remained as Keeper of Geology in Bulawayo. This was a period of great excitement for him, and one during which he built up a sound knowledge of the geology of his adopted country from first-hand experience.

He travelled far and wide but his interest gradually focused mainly on the Zambezi Valley which he grew to love. His Ph.D. in 1949 embraced the geology, climates and cultures of the Pleistocene period in Matabeleland. This was followed by a second thesis for the Diploma of Membership of Imperial College, in 1953, on the Karoo System of the Zambezi basin. Numerous scientific papers flowed from his pen and he soon became an accepted authority, inside and outside Zimbabwe, on the Karoo System of the country. Also, from his day-to-day dealings with prospectors and mining companies, an integral part of his museum advisory work, he soon acquired a sound knowledge of the country's mineral wealth.

By 1960, therefore, when plans were laid for a department of geology at the recently established University College of Rhodesia and Nyasaland, Geoffrey Bond was the obvious man to start it off. An enthusiastic cricketer all his life the idea of 'opening bat' probably held no qualms for him and on 1 November 1960 he was appointed as the inaugural Professor and Head of the Department of Geology. On the same day he married Marina Levitzky from Bulawayo, achieving, as he put it, a unique double.

Within a few years his leadership and guidance built up a small but enthusiastic and happy Department that, over the years, successfully produced not specialists but 'thinking men' sufficiently well versed in all branches of their science to face with confidence any aspect of the local geological and mining scene. On the research side the Department produced results that have gained international recognition. Professor Bond contributed to this achievement by being the author of some 60 publications.

In July 1974 Geoffrey Bond combined the Headship with that of part-time Vice-Principal of the University and in 1977 retired as Professor of Geology to become full-time Vice-Principal and Deputy Vice-Chancellor. From June 1981 until his final retirement in June 1982 he served as Vice-Principal. His contribution to the development and progress of the University was immense, and in recognition of this the University conferred on him the title of Professor Emeritus on his retirement.

Geoffrey Bond's activities, however, extended well beyond the University confines. He gave long and distinguished service on the Natural Resources Board, and on the Board of Museums and Monuments. He was a Director of Rio Tinto Mining Zimbabwe Limited. A radio ham for many years he also had a keen interest in broadcasting and will be remembered for his popular weekly programmes which brought science into many a home. Professor Bond will also be remembered for his part in the establishment of the Southern Rhodesian branch of The Geological Society of South Africa, which, in 1981, became the Geological Society of Zimbabwe. For many years he served on the Council of The Geological Society of South Africa and was elected President in 1971. In 1983, only three months prior to his death, Geoffrey Bond was presented with the Draper Memorial Medal of The Geological Society of South Africa. Earlier he had distinguished himself by being awarded the Murchison Fund of the Geological Society of London and the Order of the Legion of Merit for services rendered to Rhodesia.

On his retirement as Vice-Principal in 1982 Geoffrey Bond was appointed Honorary Research Fellow in the Department of Geology. The return 'home' of their friend and mentor was welcomed by all. This return was real and within a very short time he was once again back to active research and work on his beloved Zambezi Valley.

Geoffrey Bond was a man of many parts who distinguished himself in many fields. He led an active and full life. In all that he did, the support of his wife and family, which was always there, was of vital importance to him. He was a distinguished scholar, an inspiring teacher and a wise administrator. Perhaps above all he was a kindly man who was concerned about others and was always ready to ease a tense situation with an amusing anecdote. His death is a tremendous loss to Zimbabwe.

Originally published as
J.F. Wilson, 'Obituary: Geoffrey Bond, 1912–1983'
Transactions of the Geological Society of South Africa (1983) 86: 161–3
and reproduced with the permission of the Geological Society of South Africa

Appendix 3

Geoffrey Bond's Publications

Based on a compilation by Tim Broderick, September 2020,
who was greatly assisted by
C. C. Smith and H. E. van der Heyde,
'Rhodesian geology: A bibliography and brief index to 1968',
Occasional Papers of the National Museum of Southern Rhodesia (1971), 4(31B)

Single-authored works

1941

Species and variation in British and Belgian carboniferous Schizophoriidae. *Proceedings of the Geologists' Association* [London], 52(4): 285–303.

1943

Solution etching of detrital staurolite. *Geological Magazine*, 80(4): 155–6.

1946

A Lower Beaufort (Karroo) invertebrate fauna from Southern Rhodesia. *Transactions of the Royal Society of South Africa*, 31(2): 125–31.

The Pleistocene succession near Bulawayo. *Occasional Papers of the National Museum of Southern Rhodesia*, 2(12): 104–15.

1947

Notes on some Karroo lamellibranches from the Sebungwe District, Southern Rhodesia. *Occasional Papers of the National Museum of Southern Rhodesia*, 2(13): 209–12.

Radio-active minerals may be found in Colony. *Rhodesian Mining Review*, 144: 15, 17.

1948

The direction and origin of the Kalahari sand of Southern Rhodesia. *Geological Magazine*, 85(5): 305–13.

Outgrowths on zircon from Southern Rhodesia. *Geological Magazine*, 85(1): 35–40.

Rhodesian Stone Age Man and his raw materials. *South African Archaeological Bulletin*, 3(11): 56–60.

Will deep drilling aid Rhodesia in mine exploration? *Rhodesian Mining Review*, 148: 18–19.

1949

The Geology, Climates and Cultures of the Pleistocene Period in Matabeleland, Southern Rhodesia. (Unpublished Ph.D. thesis, University of London.)

1950

Beryllium – an up-and-coming metal. *Rhodesian Mining Review,* 15(4): 13.

The dating of deposits of 'Kalahari' sand. *In:* J. Desmond Clark, *The Stone Age Cultures of Northern Rhodesia: With Particular Reference to the Cultural and Climatic Succession in the Upper Zambezi Valley and its Tributaries* (Claremont: South African Archaeological Society), Appendix D, 151–2.

The lower carboniferous reef limestones of Cracoe, Yorkshire. *Geological Society London Quarterly Journal*, 105: 157–88.

The lower carboniferous reef limestones of northern England. *Journal of Geology*, 58(4): 313–29.

1950 (cont.)

Note on the geology of Khami Magosian sites. *In:* C.K. Cooke, The Middle Stone Age site at Khami, Southern Rhodesia: A further examination. *South African Archaeological Bulletin*, 5(18): 68.

1951

Field test for columbite. *Rhodesian Mining Review*, 16(2): 25.

Museum geology in Southern Rhodesia. *Bulletin of the South African Museum Association*, 5(4): 98–101.

Structures and metamorphism in the Chimanimani Mountains, Southern Rhodesia. *Transactions of the Geological Society of South Africa*, 54: 69–83.

The Victoria Falls. *South African Railways and Harbours Magazine* (March), 141–3.

1952

Bulawayo to Victoria Falls via the Game Reserve. *Rhodesian Motorist* (September), 13–18, 34.

The deformation of banded ironstones in Southern Rhodesia. *Economic Geology*, 47(8): 823–4.

Evidence of glaciation in the lower part of the Karroo System in Southern Rhodesia. *Transactions of the Geological Society of South Africa*, 55: 1–11.

The finest of all gems [Diamond]. *Rhodesian Mining Review*, 17(4): 23–4.

The Karroo System in Southern Rhodesia. *In:* C. Teichert (ed.), *Symposium sur les séries de Gondwana: 19. congrès géologique international* (Algiers), 209–23.

Looking for scheelite? *Rhodesian Mining Review*, 17(1): 23, 25.

The origin of the Victoria Falls. *In:* J. Desmond Clark (ed.), *The Victoria Falls: A Handbook to the Victoria Falls, the Batoka Gorge, and Part of the Upper Zambesi River* ([Lusaka]: Commission for the Preservation of Natural and Historical Monuments and Relics, Northern Rhodesia, 49–56.

Worth looking for columbium and tantalum. *Rhodesian Mining Review*, 17(12): 21, 23, 25.

1953

An account of the Karroo System in the Middle Zambezi Basin. (Unpublished thesis submitted for the Diploma of Membership of the Imperial College of Science and Technology, London.)

Notes on the geology of the head-waters of the Deka River. *Occasional Papers of the National Museum of Southern Rhodesia*, 2(18): 521–8.

Notes on the Stone Age in the Eastern Sebungwe, Southern Rhodesia. *Occasional Papers of the National Museum of Southern Rhodesia* 2(18): 518–20.

The origin of thermal and mineral waters in the Middle Zambezi Valley and adjoining territory. *Transactions of the Geological Society of South Africa*, 56: 131–48.

1954

Lamellibranchia and plants from the Lower Karroo beds of Northern Rhodesia. *Geological Magazine*, 91(3): 189-92.

Mining - and the National Museum of Southern Rhodesia. *Rhodesian Mining Review*, 19(8): 39-42.

Surface textures of sand grains from the Victoria Falls area. *Journal of Sedimentary Petrology*, 24(3): 191-5.

1955

Geology Report. *Rhodesian Schools' Exploration Society, West Sebungwe Expedition, 7-19 May 1955*, 81-5 [cyclostyled].

The Madumabisa (Karroo) Shales in the Middle Zambezi region. *Transactions of the Geological Society of South Africa*, 58: 71-99.

A note on dinosaur remains from the Forest Sandstone (Upper Karroo). *Occasional Papers of the National Museum of Southern Rhodesia*, 2(20): 795-800.

Quaternary geology and correlation. *In:* G. Bond, R.H.F. Summers, C.K. Cooke and K.R. Robinson, *Excursion Guide Book for Southern Rhodesia - Third Pan-African Congress on Prehistory, Livingstone, 1955*, 1-3.

1956

The geology of the hills. *In:* Robert Tredgold (ed.), *The Matopos* ([Lusaka]: Federal Department of Printing and Stationery), 34-8.

Geology Report. *Rhodesian Schools' Exploration Society, Upper Kariba Expedition, 3-19 May 1956*, 71-7 [cyclostyled].

A preliminary account of the Pleistocene geology of the Plateau tin fields region of northern Nigeria. *In: Proceedings of the International West Africa Conference 1949, Ibadan* (Lagos: Nigerian Museum), 187-202.

1957

The geology of the Khami Stone Age sites. *Occasional Papers of the National Museums of Southern Rhodesia* A, 3 (21A): 44-55.

Mineral Wealth. *Round Table, Industrial Bulawayo* [publicity brochure].

Quaternary sands at the Victoria Falls. *In:* J. Desmond Clark (ed.), *Third Pan-African Congress on Prehistory, Livingstone, 1955* (London: Chatto and Windus), 115-22.

1960

The fossil history of Rhodesia. *Rhodesia Mining Engineering*, 25(13): 37-9.

1962

Past Climates in Central Africa: An Inaugural Lecture given in the University College of Rhodesia and Nyasaland (London: Oxford University Press).

Pleistocene research in Southern Rhodesia. *In:* G. Mortelmans and J. Nenquin (eds.), *Actes du IVe Congrès Panafricain de Préhistoire et de l'étude du Quaternaire* (Tervuren: Musée Royal de l'Afrique Centrale), Section I, 141-50.

1963

Environments of east and southern Africa since the mid-Tertiary. *South African Journal of Science*, 59(7), 347–52.

Pegmatites in Southern Rhodesia: Proceedings of a symposium arranged by the Southern Rhodesia Section of the Institution of Mining and Metallurgy in collaboration with the Southern Rhodesia Branch of the Geological Society of South Africa held at the University College of Rhodesia and Nyasaland, Salisbury, on 22nd and 23rd November 1962 (Salisbury: Institution of Mining and Metallurgy, Southern Rhodesia Section).

Pleistocene environments in southern Africa. *In:* Clark F. Howell and François Bourlière (eds), *African Ecology and Human Evolution* (Chicago: Aldine Press, Viking Fund Publications in Anthropology, 36), 308–34.

1964

The origin of the Victoria Falls. *In:* B.M Fagan (ed.), *The Victoria Falls: A Handbook to the Victoria Falls, the Batoka Gorge, and Part of the Upper Zambesi River (Second Edition)* ([Lusaka]: Commission for the Preservation of Natural and Historical Monuments and Relics, Northern Rhodesia), 45–54.

A new phyllopod from the Karroo System of Northern Rhodesia. *Arnoldia (Rhodesia)*, 1(15): 1–4.

1965

Preliminary observations on the development of shoreline features on Lake Kariba. *In: Record of a Kariba Research Symposium held at the Fisheries Research Institute on 8–9 June 1965* (Kariba: Lake Kariba Fisheries Research Institute), 33–42.

Quantitative approaches to rainfall and temperature changes in the Quaternary of southern Africa. *In:* H.E. Wright, Jr and David G. Frey (eds.), *International Studies on the Quaternary: Papers Prepared on the Occasion of the VII Congress of the International Association for Quaternary Research, Boulder, Colorado, 1965* (New York: Geological Society of America, Special Papers, 84).

Some new fossil localities in the Karroo System of Rhodesia. *Arnoldia (Rhodesia)*, 2(11): 1–4.

1966

Obituary: Frederick Philip Mennell, F.G.S., M.I.M.M. *Proceedings and Transactions of the Rhodesia Scientific Association*, 51: 14–16.

1967

A review of Karroo sedimentation and lithology in Southern Rhodesia. *Reviews prepared for the First Symposium on Gondwana Stratigraphy, Mar del Plata, Argentina, 1967* (Harlem: International Union of Geological Sciences), 173–95.

1967 (cont.)

River valley morphology, stratigraphy, and palaeoclimatology in southern Africa. *In:* Walter W. Bishop and J. Desmond Clark (eds), *Background to Evolution in Africa* (Chicago: University of Chicago Press), 303-12.

1968

The flora of the Black Shale and Coal Group (K2) in the mid-Zambezi basin. *Arnoldia (Rhodesia)*, 3(38): 1-2.

Role of the University College of Rhodesia. *Rhodesian Mining: Supplement to Rhodesian Property & Finance* (April), 34-5.

1969

From desert to woodland savanna: The natural reclamation of Pleistocene sand dunes in the Wankie National Park. *Rhodesia Science News*, 3(5): 127-8.

The geology of the Kalambo Falls prehistoric site. *In:* J. Desmond Clark, *The Kalambo Prehistoric Site* (Cambridge: Cambridge University Press), Appendix A, 197-214.

Milestones in Rhodesian Palaeontology. *Detritus*, 4: 16-18.

Summing up: Achievements and future prospects for Quaternary studies in southern Africa: Geological aspects. South *African Archaeological Bulletin*, 24(95/96): 209-10.

The task of the geologist in site investigation. *The Rhodesian Engineer*, 7(3): 826-30.

1970

The Dwyka Series in Rhodesia. *Proceedings of the Geologists' Association*, 81(3): 463-72.

Research for mining in Rhodesia. Society News - Rhodesian Scientific Association. *Rhodesia Science News*, 4(12): 417-18.

1971

The Madziwadzido fossil locality revisited. *Detritus*, 6: 14-15.

The Problem of the Kalahari Sands. *Mining in Rhodesia*, 33-4.

1972

Milestones in Rhodesian palaeontology (1901-1971). Presidential Address. *Transactions of the Geological Society of South Africa*, 75(2): 149-58.

The crust of the Earth. In: *Gems and Minerals of Rhodesia: A Guide to the Crust of the Earth, Geology Prospecting, Lapidary, Art and Birthstones* (Salisbury: Rhodesian Gem and Mineral Society), 11-17.

1973

Geology and the development of Rhodesia. *Rhodesia Science News*, 7(6): 141-3.

[Book review of] Karl W. Butzer, *Environment and Archeology, an Ecological Approach to Prehistory* (Methuen, London, 1972, 2nd ed.) *Palaeogeography, Palaeoclimatology, Palaeoecology*, 13(2), 158-9.

1973 (cont.)

The Palaeontology of Rhodesia, with a section on the Palynology of the Middle Zambezi Basin, by R.M.S. Falcon. *Rhodesia Geological Survey Bulletin*, 70.

1974

The stratigraphic distribution of plant and animal macro-fossils in the Karoo System of the mid-Zambezi region. *Arnoldia (Rhodesia)*, 7(3): 1–6.

1975

The geology and formation of the Victoria Falls. *In:* D.W. Phillipson (ed.), *Mosi-oa-Tunya: A Handbook to the Victoria Falls Region* (Salisbury: Longman Rhodesia), 19–27.

1976

Plants and time - a geologist's view. *Excelsa*, 6: 21–5.

1980

Science broadcasting in Zimbabwe - the lighter side. *Zimbabwe Science News*, 14(6): 131–2, 134.

1981

Late Palaeozoic (Dwyka) glaciation in the Middle Zambezi region. *In:* M.J. Hambrey and W.B. Harland (ed.), *Earth's pre-Pleistocene Glacial Record* (Cambridge: Cambridge University Press), 55–7.

Late Palaeozoic (Dwyka) glaciation in the Sabi–Limpopo region, Zimbabwe. *In:* ibid., 58–60.

A possible Late Precambrian tillite from the Urungwe District, Zimbabwe. *In:* ibid., 178–9.

1983

The Stone Age in the Middle Zambezi Valley. *Zimbabwean Prehistory*, 19 (December): 2–6.

1990

The geology and formation of the Victoria Falls. *In:* D.W. Phillipson (ed.), *Mosi-oa-Tunya: A Handbook to the Victoria Falls Region* (Salisbury: Longman Rhodesia, Second edition), 19–25.

2017

Verse & Worse: From Bad to Worse (Harare: M. Bond, privately published).

Collaborative works

1948

Jones, N. and Bond, G. The Lydiate pebble industry. *Occasional Papers of the National Museum of Southern Rhodesia*, 2(14): 251–8.

1951

Bond, G. and Summers, R.F.H. The Quaternary succession and archaeology at Chelmer, near Bulawayo, Southern Rhodesia. *South African Journal of Science*, 47(7): 200–4.

1954

Bond, G. and Clark, J.D. The Quaternary sequence in the Middle Zambezi Valley. *South African Archaeological Bulletin*, 9(36): 115–30.

Bond, G. and Summers, R. A Late Stillbay hunting-camp on the Nata River, Bechuanaland Protectorate. *South African Archaeological Bulletin*, 9(35): 89–95.

1955

Bond, G., Summers, R.H.F., Cooke, C.K. and Robinson, K.R. *Excursion Guide Book for Southern Rhodesia: Third Pan-African Congress on Prehistory, Livingstone, 1955*.

Tyndale-Biscoe, R. and Bond, G. The Muva-Frontier System correlation. *First Meeting, C.C.T.A. Southern Regional committee for Geology C.C.T.A.*, Salisbury, 41–48; discussion, 48–53.

1956

Bond, G., Maufe, H.B. and Tyndale-Biscoe, R. Stratigraphy of Southern Rhodesia. *In:* R. Furon (ed.), *Lexique Stratigraphique International, Vol.IV, Afrique, Fascicule 9c, Rhodesie du Sud* (Paris: Centre National de la Recherche Scientifique), 31–75.

1960

Sutton, E.R. and Bond, G. The Karroo succession on the lower Bubye River, Southern Rhodesia. *In:* Boughney, A.S. and Robins, P.A. (eds.), *Proceedings of the First Federal Science Congress, Salisbury, Rhodesia, May 18–22nd 1960* (Salisbury: Mardon Printers), 45–52.

1962

Bond, G. and Anderson, R.B. The significance of pegmatites in Southern Rhodesian mining. *Chamber of Mines Journal*, 4(11): 30–2.

Bond, G. and Macdonald, D. The Karroo succession at Sawmills, Southern Rhodesia. *Occasional Papers of the National Museum of Southern Rhodesia* 3(26B): 623–30.

1964

Bond, G. and Bliss, N.W. Recent discoveries of stromatolites in the dolomites of the Lomagundi System, Southern Rhodesia. Synopsis. In: *Proceedings of the 7th Annual Congress of the Geological Society of South Africa* (Salisbury, 1964), 63.

1965

Bond, G. and Stocklmayer, V.R.C. Heavy mineral residues from early and late Karroo sandstones in parts of the Zambezi basin. *Arnoldia (Rhodesia)*, 1(34): 1–7.

Bond, G. and Stowe, C.W. The geology of Rhodesia. *In:* [M.O. Collins (ed.)], *Rhodesia: Its Natural Resources and Economic Development* (Salisbury, M.O. Collins), 40–1.

1967

Bond, G. and Stocklmayer, V.R.C. Possible ice-margin fluctuations in the Dwyka Series in Rhodesia. *Palaeogeography, Palaeoclimatology, Palaeoecology*, 3(4): 433–46.

1968

Flint, R.F. and Bond, G. Pleistocene sand ridges and pans in western Rhodesia. *Geological Society of America Bulletin,* 79(3): 299–313.

1969

Bond, G., Raath, M.A. and Wilborg, M. Queen Victoria Museum notes – Palaeontology. *Rhodesia Science News*, 3(12): 377.

1970

Bond, G. and Bromley, K. Sediments with the remains of dinosaurs near Gokwe, Rhodesia. *Palaeogeography, Palaeoclimatology, Palaeoecology*, 8: 313–27.

Bond, G., Wilson, J.F. and Raath, M.A. Upper Karroo pillow lava and a new sauropod horizon in Rhodesia. *Nature*, 227: 1339.

Raath, M.A., Smith, C.C. and Bond, G. A new Upper Karroo dinosaur fossil locality on the lower Angwa River, Sipolilo District, Rhodesia. *Arnoldia (Rhodesia)*, 4(35): 1–9.

1972

Nesbitt, E.A. and Bond, G. Fossil chelonians from the Gokwe Formation, Rhodesia. *Arnoldia (Rhodesia)*, 5(37): 1–7.

1973

Bond, G., Wilson, J.F. and Winnall, N.J. Age of the Huntsman Limestone (Bulawayan) stromatolites. *Nature*, 244: 275–6.

1974

Bond, G. and Fernandes, T.R.C. Scanning electron microscopy applied to quartz grains from Kalahari type sands. *Transactions of the Geological Society of South Africa*, 77(2): 191–9.

Index

Made in United States
Orlando, FL
05 November 2026

90231493R00195